Experience the California Coast

A Guide to Beaches and Parks in Northern California

COUNTIES INCLUDED

DEL NORTE · HUMBOLDT · MENDOCINO · SONOMA · MARIN

State of California

Arnold Schwarzenegger, *Governor*

California Coastal Commission

Peter Douglas, *Executive Director* Susan Hansch, *Chief Deputy Director*

Steve Scholl
Editor and Principal Writer

Erin Caughman
Designer and Co-Editor

Jonathan Van Coops
Mapping and GIS Program Manager

Gregory M. Benoit
Principal Cartographer

Caitlin Bean, Biology
John Dixon, Ph.D., Biology
Lesley Ewing, P.E., Coastal Engineering
Mark Johnsson, Ph.D., Geology
Chris Parry, Public Education
Contributing Writers

Jo Ginsberg
Consulting Editor

Artwork by Tom Killion

Linda Locklin
Coastal Access Program Manager

University of California Press
Berkeley Los Angeles London

St. Louis Community College
at Meramec
LIBRARY

University of California Press
Berkeley and Los Angeles, California

University of California Press, Ltd.
London, England

© 2005 by the State of California, California Coastal Commission.
All rights reserved.
No part of this book may be used or reproduced in any manner whatsoever without written permission except in the case of brief quotations embodied in articles and reviews. For information, write to the Executive Director, California Coastal Commission, 45 Fremont Street, Suite 2000, San Francisco, California 94105.

Tom Killion's woodcut prints from his *Coast of California* series are used by permission of the artist, who retains sole copyright to them.

Library of Congress Cataloging-in-Publication Data

Experience the California coast : a guide to beaches and parks in northern California : counties included : Del Norte, Humboldt, Mendocino, Sonoma, Marin / [compiled by the] California Coastal Commission.

p. cm.

Includes bibliographical references and index.

ISBN 0-520-24540-7 (cloth : alk. paper)

1. California, Northern—Guidebooks. 2. Pacific Coast (Calif.)—Guidebooks. 3. Beaches—California, Northern—Guidebooks. 4. Parks—California, Northern—Guidebooks. 5. Beaches—California—Pacific Coast—Guidebooks. 6. Parks—California—Pacific Coast—Guidebooks. I. California Coastal Commission.

F867.5.E96 2005

917.94 ' 10454—dc22

Manufactured in China

15 14 13 12 11 10 09 08 07 06 05

10 9 8 7 6 5 4 3 2 1

The paper in this publication meets the minimum requirements of ANSI/NISO Z39.48-1992 (R 1997) (Permanence of Paper).

Contents

Features

Mendocino coast, north of Elk

© 2004, Tom Killion

Introduction

ALTHOUGH KNOWN TO Californians as the North Coast, the shoreline between the Golden Gate and the Oregon border lies squarely at the center of the continent's Pacific coast. The range of environments available to coastal visitors reflects this middle location. Soft white sand and coarse pebbles, Sitka spruce and California poppies, roadless areas and campgrounds, redwood glades and pygmy forests—all find their place on California's North Coast.

The climate of coastal northern California is moderate, with a strong marine influence. Average afternoon temperatures typically fluctuate only about 10 degrees Fahrenheit between winter and summer, and in any season the spread between day and night may be only 15 degrees or less. Even though sun and fog may play hide-and-seek on summer afternoons, microclimate is everything; there are plenty of coves and sheltered beaches where building sandcastles is the logical pursuit, while protected waters in rivers and lagoons draw swimmers and kayakers. Wetsuited surfers ride the waves, while clam diggers burrow in mudflats, and hikers explore miles of trails. This guide is an introduction to a nearly endless range of coastal experiences available to visitors.

The California Coastal Commission, along with sister state agencies including the State Coastal Conservancy, the Department of Parks and Recreation, and the Department of Fish and Game, is charged with conserving, enhancing, and making available to the public the beaches, accessways, and resources of the coast. The Coastal Commission's responsibilities include providing the public with an informative and educational guide to coastal resources and maintaining an inventory of paths, trails, and other shoreline accessways that are available to the public. Along with the previously published *California Coastal Resource Guide* and the *California Coastal Access Guide*, this book furthers those purposes.

This guide tells you where to go on the North Coast, how to get there, what facilities and coastal resources you will find, and what you might do at each location. It is meant for all coastal visitors—picnickers, hikers, campers, surfers, divers, birders, boaters, anglers—and is intended to introduce the richness and diversity of the California coast. This book includes a comprehensive list of beaches, parks, and paths to the shoreline, some 350 in all. Most are publicly owned or controlled, while others are privately managed but available to all users. Use your public tidelands and beaches, but respect private property; do not trespass.

The book also lists commercial recreational outfitters and guides, including fishing boat services, kayak rentals, and equestrian facilities. An effort has been made to be as comprehensive as possible in describing these enterprises, but businesses may change. Call ahead to make sure the recreational offerings you seek are available. The editors welcome additions for inclusion in future editions (see p. 300).

A coastal visit has more depth with an overnight stay, and this guide lists hostels, campgrounds in state and local parks, and, as space permits, private campground facilities. Campsites in public or private parks include family camps, group camps, sites with RV hookups, walk-in environmental campsites, hike or bike sites, and enroute (overflow) spaces. Many can be reserved in advance, while others are available to those who come first. In some areas, such as along the lower Klamath River in Del Norte County, private campgrounds and RV facilities are too numerous to be listed here individually, and visitors are instead directed to clearinghouses such as the local chambers of commerce. Information

Doran Beach Regional Park, Sonoma County

about market-rate hotels, inns, and eating establishments is available in numerous other guidebooks.

Enjoy your visits to California's magnificent North Coast. Keep safe by observing posted restrictions along hazardous stretches of shoreline. Keep in mind that sleeper waves are a constant factor on the Northern California coast; when strolling the beach or checking out tidepools, do not turn your back on the ocean, and be aware that occasional out-sized waves may wash over what look like safe spots on rocks and bluffs.

Natural conditions along the California coast are always changing, and the width of beaches and shape of bluffs can be altered by the seasonal movement of sand or by erosion. Coastal access and recreation facilities can be damaged by these forces, and from time to time, trails, stairways, parking areas, and other facilities may be closed for repairs. When planning any trip to the coast, but especially right after a storm, it is advisable to check ahead to make sure that the coastal area you choose to visit is currently accessible and usable. Also be aware that some fa-

For general information on state parks, including a list of camping and day-use fees and campgrounds available without a reservation, see www.parks.ca.gov.

For state park camping reservations, call: 1-800-445-7275 (available 24 hours), or see www.reserveamerica.com.

For other camping opportunities, see individual entries that follow.

For information on Hostelling International's facilities, see www.hiayh.org.

cilities, such as park visitor centers, may be run by volunteers and are open only limited hours; call ahead to check open times. Facilities such as running water are limited or not available at some parks and shoreline accessways; it is a good idea on a coastal trip to bring water, food, waterless hand cleaner, and perhaps one more layer of clothing than you expect to need. Much of the area covered by this guidebook is rural in nature, and only very limited public transit options are listed here.

Dogs enjoy coastal outings just as their human companions do, but their naturally inquisitive nature can create hazards for coastal wildlife. In state parks, dogs must be kept on leashes no longer than six feet and in a tent or enclosed vehicle at night. Except for guide dogs, pets are not allowed in state park buildings, on trails, or on most beaches. Although allowed in some county beach parks, dogs may be subject to leash requirements. See individual site descriptions for more information. Please observe signs regarding dogs on trails and beaches and in parks.

Snowy plovers are a threatened species of sparrow-sized shorebirds whose habitat includes beaches and dunes that are also attractive to humans. During the breeding season in spring to late summer, snowy plovers lay well-camouflaged eggs in sand nests, where they are subject to disturbance by beach visitors or dogs, or to predation by ravens attracted by careless trash disposal by humans. The snowy plover is a key link in the interconnected web of life along the coast. Please observe signs and temporary beach closures intended to protect the habitat of this species.

In the words of the California Coastal Act, the purpose of this guide is to contribute to a better understanding by the public of the importance of coastal resources, both to the quality of life for people and to the maintenance of a healthy and productive natural environment. This book is offered with the knowledge that a wide appreciation among Californians for the coast plays an important role in the protection and restoration of coastal resources.

Stinson Beach, Marin County

Using This Guide

Each group of sites is accompanied by a map and a chart that summarizes key facilities and characteristics. The "Facilities for Disabled" chart category includes wheelchair-accessible restrooms, trails, campsites, or visitor centers; text descriptions note where restrooms are not wheelchair-accessible. The "Fee" chart category refers to a charge for either entry, parking, or overnight use at a facility. Most parks and recreational outfitters maintain websites, but URL addresses may change, use any popular search engine to locate websites for more information on many of the parks and facilities listed in this guide.

Brief introductions to some of the coast's major environments, such as beaches, rocky shoreline, and the redwood forest, are included in this volume, along with highlights of plants, animals, and birds that you may see there. For more information on the resources of the California coast, consult sources listed in the Bibliography (p. 309), especially under Suggestions for Further Reading (p. 310).

Sandy Beach	Rocky Shore	Trail	Visitor Center	Campground	Wildlife Viewing	Historic Building	Fishing or Boating	Facilities for Disabled	Restrooms	Parking	Fee

Building a sandcastle

Caring for the Coast

WITH ITS ASTONISHING beauty and bounty of recreational opportunities, California's coastline enriches our lives. We can express our appreciation for this largesse and contribute to the continuing good health of the coast by being good coastal stewards. A coastal steward understands how his or her everyday choices can impact the coast, and strives to act in ways that will have beneficial results. Here are some examples of coastal stewardship practices. For more ideas and to take the Coastal Stewardship Pledge, call 1-800-COAST-4U or visit www.coastforyou.org.

Stash Your Trash

Litter isn't just ugly, it can kill. Each year, thousands of marine animals die from encounters with human debris—by becoming entangled in waste such as fishing line and plastic strapping, or by ingesting plastic bags and other debris.

Where does beach debris come from? It may have been left by a beachgoer, but most likely it traveled from miles away, having floated there on a tide, washed up from a stream or storm drain outfall, or blown there on a strong wind. Regardless of where it came from, each piece of beach trash was, at some point, handled by a human who didn't dispose of it properly. Sometimes it's an accident—a paper cup placed in an overflowing garbage can falls on the ground, gets washed down the gutter, through the storm drain and out onto the beach. Sometimes it's intentional, as when someone stubs a cigarette butt out on a city street.

Be a coastal steward by not littering, and by practicing the three "Rs"—**reduce** the waste you generate, purchase **reusable** items, and **recycle** trash when possible. Volunteer for a beach cleanup activity such as Coastal Cleanup Day each September or the year-round Adopt-A-Beach Program.

Volunteers participating in a beach cleanup

Walking on bare rock among tidepools

Tidepool Etiquette

North Coast tidepools offer the opportunity to see fascinating marine creatures at close range. Tidepool plants and animals, however, are very sensitive to human contact. When visiting tidepools:

- Watch where you step. Step only on bare rock or sand.
- Don't touch any living organisms. A coating of slime protects most tidepool animals. Touching them with dry hands can damage them.
- Don't prod or poke tidepool animals with a stick. Don't attempt to pry animals off of rocks.
- Leave everything as you found it. Collecting tidepool organisms is illegal in most locations and will kill them. Cutting eelgrass, surfgrass, and sea palm is prohibited.

Watching Wildlife

Observing wild animals in their natural environment is a rare treat that can be experienced on California's North Coast. To ensure that the encounter results in no harm to either the animal or the human observer, we recommend following the guidelines noted below (adapted from the Alliance for Marine Parks and Aquariums and the Farallones Marine Sanctuary Association). For additional information, visit www.coastforyou.org and click on How Can I Help.

- **Keep your distance**. Maintain enough distance so that the animal is not aware of you. One hundred yards is a good rule of thumb, although some animals require 200 yards or more. If your presence causes an animal to change its behavior—even if it just looks at you—you should move away immediately. Use binoculars or zoom lenses for a close-up look.

- **Watch quietly and limit the time spent observing animals.** Encounters with people can be stressful to animals. Half an hour is reasonable.
- **Stay clear of mothers with young.** Nests, dens, and rookeries are especially vulnerable to human disturbance.
- **Resist the temptation to "save" animals.** If an animal appears sick, get help from a professional. In Del Norte and Humboldt Counties, contact the North Coast Marine Mammal Rehabilitation Center: 707-465-MAML. From Mendocino to San Luis Obispo County, contact the Marine Mammal Center: 415-289-SEAL.
- **Never surround an animal.** Avoid approaching wildlife directly and always leave an escape route.
- **Keep pets on a leash or leave them at home**.
- **Never feed wild animals.**
- **Report illegal poaching (or polluting) to the authorities: call 1-888-DFG-CALTIP.** The program is confidential and you may be eligible for a reward.

Sensible Seafood Choices

Increasing consumer demand for seafood has led to overfishing. Some fishing practices destroy habitat and harm non-target fish and animals. Use your purchasing power to support healthy oceans by selecting seafoods that are harvested in a sustainable and environmentally responsible manner. This website offers a pocket guide to sensible seafood choices: www.montereybayaquarium.org/cr/seafoodwatch.asp

Non-point Source Pollution

Another way that people can affect the health of the coast is through non-point source pollution. Non-point source pollutants are carried to the ocean by stormwater runoff. Individuals can avoid contributing to this problem by taking simple actions—for example, picking up after the dog, using least-toxic gardening products, and maintaining the car so there are no oil leaks.

Whale Tail License Plate

California drivers can help the coast by purchasing a Whale Tail License Plate. The plate funds coastal access trails, beach cleanups, and marine education throughout California, including grants to local groups. For information, call 1-800-COAST-4U, or visit www.ecoplates.com.

Map Legend

TRANSPORTATION

- Major Road
- California State Highway
- United States Highway
- Interstate

SHORELINE AND HYDROGRAPHY

- Shoreline
- Rivers and Streams
- Pacific Ocean, Lakes, and Ponds

TRAILS AND BIKE WAYS

- Hiking Trail
- Hiking Trail Along State Highway
- Hiking Trail Along Major Road
- Pacific Coast Bicentennial Bike Route
- Bike Route Along State Highway
- Bike Route Along Major Road

TOPOGRAPHY AND BATHYMETRY

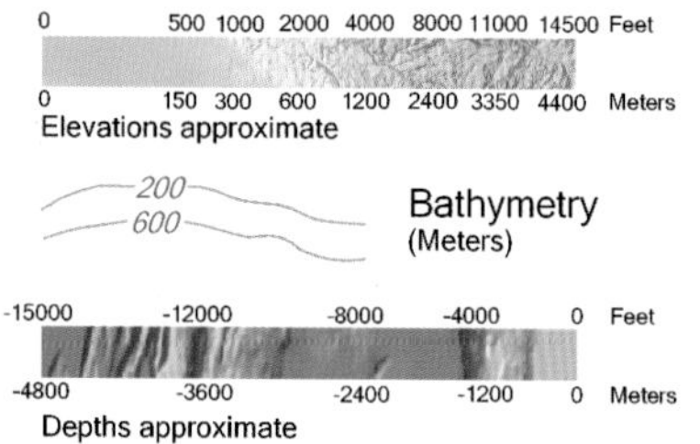

BOUNDARIES

- County
- State
- International

Public Land

NORTH ARROW AND BAR SCALE

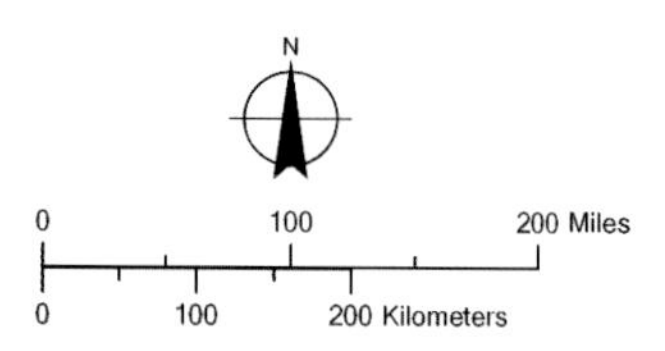

Data and Information Sources:

California Coastal Commission
California Department of Fish and Game
California Department of Parks and Recreation
California Spatial Information Library
90 meter Hillshade and USGS 90 meter DEMs provided by Greeninfo Network

Page opposite: Smith River, Jedediah Smith Redwoods State Park, Del Norte County

Del Norte County

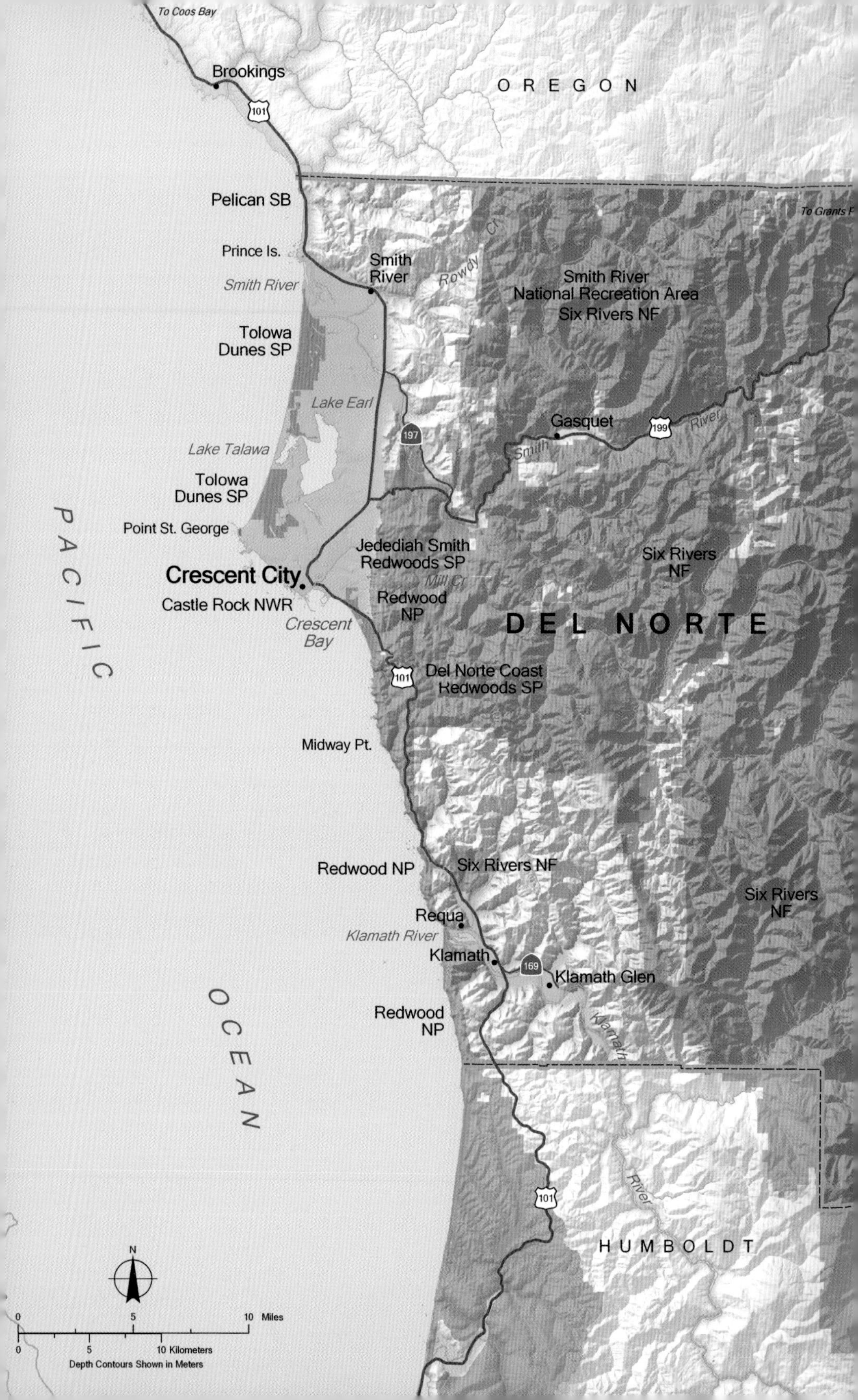

To Coos Bay
Brookings
101
OREGON
Pelican SB
To Grants P
Prince Is.
Smith River
Smith River
Rowdy Cr
Smith River National Recreation Area
Six Rivers NF
Tolowa Dunes SP
Lake Earl
197
Gasquet
199
Smith
River
Lake Talawa
Tolowa Dunes SP
PACIFIC
Point St. George
Jedediah Smith Redwoods SP
Six Rivers NF
Crescent City
Mill Cr
Castle Rock NWR
Redwood NP
Crescent Bay
DEL NORTE
101
Del Norte Coast Redwoods SP
Midway Pt.
Redwood NP
Six Rivers NF
Six Rivers NF
Requa
Klamath River
Klamath
169
Klamath Glen
Redwood NP
Klamath
OCEAN
River
101
HUMBOLDT
N
0
5
10 Miles
0
5
10 Kilometers
Depth Contours Shown in Meters

Del Norte County

DEL NORTE has by far the smallest population (approximately 28,000) of any coastal county in California, but with 42 miles of coastline, sandy beaches, dunes, lakes, forests, and pristine rivers, it more than holds its own in drawing coastal visitors. Strollers and anglers enjoy long sandy beaches near the Oregon border. Scenic Lakes Earl and Talawa hold a wealth of wildlife resources, and windswept Point St. George and nearby Castle Rock are significant seabird nesting areas. Crescent City is the headquarters of Redwood National Park and is a good starting point for visits throughout the county. South of town, rocky coast and redwood forest extend to the mouth of the Klamath River, where boating and fishing are popular.

On the Del Norte coast, cool temperatures prevail (average highs range between 55 and 63 degrees year-round) and fogs are frequent, but in the summertime sunny inland parks are 10 or 20 degrees warmer, drawing picnickers, swimmers, and kayakers to the Smith River. Redwood National and State Parks, including Del Norte Coast Redwoods State Park, Jedediah Smith Redwoods State Park, and other state and county parks provide ample facilities for sightseeing, hiking, and camping.

Del Norte County's Aleutian Goose Festival, held annually in late March, centers on the tens of thousands of geese, the world's entire population of this subspecies of Canada goose, which visit the county's coast every spring. Over 400 species of birds have been recorded in Del Norte County, exceeding the count in many entire states. The festival includes dozens of workshops and excursions to local parks, offering a look at local geology, history, wildflower identification, Native American culture, and more. For information, call: 707-465-0888.

The headquarters of Redwood National and State Parks, which extend south into Humboldt County, is at 1111 2nd St., Crescent City 95531; call: 707-464-6101.

Information on lodging and recreation throughout the county can be obtained from the Crescent City/Del Norte County Chamber of Commerce at 1-800-343-8300. For information on lodging and activities in the Klamath area, contact the Klamath Chamber of Commerce at 1-800-200-2335 or 707-482-7165. Public transit information is available from the Del Norte Senior Center, 810 H Street, Crescent City 95331, or call: 707-464-3069. For DIAL-A-RIDE, a service that carries passengers for an 8–10-mile-area around Crescent City, call: 707-464-9314 or 707-464-4314.

Fishing trips for salmon or steelhead on the Smith and Klamath rivers:

Cast Guide Service, 707-487-2278.

Greg Nicol Guide Service, 541-469-7321.

Phil's Smiling Salmon Guide Service, 707-487-0260.

Dan Carter's Guide Service, 707-458-3527.

Early Fishing Guide Service, 541-469-0525.

Klamath Wild River Fishing, 707-498-4491.

Lunker Fish Trips (also offering raft and bicycle rentals), 1-800-248-4704.

Ocean fishing, birding, and whale-watching trips:

Tally Ho-II Ocean Charters, Crescent City, 707-464-1236.

Diving equipment and kayaks:

Pacific Quest Dive Center, Crescent City, 707-464-8753.

Surfboards, skateboards, and other recreational equipment:

Noll Surf & Skate, Crescent City, 707-465-4400.

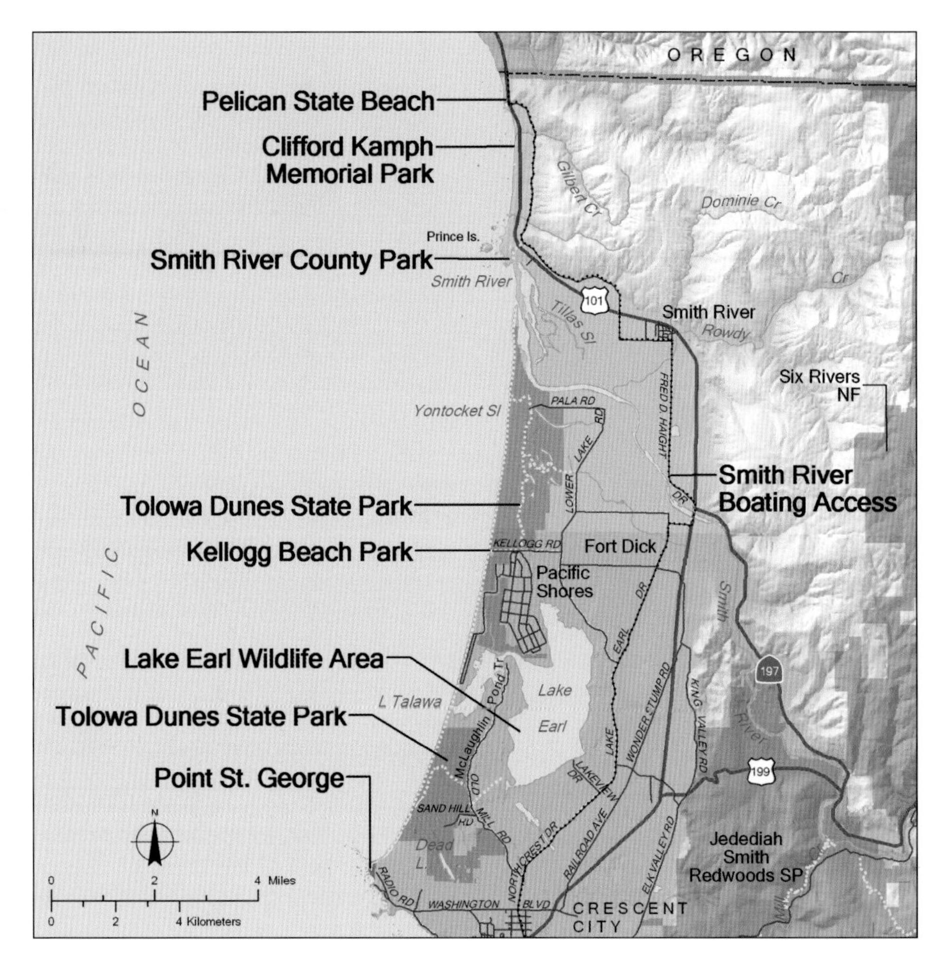

Clifford Kamph Memorial Park

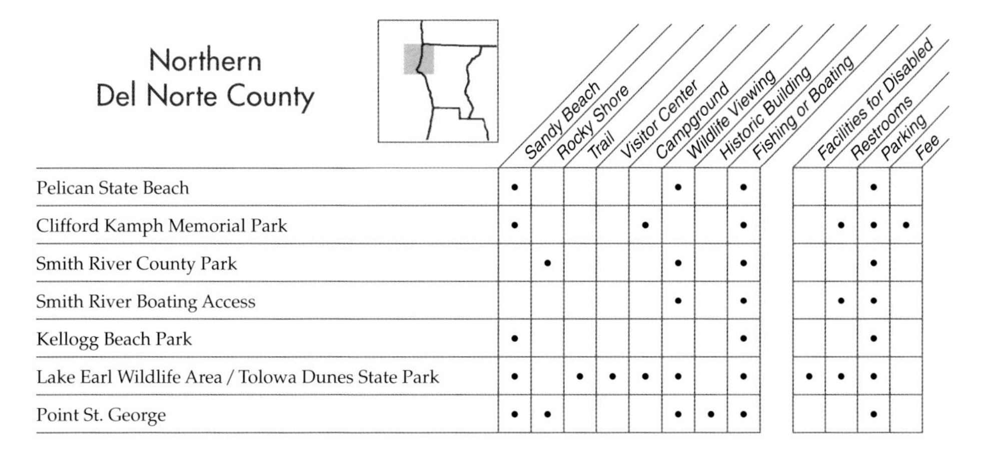

Northern Del Norte County

	Sandy Beach	Rocky Shore	Trail	Visitor Center	Campground	Wildlife Viewing	Historic Building	Fishing or Boating	Facilities for Disabled	Restrooms	Parking	Fee
Pelican State Beach	•					•		•			•	
Clifford Kamph Memorial Park	•				•			•		•	•	•
Smith River County Park		•				•		•			•	
Smith River Boating Access						•		•		•	•	
Kellogg Beach Park	•							•			•	
Lake Earl Wildlife Area / Tolowa Dunes State Park	•		•	•	•	•		•	•	•	•	
Point St. George	•	•				•	•	•			•	

PELICAN STATE BEACH: *W. of Hwy. 101, .5 mi. S. of the Oregon border.* California's northernmost state beach is good for beachcombing and surf fishing. A 250-yard-long stretch of sand has abundant driftwood and is backed by grass-covered dunes. From Hwy. 101, turn west on White Rock Loop; a short path leads down a low bluff. No facilities. Nearshore fish include cabezon, lingcod, and several species of rockfish, while surfperch, Pacific sanddab, and starry flounder are found in the surf zone.

CLIFFORD KAMPH MEMORIAL PARK: *W. of Hwy. 101, 2 mi. S. of the Oregon border.* Located on a low bluff, this small park has a picnic area overlooking the sea and a short, gentle path over a creek to a wide, sandy beach with ample driftwood. There are nine tent-only campsites, one of which is on the beach, with picnic tables, barbecue grills, and fire pits. Running water available; firewood for sale by camp host. No fee for day use; call: 707-464-7230.

Directly offshore is Cone Rock, which provides nesting habitat for pelagic cormorants and pigeon guillemots. Visible to the south is Prince Island, which, along with a handful of other islets off Del Norte County, provides nearly half of the seabird nesting sites on coastal rocks in California.

Prince Island is held in trust for the Indians of the Smith River Rancheria, located adjacent to the Smith River. The island provides nesting sites for many species of birds, including black oystercatchers, Leach's and ashy storm-petrels, and tufted puffins, and it is one of only three California breeding sites for the rhinoceros auklet.

SMITH RIVER COUNTY PARK: *End, Mouth of Smith River Rd., Smith River.* A sand and pebble beach at the river mouth is used for fishing and wildlife viewing. There is parking at the street end, and 20 steps lead to a rocky beach at the water's edge. A bend in the river provides an expansive view upstream and out to sea. To the west is a rocky headland called Pyramid Point. No facilities.

Waterfowl, especially diving ducks, are the most abundant species in the delta in winter. In the nesting season, green-backed herons, American bitterns, mallards, and cinnamon teals may be present. In summer, there is a great blue heron rookery in the delta, and brown pelicans can be seen flying low over the river.

The clear waters of the Smith River support many sport fish species, including coho and king salmon and steelhead trout. Also present are Pacific herring, northern anchovy, surfsmelt, and starry flounder. There is a Dungeness crab nursery in the estuary, and gaper, littleneck, Washington, and soft-shelled clams are found here.

SMITH RIVER BOATING ACCESS: *Fred Haight Dr., W. of Hwy. 101, 3 mi. S. of Town of Smith River.* Access to the lower reach of the Smith River for boating and fishing, with paved boat ramp, parking, and restrooms.

KELLOGG BEACH PARK: *W. end of Kellogg Rd., 9 mi. S. of Smith River.* A broad sandy ocean beach, backed by grass-covered dunes, extends north and south of the end of Kellogg Rd. There is a parking lot at the edge of the dunes; no facilities. The sandy beach runs for miles to the north and south, where the beach adjoins Tolowa Dunes State Park.

LAKE EARL WILDLIFE AREA/TOLOWA DUNES STATE PARK: *W. of Lake Earl Dr. and Lower Lake Rd., 2 mi. N. of Crescent City.* A 5,500-acre wildlife area and adjacent 5,000-acre park located on Lakes Earl and Talawa, with few developed facilities. Only a few roads approach the lakes, but there are trails, a boat launch, and a small camping area within the wildlife area/state park complex.

Start a visit at the Wildlife Center at Department of Fish and Game headquarters at 2591 Old Mill Rd., off Northcrest Dr. The visitor center has trail maps and information on summer programs conducted by the North Coast Redwood Interpretive Association, including Sunday afternoon biologist-led walks through the dunes and other habitats. Call: 707-464-6101 ext. 5300, or 707-465-6191. There are restrooms, a three-quarter-mile self-guided nature walk, and access to other hiking and bicycle trails. Dogs must be leashed. For information on the Lake Earl Wildlife Area, call: 707-464-2523.

From the Wildlife Center, follow the graveled extension of Old Mill Rd. on foot or bicycle past a vehicle gate to McLaughlin Pond, about two miles distant. Beavers and river otters may be spotted here, along with ducklings paddling on the pond in season. The Cadre Point Trail turns east off the McLaughlin Pond trail and leads one mile through woodlands and swamps; watch for coastal blacktail deer, brush rabbits, woodpeckers, hawks, and songbirds. At the end of the trail, a picnic table is situated on a point overlooking Lake Earl, where you may see great blue herons, grebes, swans, and raptors including bald eagles and peregrine falcons. On the east side of Lake Earl, the 1.5-mile-long Bush Creek Trail starts at a parking area on Lake Earl Dr., near Elk Valley Cross Rd. On the north side of Lake Earl, there is lake access at Teal Point, reached by a public road in the Pacific Shores subdivision south of Kellogg Rd.

Lakes Earl and Talawa are estuarine freshwater lagoons connected to each other and periodically open to the Pacific Ocean. Cutthroat trout and starry flounder are found in the lakes. A boat launch on the shore of

Kellogg Beach Park

Lake Earl

Lake Earl can be accessed from Lake Earl Dr., which is the northern extension of Northcrest Dr.; turn west onto Lakeview Dr. The lakes are very shallow and can accommodate only small, shallow-draft boats. Motorboats are prohibited during waterfowl hunting season.

Trails to the ocean beach include one leading a mile west through the dunes from Sand Hill Rd., which turns west off Old Mill Rd. south of the Wildlife Center. From the beach, sea lions, harbor seals, and migrating gray whales may be seen. Ocean beach access is also available north of Lakes Earl and Talawa. In Tolowa Dunes State Park, west of Yontocket Slough off Pala Rd., there is a picnic area with parking, restrooms, and trails leading west to the beach. Also in the state park are six walk-in environmental campsites one-half mile from the beach off Kellogg Rd. Fire rings, stoves, and chemical toilets; no water available, no reservations. Fee charged; pay at the Redwood National and State Parks visitor center at Second and K Streets in Crescent City or at Jedediah Smith State Park. Call: 707-464-6101.

POINT ST. GEORGE: *End of Radio Rd.* Radio Rd., the western extension of Washington Blvd., dead-ends at a parking lot at Point St. George; no facilities. A gradually sloping gravel trail leads north one-fifth mile to the north side of the point. Low tide exposes a rocky reef, which extends under the sea several miles offshore and includes the "Dragon's Teeth," rocks sometimes exposed by low water. The St. George Reef lighthouse is located on one of these rocks; on a clear day, the lighthouse can be seen some four miles away. To the northeast of Point St. George, miles of curving sandy beach along Pelican Bay invite beachcombing; the coast of Oregon is visible across the bay. Shorebirds feed along the beach, and pelicans and cormorants can be seen flying overhead. A former Coast Guard Station, now a private residence, is located near the parking area; do not trespass.

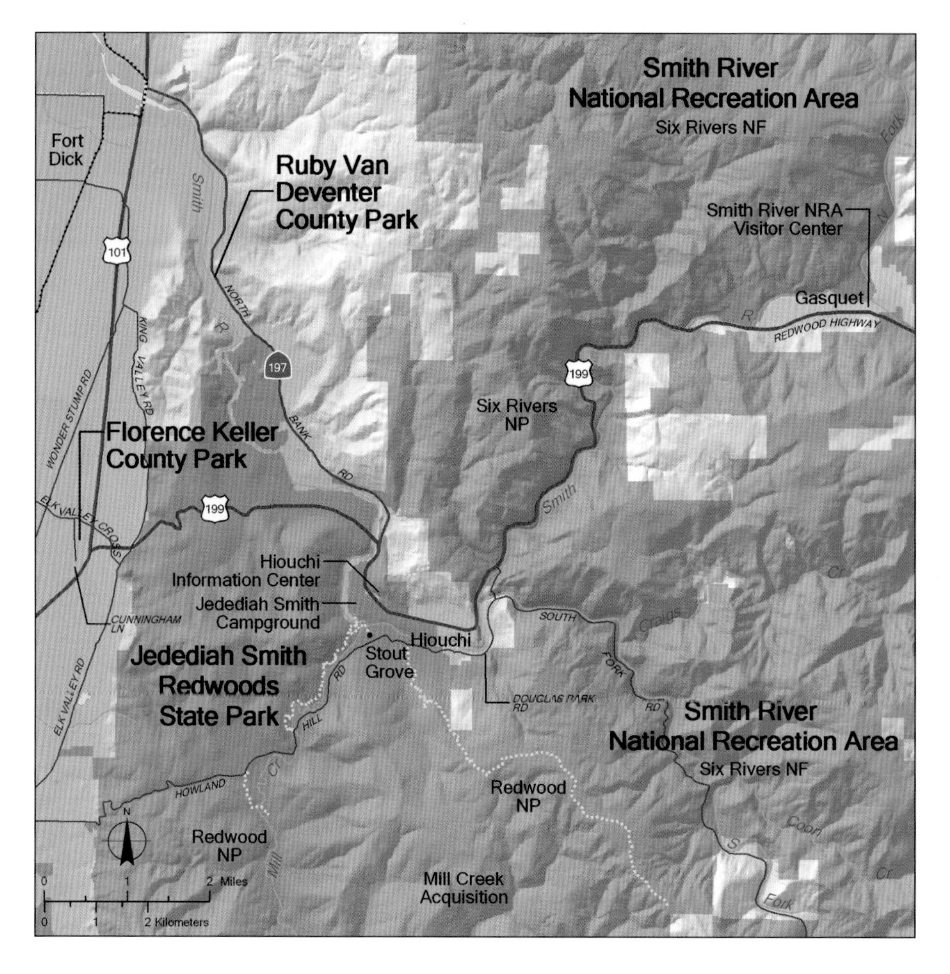

The Smith River is known for its clarity. Even after a heavy rain, the water may clear within hours, regaining its beautiful jade-green color. This is due in part to the tendency of rocks underlying the river, including the mineral serpentine, to break down into flakes rather than fine silt. Fishing on this undammed river is good, too; steelhead of up to 20 pounds may be caught, and the state's record steelhead, weighing 27 pounds 4 ounces was caught in the Smith River. King salmon are also fished during the winter months (the state's second largest, weighing 86 pounds, was caught in the Smith River). During the summer, fishing for cut-throat trout is popular.

Smith River Area

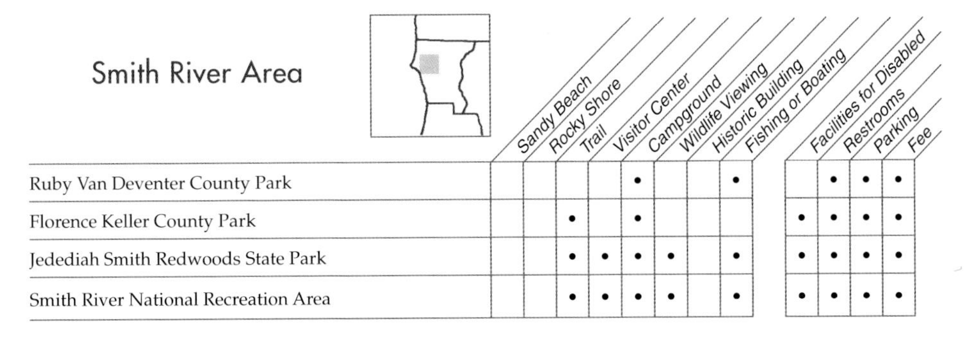

	Sandy Beach	Rocky Shore	Trail	Visitor Center	Campground	Wildlife Viewing	Historic Building	Fishing or Boating	Facilities for Disabled	Restrooms	Parking	Fee
Ruby Van Deventer County Park					•			•		•	•	•
Florence Keller County Park			•		•				•	•	•	•
Jedediah Smith Redwoods State Park			•	•	•	•		•	•	•	•	•
Smith River National Recreation Area			•	•	•	•		•	•	•	•	•

RUBY VAN DEVENTER COUNTY PARK: *North Bank Rd. (Hwy. 197), 2.5 mi. E. of Hwy. 101.* A small campground is nestled in a redwood forest next to the Smith River. The river offers swimming, kayaking, and seasonal trout and salmon fishing. There are 18 campsites with picnic tables, barbecue grills, and firepits, and a group picnic area. Restrooms and running water; fee for camping. Pets welcome if leashed. Call: 707-464-7230.

FLORENCE KELLER COUNTY PARK: *Cunningham Ln., off Elk Valley Cross Rd., W. of Hwy. 101, 3 mi. N. of Crescent City.* A campground accommodating tent campers and RVs up to 32 feet is located in a redwood forest near Hwy. 101. There are 50 campsites with picnic tables and firepits; running water is available. Group picnic sites also available; dogs permitted on leash. There are several easy walking trails, one of which is wheelchair accessible. Call: 707-464-7230.

JEDEDIAH SMITH REDWOODS STATE PARK: *Hwy. 199, 5 mi. E. of Hwy. 101.* Jedediah Smith Redwoods State Park is the most northerly of California's magnificent coastal redwood parks. The park is named for the explorer who was the first non-native to travel overland from the Mississippi River across the Sierra Nevada range to the Pacific Coast. Some of the tallest trees on earth grow here, and the park's 10,000 acres also encompass six miles of the pristine Smith River. There are opportunities for fishing, boating, camping, exploring, and, in summer, enjoying warm daytime temperatures, compared to the nearby coast.

The park visitor center is located off Hwy. 199, and nearby are picnic areas, a pebble river beach, and a nature trail through the redwood forest. The campground has 106 sites, each with picnic table, fire ring, and cupboard; restrooms and hot showers are available. Campfire programs are offered at the campground. Some campsites can accommodate trailers or RVs up to 35 feet; there are no hookups. To reserve campsites, call: 1-800-444-7275. Primitive campsites for hike-in, bike-in, or equestrian use are available outside park boundaries along the Little Bald Hills Trail, which leads from the Stout Grove.

Smith River

In accordance with the Tolowan account of genesis, the Tolowa people and their ancestors have always lived in much of present-day Del Norte and Curry Counties, from just north of the Klamath River into southwest Oregon. Tolowa villages with structures made of split redwood and cedar planks were located along the coast, on the shores of the lagoons now known as Lakes Earl and Talawa, and inland along the Smith, Chetco, Pistol, Rogue, Sixes, and Applegate Rivers and their tributaries. The people lived on the bounty of the land and sea, which varied with the season; their foods included salmon, elk, deer, shellfish, berries, acorns, and others. The Aleutian Canada goose is a part of the Tolowan creation account, in which the sounds of the returning geese each year are a signal of the coming spring. During the March staging period, the geese feed in pastures in Del Norte County to prepare for their nonstop journey over thousands of miles of ocean to their summer breeding sites on islands off Alaska, Russia, and Japan.

As settlers arrived in Del Norte County in the early 1850s in search of gold or farmland, they displaced the Tolowa from their ancestral homes. In 1853 the first in a series of massacres took place at the Tolowan axis mundi ("center of the earth") of Yontocket, where hundreds of Tolowa people were murdered as they gathered for their annual world renewal ceremony, known as Nee-dash. Mass killing of the Indians and the effects of disease drove the population of the Tolowa people to a low of barely one hundred persons, while many Tolowa were banished to distant reservations.

In recent years, the Tolowa people have come back from near-extinction. The Smith River Rancheria includes nearly one thousand members, and the tribal government today operates a health clinic, a Headstart program for children, a daycare program, and a housing and meals program for the elderly and disabled; there are also plans for development of a community center. With their increasing numbers, the Tolowa people have been able to resume the sacred Nee-dash ceremony, including prayers for the return of the Aleutian Canada goose, which in 1967 was listed as an endangered species.

Like the Tolowa people, the Aleutian Canada goose has also made a remarkable comeback. In the 19th century, Russian fur trappers introduced nonnative foxes to the breeding islands of the Aleutian goose in Alaska. The foxes fed on the geese, causing their numbers to dwindle to a few hundred. However, efforts to restore their island breeding habitat have been successful, and the Aleutian Canada goose has rebounded such that in 1990, the goose was removed from the Endangered Species List, with numbers now in the tens of thousands.

Kayaking in Jedediah Smith Redwoods State Park

The Stout Grove, with redwood trees up to 340 feet in height, is located one-half mile by trail from a parking area on scenic gravel Howland Hill Rd., which winds along Mill Creek. Wildlife in Jedediah Smith Redwoods State Park includes river otters, beavers, and sometimes bears, along with deer, squirrels, and raccoons. The marbled murrelet is a small bird, no more than 10 inches long, that frequents the forest but remains virtually unseen. The murrelet ordinarily never touches land; during the nesting season it spends its nights on a nest near the top of an old-growth redwood tree and its days diving for fish in the open ocean. Other birds more likely to be seen by visitors include several species of woodpeckers, belted kingfishers, and owls.

Some park facilities are wheelchair accessible, including picnic areas, some campsites, the visitor center, and the campfire center. Dogs are not allowed on trails. For park information, call: 707-464-6101.

SMITH RIVER NATIONAL RECREATION AREA: *Hwy. 199, 14 mi. E. of Hwy. 101.* Adjoining Jedediah Smith Redwoods State Park on the east is the 300,000-acre Smith River National Recreation Area, where visitors enjoy river rafting, fishing, and hiking. There are over 65 miles of trails, such as the Myrtle Creek Trail, which leaves Hwy. 199 one mile east of the Jedediah Smith park boundary and passes through serpentine soils and an unusual native plant assemblage including the insectivorous *Darlingtonia*, or California pitcher plant. The recreation area includes four campgrounds: Big Flat, Grassy Flat, Panther Flat, and Patrick Creek; for reservations, call: 1-800-280-2267. For the interpretive center and headquarters, call: 707-457-3131.

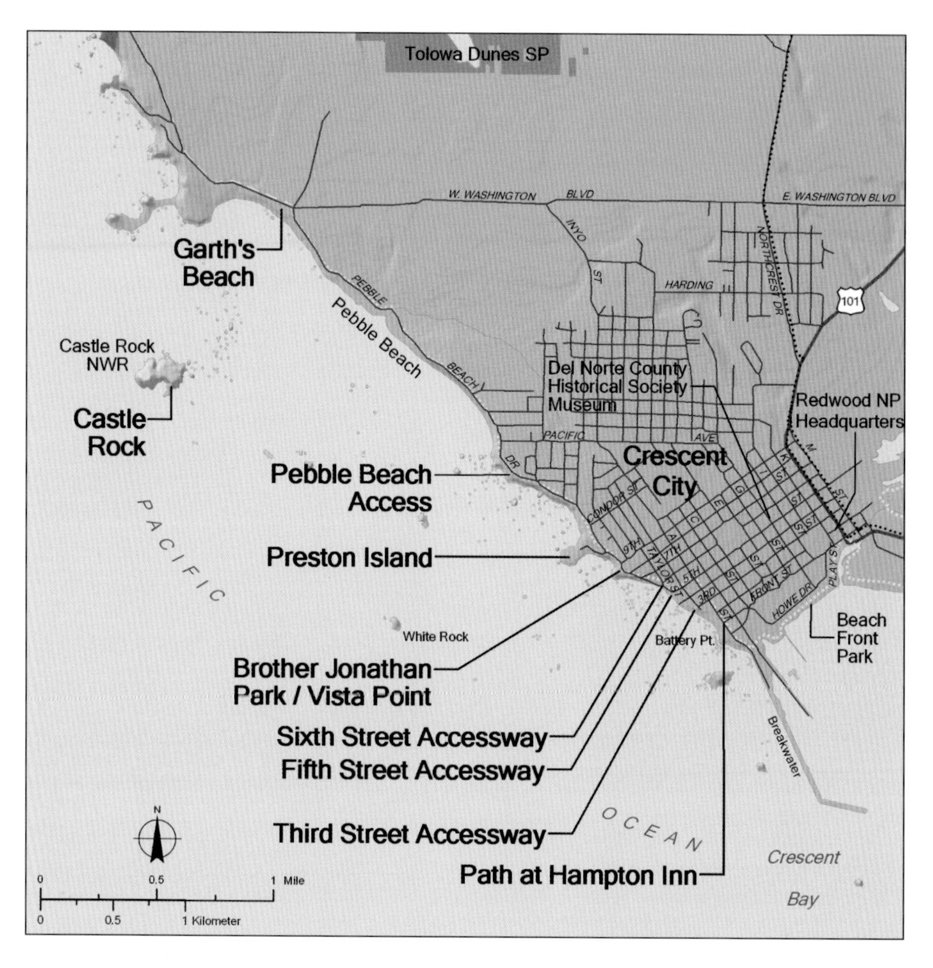

Path at Hampton Inn

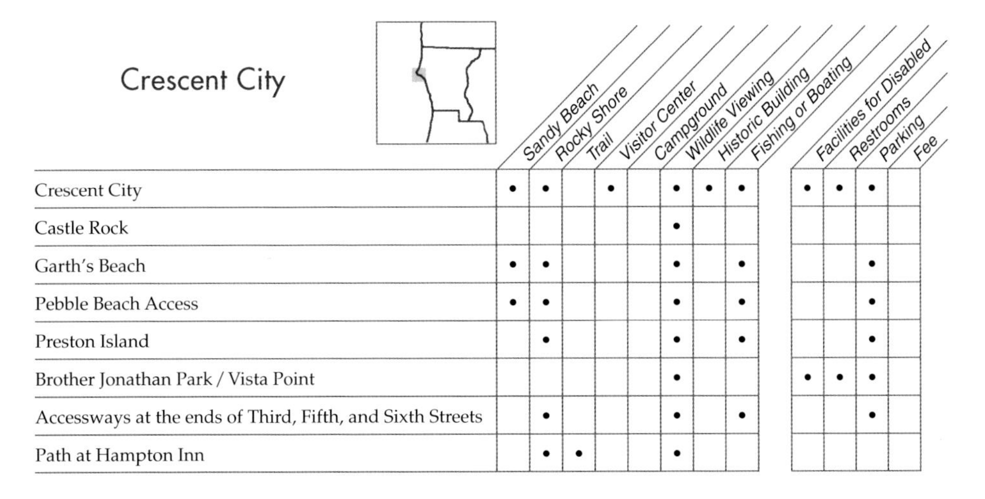

Crescent City

	Sandy Beach	Rocky Shore	Trail	Visitor Center	Campground	Wildlife Viewing	Historic Building	Fishing or Boating	Facilities for Disabled	Restrooms	Parking	Fee
Crescent City	•	•		•		•	•	•	•	•	•	
Castle Rock						•						
Garth's Beach	•	•				•		•			•	
Pebble Beach Access	•	•				•		•			•	
Preston Island		•				•		•			•	
Brother Jonathan Park / Vista Point						•			•	•	•	
Accessways at the ends of Third, Fifth, and Sixth Streets		•				•		•			•	
Path at Hampton Inn		•	•			•						

CRESCENT CITY: *Hwy. 101, 79 mi. N. of Eureka.* Many North Coast towns enjoy close proximity to the sea and its wildlife, but few can rival the wealth of resources at the doorstep of Crescent City. California's second largest seabird colony occurs on Castle Rock, less than a mile across the water from city limits. A short stroll from downtown offers a visitor easy sightings of an osprey perched on a snag, harbor seals on the beach, or whales migrating offshore. The largest coastal lagoon complex in the western contiguous United States lies a few minutes' drive north of town, and old-growth redwood forests and California's largest undammed river are located nearby.

A gold rush town, Crescent City boomed within the space of a few years in the 1850s after gold was discovered in the Trinity River and the south fork of the Smith River. Lumbering later became Crescent City's dominant enterprise, and the construction of a deepwater harbor supported the fishing industry. Crescent City derived its name from the crescent-shaped bay around which the community was built. This is one of many crescent bays in California formed by waves coming from the northwest that refract around a resistant rock outcrop, such as Battery Point, and erode the less resistant land to the southeast.

Few old commercial buildings remain in the town, despite its 19th century origin, due to a tsunami that took the lives of 12 persons and destroyed much of the waterfront area. In the early morning hours of March 28, 1964, a succession of seismic sea waves caused by a huge earthquake centered off Alaska swept inland as far as 5th Street, some 1,600 feet from the harbor's edge. Redwood logs that had washed up on nearby beaches, along with automobiles and other debris, battered buildings and ended up in heaps on the streets after the waters receded. A 25-ton concrete tetrapod, similar to those used in the harbor breakwater, was moved off its display base downtown by water and debris forces. Today the waterfront area of the rebuilt town is dominated by the expansive lawn of Beach Front Park.

The Del Norte County Historical Society Museum at 577 H St. displays photographs from the tsunami, relics of pioneer settlement, a Fresnel lens from the St. George Reef lighthouse, and items recovered from the wreck of the *Brother Jonathan*. There are also articles made by the Tolowa, Yurok, Hupa, and Karuk peoples of the Crescent City area, including an excellent collection of 19th and 20th century baskets and caps finely woven of beargrass, maidenhair fern, and willow. Call: 707-464-3922.

CASTLE ROCK: *Offshore, W. of Pebble Beach Dr., N. of Pacific Avenue.* The largest of the islets near Crescent City offers a relatively close-to-shore viewing opportunity for wildlife, including ocean-dwelling species that are rarely seen from the mainland. A National Wildlife Refuge, steep-sided Castle Rock supports California's largest breeding colony of common murres, reaching over 100,000 birds in the late spring and early summer. Other seabirds are plentiful, including the Leach's storm-petrel, Cassin's auklet, and Brandt's cormorant. Viewing some pelagic birds requires a spotting scope, while other species, such as pigeon guillemots with their distinctive coloring, can be seen from shore with binoculars. Even with the best equipment it is hard to spot some birds on Castle Rock, such as nesting fork-tailed storm-petrels, which remain in burrows during the day and venture out only at night.

GARTH'S BEACH: *Radio Rd., .3 mi. N.W. of Washington Blvd./Pebble Beach Dr. intersection.* A small parking area is located adjacent to the road; no facilities. A gently sloping path with a few steps leads to a curving sandy beach with rocks and tidepools at the north end and low white cliffs at the south. Point St. George and Castle Rock shelter the beach, reducing wave action. In winter and early spring, look for strikingly colored harlequin ducks among the rocks and tidepools. From March to mid-April, the dominant bird is the Aleutian Canada goose. Once on the endangered species list, these birds have grown in such numbers that some 40,000 of them use Castle Rock as their main staging grounds, where they roost overnight. At dawn tens of thousands of birds lift off in a noisy show and head inland across the beach to feed in pastures along the Smith River. Most of Point St. George has been acquired by the County of Del Norte, assisted by the State Coastal Conservancy, for resource protection and public access; future trails and interpretive facilities are envisioned.

PEBBLE BEACH ACCESS: *Pebble Beach Dr. near Pacific Ave., Crescent City.* Pebble Beach Dr. parallels the shore on a bluff overlooking the beach and numerous sea rocks. There are several pull-outs for parking, picnic tables, and three stairways to the beach. The stair south of the Pacific Ave. intersection descends from the parking area to a huge flat boulder, offering an elevated observation platform with prime views of wildlife activity. Bring field glasses to see roosting cormorants and pigeon guillemots on the largest rocks offshore. The sandy beach and rocky shore also draw anglers who surf fish. A State Historical Landmark on the bluff south of Pacific Ave. marks the site of a Tolowa Indian settlement that existed here until the latter part of the 19th century.

PRESTON ISLAND: *Pebble Beach Dr. and Condor St., Crescent City.* Now more of a rocky spit than an island, Preston Island is accessible via a paved road down the bluff face. Preston Island was quarried nearly to sea level for the Crescent City Harbor breakwaters. Now the pebble beach is used for beachcombing, fishing, and tidepool exploring.

Castle Rock

BROTHER JONATHAN PARK/VISTA POINT: *Pebble Beach Dr. and 9th St., Crescent City.* This grassy park on the inland side of Pebble Beach Dr. includes a playground with restrooms and baseball and basketball facilities, as well as a cemetery and a memorial to those lost at sea in the 1865 shipwreck of the steamer *Brother Jonathan*. On the seaward side of the street is a vista point with picnic tables and bench. This is a good spot for a sunset view of Crescent Lighthouse at Battery Point, flanked by offshore rocks and backed by forested mountains.

ACCESSWAYS AT THE ENDS OF THIRD, FIFTH, AND SIXTH STREETS: *W. ends of 3rd, 5th, and 6th Sts., Crescent City.* A rocky beach can be reached by paths or stairways. There is parking at Third and Fifth Streets. No facilities.

PATH AT HAMPTON INN: *Front and "A" Sts., Crescent City.* The grounds of this hotel, built in 2003, include a short public path leading around the seaward side of the building to a narrow driftwood-strewn beach. Fine views of offshore rocks.

Brother Jonathan, painting by James Bard, 1958, for the Book Club of California. San Francisco Maritime National Historical Park.

The wreck of the *Brother Jonathan* is California's worst maritime disaster. The *Brother Jonathan*, driven by steam-powered side wheels as well as sails, carried passengers and freight between San Francisco, Seattle, and other Northwest ports. She was fast; a trip between San Francisco and Portland in less than three days set records for the time. But in July 1865 after a brief stop at Crescent City, the ship, which was drastically overloaded, headed north into the face of a storm, only to strike partially submerged and uncharted rocks some four miles west of Crescent City. The *Brother Jonathan* went down with the loss of 225 lives, the U.S. Army payroll intended for troops at Northwest forts, and an unknown number of gold coins. Nineteen persons reached shore in one of the lifeboats. In 1993 the sunken ship was found, and some of the items were recovered. A selection of these can be viewed at the Del Norte County Historical Society museum in Crescent City. Following the disaster, a lighthouse was constructed at tremendous expense on St. George Reef, but the danger to keepers along with high upkeep costs led to its abandonment in 1975.

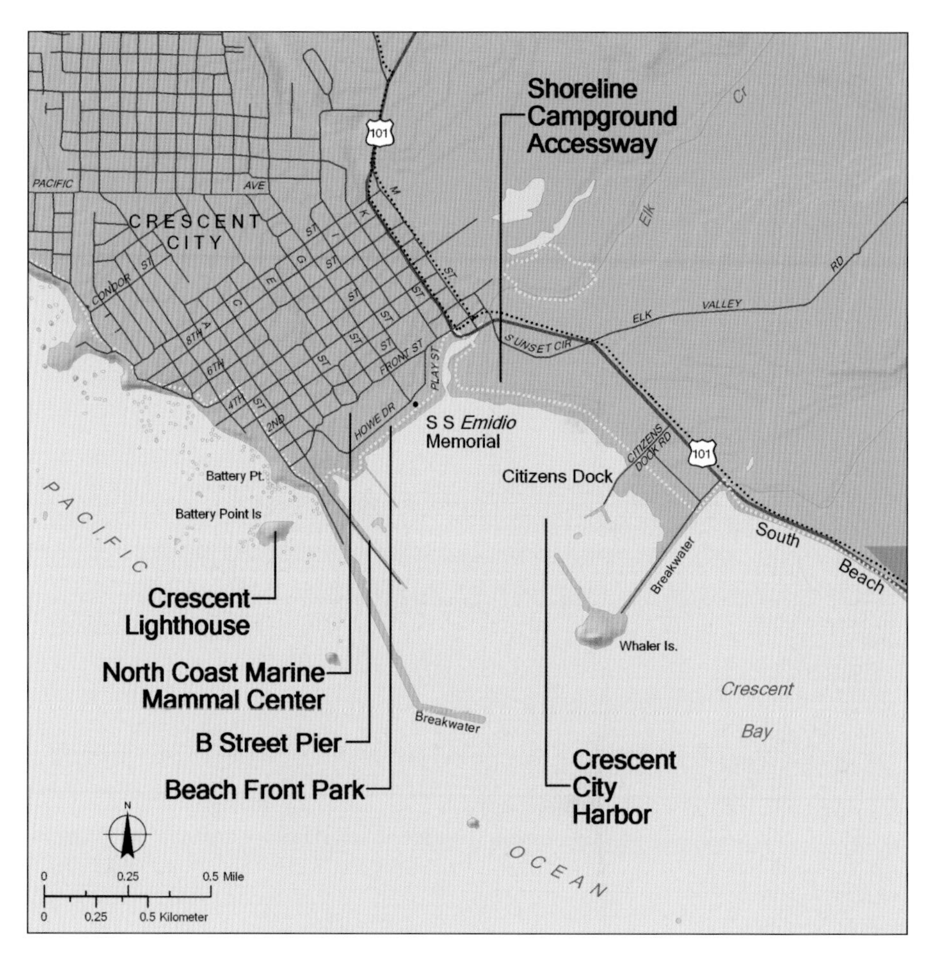

Crescent Lighthouse at Battery Point

Crescent City Harbor

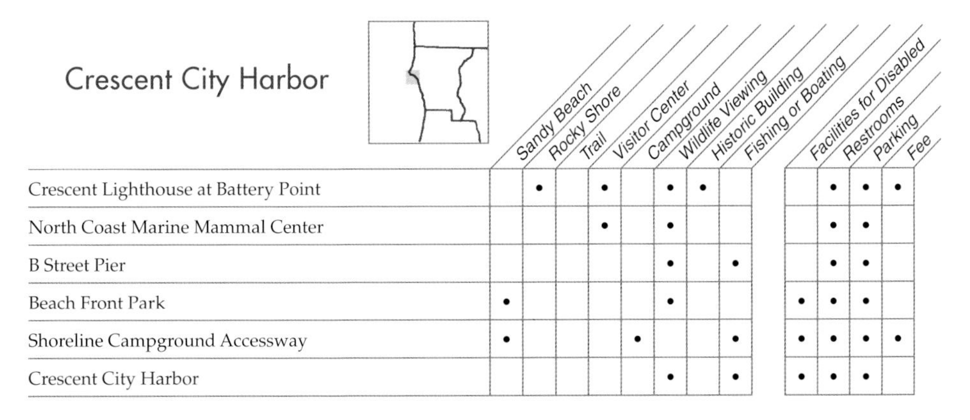

	Sandy Beach	Rocky Shore	Trail	Visitor Center	Campground	Wildlife Viewing	Historic Building	Fishing or Boating	Facilities for Disabled	Restrooms	Parking	Fee
Crescent Lighthouse at Battery Point		•		•		•	•			•	•	•
North Coast Marine Mammal Center				•		•				•	•	
B Street Pier						•		•		•	•	
Beach Front Park	•					•			•	•	•	
Shoreline Campground Accessway	•				•			•	•	•	•	•
Crescent City Harbor						•		•	•	•	•	

CRESCENT LIGHTHOUSE AT BATTERY POINT: *S. end of "A" St., Crescent City.* An offshore light station that is relatively easy to get to, the Crescent Lighthouse at Battery Point is located on a small island about 200 yards from shore. Tides permitting, the island can be reached from the parking area by a short walk across the rocky beach and causeway. Tours of the lighthouse, which was built in 1856 and was one of the first lighthouses in California, are available Apr.–Oct., Wed.–Sun.; fee charged. From the rock on which the lighthouse is located there are panoramic views of Crescent City Harbor to the east and the rocky shoreline of Pebble Beach to the northwest; on a clear day, St. George Reef Lighthouse is visible offshore, some seven miles distant.

Battery Point was the site of Ta'atun, one of the major Tolowa Indian villages at the time of white contact. Other Tolowa villages included Meslteltun on Pebble Beach, Tawiatun, north of Point St. George, and Tatintun, south of Point St. George. There are picnic tables, benches, and restrooms on Battery Point near the parking area.

NORTH COAST MARINE MAMMAL CENTER: *Howe Dr., Beach Front Park, Crescent City.* The center provides the northernmost counties in California with a rescue and rehabilitation service for injured, orphaned, or sick marine mammals. When present for rehabilitation, harbor seals or sea lions may be viewed by visitors. There is a gift shop, and education-

Tolowas ca. 1903: Mary Grimes, Clara La Fountain, Lizzie Grimes, and Bessie Stewart al materials are available. For information, call: 707-465-6265.

B STREET PIER: *Foot of B St., Crescent City.* A 900-foot-long public fishing pier is located at the south end of B Street. This is a popular spot for taking perch and Dungeness crabs. Watch the beach on the harbor side for shorebirds, gulls, and terns. Restrooms at the parking area.

BEACH FRONT PARK: *Between Front St. and Howe Dr., Crescent City.* The park comprises ten city blocks of sports fields, picnic areas, and playgrounds with views of fishing boats in Crescent City Harbor. There is a par course and putting green and a shuffleboard court. The Fred Endert indoor municipal swimming pool with water slide is located at 1000 Play St. off Front St., at the east end of the park complex, and is open to the public; fee charged. On the harbor side of Howe Dr. is a half-mile-long linear park along a sandy beach, incorporating the coastal trail and bike path. Parking is available along Howe Dr., and picnic tables are located next to the sand at the west end of the beach. Shorebirds can be viewed feeding in the flats near the beach. At the corner of Front and H Streets is a State Historical Landmark—the remains of the S.S. *Emidio,* which, on December 20, 1941, was the first ship attacked by a Japanese submarine on the California coast, about 200 miles north of San Francisco. The abandoned ship drifted northward and broke up on the rocks off Crescent City.

SHORELINE CAMPGROUND ACCESSWAY: *W. of Sunset Circle Dr., Crescent City.* A public path runs along the levee at Elk Creek, through Shoreline Campground from Sunset Circle to a sandy beach. Picnic tables, tent campsites, and 189 RV/trailer campsites with hookups are available. Restrooms, showers, and laundry. Call: 707-464-2473. To the west of the campground is the mouth of Elk Creek; inland of Hwy. 101 is the Elk Creek Nature Trail, a loop through willow and alder forests, meadows, and canals where visitors may see song sparrows, marsh wrens, and black phoebes; park at the east end of 2nd St. east of Hwy. 101.

St. George Reef Lighthouse

The St. George Reef Lighthouse, one of the loneliest in America, is perched four miles offshore on a rock barely larger than the lighthouse itself. The St. George Reef Lighthouse Preservation Society is restoring the facility, which is no longer in use. The society offers occasional helicopter flights to the public, weather permitting, from October through June. The trips depart Crescent City Airport in a four-place Raven R-44 helicopter and take six minutes to reach the lighthouse. The flights land near the base of the tower, and visitors are taken on a one-hour tour. For information, write: SGRLPS-Tours, P.O. Box 577, Crescent City, CA 95531, or call: 707-464-8299.

Crescent City Harbor

CRESCENT CITY HARBOR: *W. of Hwy. 101 and Citizens Dock Rd., Crescent City.* Both a working and recreational harbor, Crescent City Harbor includes public wharves, restaurants, and marine-related activities. The Citizens Dock is used for public fishing, and there is a boat hoist. Bait and tackle are available, and there are fish cleaning stations, restrooms, showers, and parking. Visitors can watch incoming commercial fishing boats unload their catch, including shrimp and crab, and there is a fresh fish market. There is also a Small Boat Basin with boat ramps, docks, and guest slips.

Other harbor facilities include fish processing plants, a Coast Guard station, a charter boat service, marine supplies, and a fuel dock. In front of the Harbor District office on Citizens Dock Rd., look for the markers on a telephone pole showing the water level during the 1964 tsunami. Now one of the safest harbors on the north coast, Crescent City Harbor is protected by rock jetties reinforced by more than 1,600 huge, 25-ton concrete tetrapods that are shaped like children's jacks. Supplementing the tetrapods are 680 steel-reinforced concrete units, weighing 42 tons each, known as dolosse. The dolosse are shaped like a twisted "H."

A public two-lane boat ramp, sportsboat marina, ice and fuel dock, snack bar, and the Chart Room Restaurant are located in the southern part of Crescent City Harbor. There is also a picnic area and access to the southern jetty via Whaler Island, where there is a panoramic view of harbor and ocean. Surfers can be seen on nearby South Beach. Gray whales can be spotted offshore during their migration from November through early May; birders scan the ocean for loons, grebes, and sea ducks. On the harbor side of the island, look for sea lions and harbor seals and, in winter or spring, harlequin ducks. Harbor District information: 707-464-6174.

Redwood Forest

THE MAGNIFICENT coast redwoods grow in forests along the northern and central California coast, from the Oregon border to south of Monterey. Coast redwood forests were once widespread but are now relicts, restricted to the narrow coastal fog belt. Coastal fog, much of which occurs during the dry season, condenses and drips off the redwoods' leaves, supplying up to ten inches of precipitation annually to supplement winter rains. Within its range, the redwood is extremely successful and nearly always dominates other trees that grow nearby, such as Douglas-fir, Sitka spruce, and tanbark-oak. One reason for the coast redwood's dominance is its prolific regeneration. Although coast redwoods produce large amounts of seed, most regeneration in redwood forests is attributed to the trees' ability to sprout from the trunk or root system. These root and trunk sprouts are also unusually hardy and fast growing because they draw water and nutrients from the already well-developed root system of the parent tree.

Low-intensity ground fires, which were once common in undisturbed natural forests, also favor coast redwoods. The trees have an unusually thick and non-resinous bark that is more resistant to fires than that of competing hardwoods and Douglas-fir. When periodic small fires occur, the effect is to eliminate trees other than redwoods; fire scars at the base of redwoods are evidence of previous fires that left these trees still standing.

In northern California, coast redwood forests intergrade with other forest communities, including Douglas-fir forest and mixed-evergreen forest. In the wettest coastal areas, coast redwood is often accompanied by Sitka spruce, western hemlock, and canoe cedar. In moderately moist areas, farther inland, Douglas-fir takes over. Mixed-evergreen forests occur in warmer areas; typical trees include many species of maple and oak, as well as bay and madrone.

The **coast redwood** *(Sequoia sempervirens)* is the tallest tree in the world, growing to heights of over 300 feet. It is also considered to be one of the longest-lived trees. One tree was confirmed to be at least 2,200 years old. The coast redwood has beautiful and high-quality wood, making it a very valuable timber tree. It is very high in tannins and therefore resistant to termites and wood rot. Within the last 200 years, most of the old-growth redwoods have been logged; only four percent of the original forest remains standing.

Coast redwoods

Sitka spruce (*Picea sitchensis*), the largest of all spruces, is a native evergreen tree that can grow up to 200 feet in height and 3 to 5 feet in diameter. The needles are yellowish-green to bluish-green, stiff, and very sharp. Sitka spruce was widely employed medicinally by several North American Indian tribes, who used it especially for its antiseptic qualities in the treatment of lung complaints, sores, and wounds. Having similar moisture needs as redwoods, this species grows in a narrow band along the coast where fog is abundant in the summer.

Sitka spruce

Tanoak *(Lithocarpus densiflorus),* also called "tanbark-oak," is an evergreen hardwood that has flowers like the chestnut and acorns like the oak. The common name is due to the former widespread use of its high-tannin-content bark for tanning cattle hides. The earliest major industry in Spanish-Mexican California was commerce in hides and tallow. The Native Americans in California's North Coast Range collected acorns from tanoak. The large acorns were ground, leached, and then prepared as a soup, cooked mush, or a kind of bread. The acorns also provide a valuable food source for many kinds of wildlife. This medium-sized tree grows vigorously on the moist slopes of the seaward coastal ranges. It usually occurs with conifers and other hardwoods, but sometimes forms pure, even-aged stands.

Tanoak

Huckleberry *(Vaccinium ovatum)* is a slow-growing, evergreen shrub with copper-colored new growth. It bears tiny greenish-pink flowers dangling in clusters, and by late summer the fruits of these diminutive flowers have ripened into the most delicious blue berries. Huckleberry bushes are common along roadsides and in forest clearings, and can also be found under light to moderate shade at the edges of clearings. The fruits were popular with all coastal Indian tribes, who used them in jams and preserves. Huckleberries are not cultivated commercially, however, so if you are interested in tasting their delectable fruit you will have to find them in the wild.

Huckleberry

Redwood sorrel

Redwood sorrel (*Oxalis oregana*) commonly makes an almost carpet-like layer on the redwood forest floor. The leaves are clover-like with three heart-shaped leaflets attached to a long stem. Redwood sorrel is adapted to growing in the shade; it folds up its leaves within minutes of being struck by rare shafts of direct light to minimize exposure. It produces small pink flowers in the spring and was reportedly used as a medicinal herb by Native Americans and frontiersmen. The leaves contain oxalic acid, which gives them a sharp flavor.

Banana slug

The **banana slug** (*Ariolimax columbianus*) is the second largest slug in the world, growing up to 10 inches long. Its coloring often resembles that of a banana; however, banana slugs can also be green or brown. Banana slugs live on the forest floor, where they feast upon living and decaying vegetation. Slugs are hemaphrodites, meaning that they contain both male and female organs. If there is no male around to fertilize the female's eggs, she is fully capable of doing it herself. Banana slugs move slowly, because, like all slugs, they have only one muscular foot. Around their bodies, they excrete a thick coating of slime, which keeps the skin moist, protects the slug from predators, and lubricates the terrain over which they move. Some researchers are attempting, so far unsuccessfully, to reproduce slug slime, since it is one of the best natural glues, and it may have potential uses in medicine.

Winter wren

The **winter wren** *(Troglodytes troglodytes)* is a denizen of the deeply shaded forest. This stubby-tailed, chocolate-brown mouse of a bird is an insectivore, feeding on the forest floor and banks of streams. Its scientific name means "cave dweller," due to its habit of building its nest in a crevice or hole in a tree or steep bank. While it can be very difficult to observe the winter wren as it flits from bush to bush, the song is unmistakable. Surprisingly loud and quite long, the song may seem to burst out of nowhere as you hike through the redwoods. The rich, high-pitched, rising and falling song is made up of very high tinkling trills and thin buzzes, a treat to hear in the deep woods.

Redwood National and State Parks

CALIFORNIA'S REDWOOD National and State Parks have been designated by the United Nations as a World Heritage Site, reflecting the international significance of this ecosystem that is found nowhere else on earth. Three state parks and a national park together include 112,000 acres of redwood forest, rivers, and beaches.

Jedediah Smith Redwoods State Park, Del Norte Coast Redwoods State Park, and Prairie Creek Redwoods State Park were created in the 1920s by the state of California, with the help of the Save-the-Redwoods League, a non-profit preservation group. Each of these parks protects an area of old-growth redwood forest, a resource that even by the early 20th century was fast disappearing. Timber removal continued elsewhere, however, and by the 1960s, 90 percent of the original redwood forest had been logged. The U.S. Congress acted in 1968 to create a new national park that linked three existing redwood state parks in Del Norte and Humboldt Counties and also offered some protection to the Tall Trees Grove along Redwood Creek. Later expansion in 1978 brought more of the Redwood Creek watershed into the park, including 39,000 acres where timber had already been removed. Restoration of the once-logged lands is continuing.

Redwood National and State Parks, now managed jointly, stretch some 50 miles from north to south. The parks include 37 miles of coastal frontage, extending from Crescent Beach near Crescent City south to Freshwater Lagoon in Humboldt County. Highway 101, the main north-south highway paralleling the coast, is located inland of the shore along most of the park's length. But the coast can be reached from Hwy. 101 by roads and trails at various locations, providing opportunities for visitors to experience the varied resources of the shoreline. Sites that provide access to the coast are described in individual site descriptions in the pages that follow.

Jedediah Smith Redwoods State Park

The Crescent City Information Center at 1111 2nd St., Crescent City, offers information about park resources and activities, as does the Thomas H. Kuchel Visitor Center, located on Hwy. 101 one mile south of Orick. Other visitor centers are at Prairie Creek Redwoods

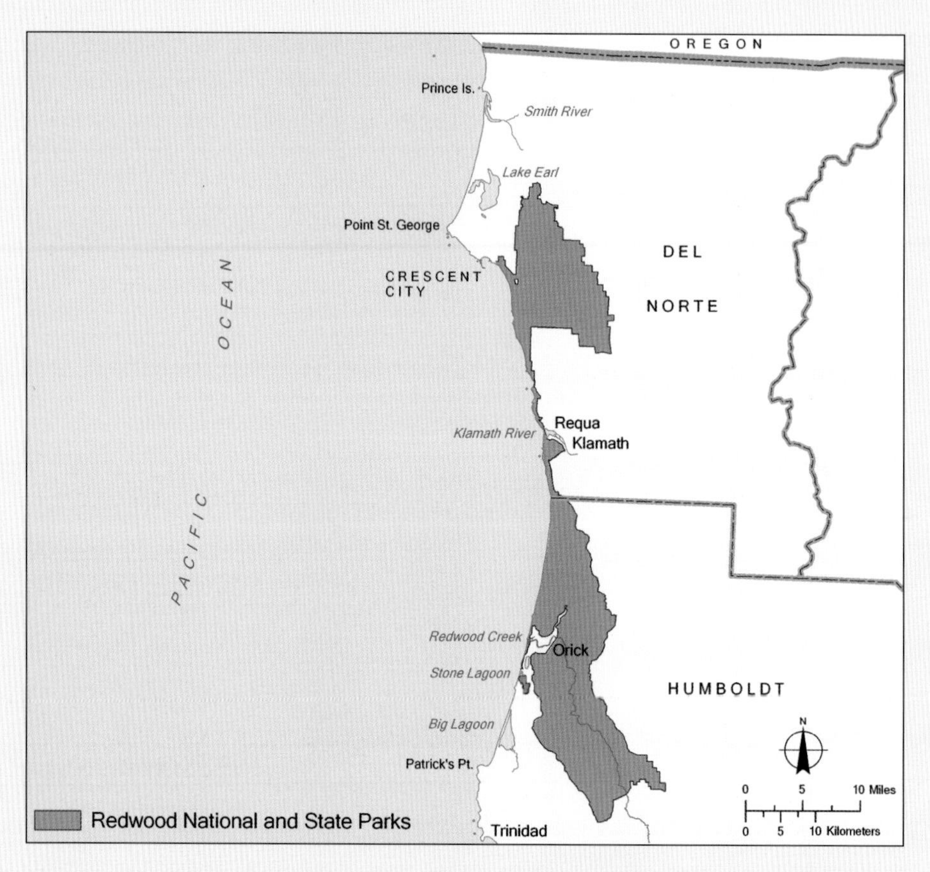

State Park on the Newton B. Drury Scenic Parkway, and at Jedediah Smith Redwoods State Park and at Hiouchi, both located on Hwy. 199 and open summers only.

Highlights of the northern part of the parks include the scenic drives on Enderts Beach Rd. and Howland Hill Rd., hiking the Coastal Trail along the roadless coast south of Enderts Beach Rd., horseback riding or bicycling on designated trails, and rafting on the Smith River. Farther south, scenic drives include the road to Fern Canyon and rugged Gold Bluffs Beach in Prairie Creek Redwoods State Park, hiking the two-mile Trillium Falls loop trail, or bicycling the Ossagon Creek Trail. Ask at the Crescent City Information Center, the Thomas H. Kuchel Visitor Center, or the Prairie Creek Visitor Center for free permits to visit the Tall Trees Grove; the hike requires three to four hours, and there is a limit of 50 cars per day. There are ranger-led campfire talks and nature walks during the summer months at several campgrounds in the parks; for information, check at a visitor center. For children 7–12 years old, the parks offer a junior ranger program. For information on all programs, call: 707-464-6101.

Camping is available at the Jedediah Smith Campground, located in an old-growth redwood forest, and Elk Prairie Campground, located next to a prairie and a redwood forest. Mill Creek Campground, also under the redwoods, is open summers only. Reservations are available for these three campgrounds; call: 1-800-444-7275. Gold Bluffs Beach Campground offers RV and tent camping on a remote stretch of coast; no reservations. Campsites are also available in nearby state parks and at privately managed

RV parks and campgrounds at Crescent City, the Klamath area, Orick, and elsewhere; check at park visitor centers or with local chambers of commerce for private campground information.

All visitor centers in the Redwood National and State Parks are wheelchair accessible. Other accessible areas are available at certain locations; see individual entries that follow. The North Coast Redwood Interpretive Association is a non-profit organization that supports interpretive and educational programs at the three state parks within Redwood National and State Parks, as well as other nearby state parks. The association operates stores for visitors at Jedediah Smith Redwoods and Humboldt Lagoons State Parks in the summer, and at Prairie Creek Redwoods State Park all year.

Restoration of the Redwood Creek Watershed

The Redwood Creek drainage in Redwood National and State Parks includes 280 square miles of mostly forested and steep slopes and the grove where the world's tallest trees grow. Redwoods depend on plenty of moisture. Heavy winter rains supplemented by summer fogs supply it; along Redwood Creek, annual precipitation averages 80 inches or more. In its natural condition, the landscape is adapted to the wet climate. Steep slopes are held in place by thick vegetation, soil, and forest duff, minimizing the amount of fine silt that washes into fast-running streams or into the ocean.

Rapid logging of land draining into Redwood Creek began in the 1940s, and within two decades, more than half the timber had been removed. Hundreds of miles of logging roads crisscrossed the slopes, exposing bare soil to the force of winter storms. Major storms in 1955 and 1964 clogged Redwood Creek with mud and sediment, 20 feet deep in places, burying pools and inhibiting the ability of juvenile salmon to find

Roosevelt elk in Prairie Creek Redwoods State Park

food or avoid predators. Although anadromous fish such as salmon and steelhead are amazingly durable, swimming from sea level upstream against the tremendous power of mountain streams, they are not very adaptable. They will lay their eggs only on coarse gravel, not on mud. The numbers of salmon and steelhead in Redwood Creek declined precipitously, removing from the ecosystem the natural "fish fertilizer" that normally results from the salmon dying following spawning. In short, the health not only of the Tall Trees Grove but also of the entire redwood forest and its wildlife and fishery resources was at stake.

In response, Redwood National and State Parks have undertaken a program to restore the watershed, for the benefit of the salmon, the trees, and ultimately the humans who use and enjoy the land. At first, park managers tried planting small trees and installing check dams to reduce erosion, but they have learned that the best approach is simply to remove unneeded roads. Logging roads and others that are no longer needed for park visitors or maintenance are scraped away using the same bulldozers that created them, and the natural contours of the land are restored. The work has created jobs for heavy equipment operators once employed in the logging industry. When natural topsoil is re-exposed and native seeds are allowed to germinate once more, the forest restores itself.

Restoration of the landscape is a slow process, and work continues. Over 200 miles of old roads were removed by 2002, and results could be measured by the movement of the plug of sediment that clogged the creek after the 1964 floods. By 1995 the bulk of that material had migrated downstream, close to Orick, while the creek channel in the upper watershed was back to its former condition, with many deep pools. But the watershed has not fully recovered. A moderate flood in 1997 again filled many pools with sediment. Some 150 miles of old logging roads remain within the parks on former

Logging road across Emerald Creek, 1981, just prior to restoration

Emerald Creek, tributary of Redwood Creek, in 1987, six years after restoration efforts

commercial timberlands. And upstream from the park, some 1,200 miles of roads exist on private lands within the Redwood Creek drainage. Park managers are working with landowners to control and prevent erosion from logging, which continues in this area.

Redwood National and State Parks are different from many other parks because they include not only ecosystems in essentially unmodified form, but also thousands of acres of land that was heavily disturbed by commercial logging activity. Progress in restoring the watershed of Redwood Creek has led to improvements in habitat in the redwood forest, and also to improved techniques for restoring the land elsewhere on California's north coast and beyond. Restoration has also begun on formerly logged lands in the Mill Creek drainage in Del Norte County, where some 25,000 acres of timberland has been acquired by the Save-the-Redwoods League and the California Department of Parks and Recreation for potential addition to the Redwood National and State Parks. Visitors can see the results of restoration efforts in the Redwood Creek drainage at Dolason Prairie on Bald Hills Road. Other restoration sites in the redwood parks that are easy to access include the Elk Meadow Day Use Area near Orick, the Ah Pah Interpretive Trail (formerly Ah Pah Rd.) at milepost 133.50 on the Newton B. Drury Scenic Parkway in Prairie Creek Redwoods State Park, and Little Bald Hills Trail (formerly a road) off Howland Hill Rd. in Jedediah Smith Redwoods State Park.

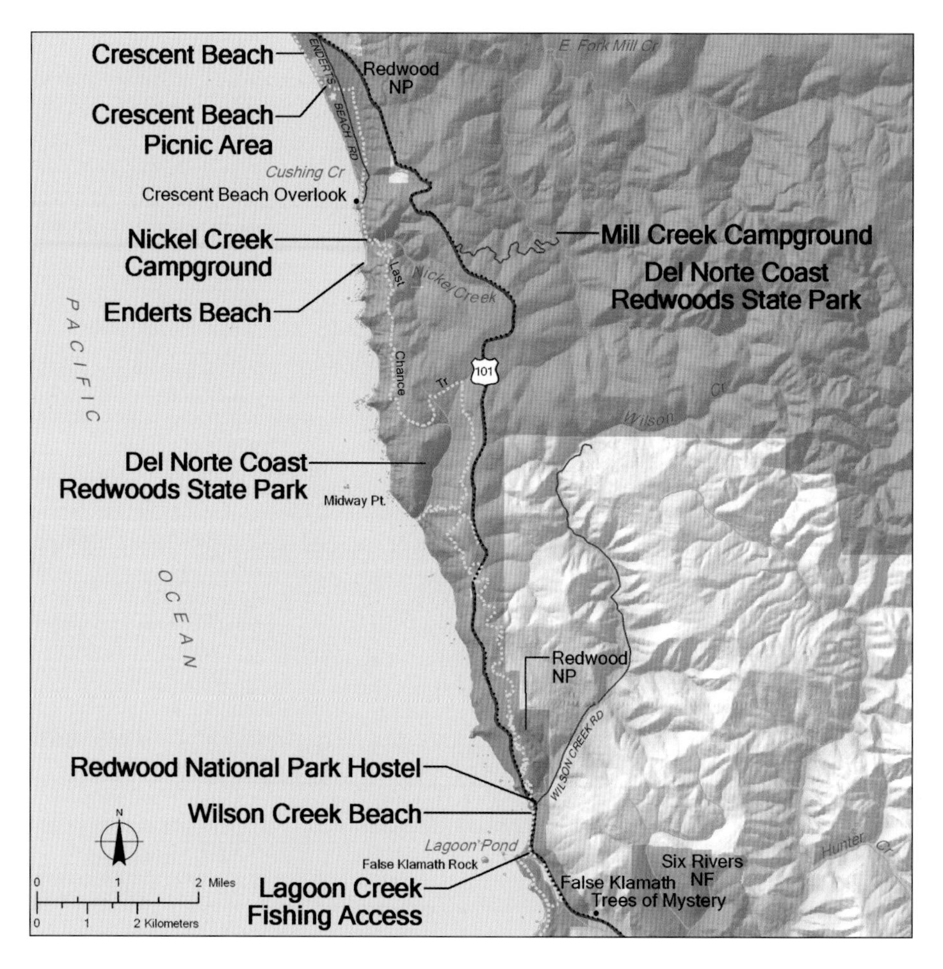

Crescent Beach

Crescent Beach to Lagoon Creek

	Sandy Beach	Rocky Shore	Trail	Visitor Center	Campground	Wildlife Viewing	Historic Building	Fishing or Boating	Facilities for Disabled	Restrooms	Parking	Fee
Crescent Beach	•							•		•	•	
Crescent Beach Picnic Areas	•		•					•	•	•	•	
Nickel Creek Campground			•		•					•	•	
Enderts Beach	•	•	•					•			•	
Mill Creek Campground			•		•	•		•	•	•	•	•
Del Norte Coast Redwoods State Park	•	•	•		•	•		•	•	•	•	•
Redwood National Park Hostel						•	•		•	•	•	•
Wilson Creek Beach	•		•			•		•	•	•	•	
Lagoon Creek Fishing Access			•			•		•	•	•	•	

CRESCENT BEACH: *W. of Hwy. 101, 1 mi. S. of Crescent City.* A long, flat sandy beach used by strollers, kite flyers, and equestrians curves nearly three miles from Crescent City Harbor to the mouth of Cushing Creek. South Beach, the end closest to the harbor, is Del Norte County's pre-eminent surf break. An annual longboard contest is held here. Wetsuits are typical; average water temperatures range from 51 to 58 degrees. Pull-outs along Hwy. 101 have no facilities, but provide direct foot access to South Beach.

CRESCENT BEACH PICNIC AREAS: *Enderts Beach Rd., .5 mi. S. of Hwy. 101.* Crescent Beach Picnic Area has barbecue pits and tables, some wheelchair accessible, in a grassy area between the parking lot and the beach. Another picnic area is located at the Crescent Beach Overlook, on a bluff with an elevated view of the sea, at the south end of Enderts Beach Rd. (which is old Hwy. 101). A two-mile-long trail links the Crescent Beach Picnic Area with the Crescent Beach Overlook. Along the way, listen for Pacific treefrogs and watch for alligator lizards and garter snakes. Dogs on leash allowed on trail.

NICKEL CREEK CAMPGROUND: *.5 mi. S. of end of Enderts Beach Rd.* Five primitive campsites are located on a level area overlooking scenic Enderts Beach. A thick canopy of vegetation shelters the campground, and blackberries and ferns grow along Nickel Creek. Campsites have picnic tables, barbecue grills, and bear lockers; restrooms are available. Stream water must be purified for drinking. No dogs allowed. The Last Chance Trail, which links the campground to the end of Enderts Beach Rd., also extends south six miles into Del Norte Coast Redwoods State Park. For information, call: 707-464-6101.

ENDERTS BEACH: *.5 mi. S. of end of Enderts Beach Rd.* From the parking lot at the end of Enderts Beach Rd., Last Chance Trail brings hikers to Enderts Beach, a sandy isolated strand bounded by rock outcroppings and backed by a forested bluff. The trail is steep in places, and offers majestic views of the coastline. Dense riparian vegetation borders Nickel Creek, and seasonal wildflowers abound on the bluffs. Rock fishing, smelt netting, and surf casting are popular; the Tolowa Indians once used the beach for fishing and seaweed gathering. Tidepools along the shore are inhabited by sea anemones, California mussels, black turban snails, and sea stars, including the unusual sun star and many-rayed star. Although Crescent City is visible in the distance across the water, this beach offers a sense of wildness and seclusion. No facilities, but restrooms available at nearby Nickel Creek Campground. For information, call: 707-464-6101.

MILL CREEK CAMPGROUND: *2.5 mi. E. of Hwy. 101, 7 mi. S. of Crescent City.* Among second-growth redwoods, in a valley sheltered from ocean winds, are 145 summer-only campsites with fire rings, picnic tables, and bear-proof lockers. Campsites are well spaced; some sites can accommodate trailers up to 27 feet although there are no hookups; RVs are limited to 31 feet, and there is an RV dump station. Hot showers, restrooms, and a campfire center are provided, and there are junior ranger programs and guided walks. Also available are hike or bike campsites; no reservations. There are trails for hiking, biking, and nature study, such as the 5.5-mile-long Mill Creek Trail, and there is fishing in Mill Creek. The campground is open May 1–Sept. 30. For campground information, call: 707-464-6101; for reservations, call: 1-800-444-7275.

Redwoods

DEL NORTE COAST REDWOODS STATE PARK: *E. and W. of Hwy. 101, 7 mi. S. of Crescent City.* This 6,400-acre park with substantial old-growth redwood forest has eight miles of wild shoreline, much of it steep and accessible only by trails. Atypical of California's coastal forest, the redwood trees here grow virtually down to the sea's edge. The vegetation understory includes rhododendrons and azaleas, many species of ferns, wild berries, and colorful wildflowers. Deer and foxes are numerous, and an occasional bobcat, mountain lion, or black bear is seen. The park extends from the south end of Enderts Beach to the mouth of Lagoon Creek and includes beaches and the Mill Creek Campground.

The challenging five-mile-long Damnation Creek Trail, which begins at Hwy. 101 mile marker 16, winds through dense old-growth redwoods and Douglas-fir and spruce groves, to a sea cove with a small beach. This trail was used by the Yurok people, who gathered shellfish and seaweed at the beach. For information, call: 707-464-6101.

REDWOOD NATIONAL PARK HOSTEL: *14480 Hwy. 101 at Wilson Creek Rd., 7 mi. N. of Klamath.* The hostel has 30 beds in the hundred-year-old DeMartin home with a view of the sea. Whale watching in season from outdoor decks; displays about the area and its original Native American inhabitants. Wheelchair accessible; open year-round 8 AM–10 AM and 5 PM–9 PM. Operated by Hostelling International. There is access to Wilson Creek Beach, 100 yards from the hostel. For information, call: 707-482-8265.

WILSON CREEK BEACH: *W. of Hwy. 101, 5.5 mi. N. of Klamath.* A 100-yard-long sandy beach that is part of Del Norte Coast Redwoods State Park is located west of Hwy. 101. The parking area entrance is 50 yards south of Wilson Creek Bridge; look sharp, as traffic moves fast here. The beach offers tidepools and abundant driftwood; cormorants roost seasonally on offshore rocks. False Klamath Rock, about 2,000 feet offshore from the south end of the beach, is second in importance only to Castle Rock as a northern California seabird breeding site. The Yurok dug for the edible *Brodiaea* bulbs, known as "Indian potatoes," that grew here, and called the rock *olrgr,* which is Yurok for "digging place." At the southern end of the beach is the Yurok Loop Trail, which leads south to Lagoon Creek Fishing Access and beyond to the Coastal Trail.

LAGOON CREEK FISHING ACCESS: *W. of Hwy. 101, 5 mi. N. of Klamath.* A freshwater lagoon stocked with trout offers picnic areas, information boards on Redwood National and State Parks, and restrooms. The lagoon is bordered by dense riparian vegetation, and the sounds of songbirds fill the air. At the north end of the parking area is the self-guided mile-long Yurok Loop nature trail, which passes through dense alder forests to the ocean bluffs along a route used for centuries by Native Americans. The Yurok Loop is a well-maintained trail that is only moderately challenging. A spur trail less than one-half mile long leads to Hidden Beach. From Lagoon Creek there is also access to a four-mile-long segment of the Coastal Trail that follows the bluffs south to Requa Overlook at the mouth of the Klamath River.

Lagoon Creek

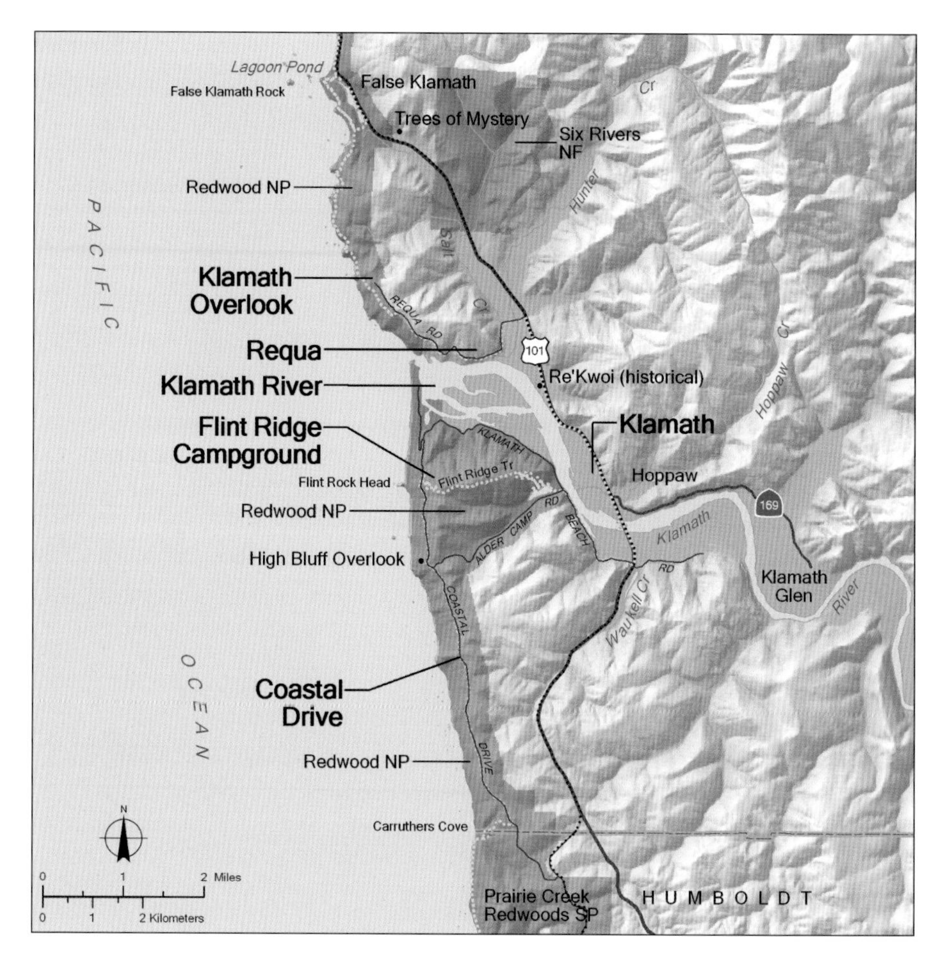

Flint Rock Head

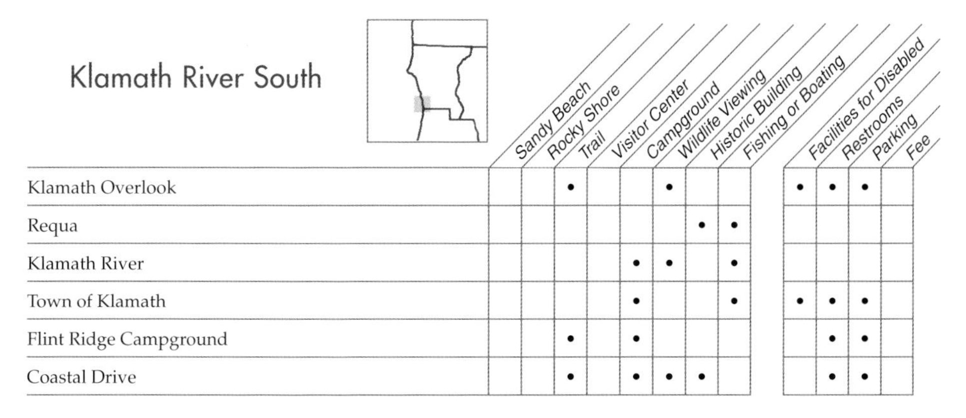

Klamath River South

	Sandy Beach	Rocky Shore	Trail	Visitor Center	Campground	Wildlife Viewing	Historic Building	Fishing or Boating	Facilities for Disabled	Restrooms	Parking	Fee
Klamath Overlook			•			•			•	•	•	
Requa							•	•				
Klamath River					•	•		•				
Town of Klamath					•			•	•	•	•	
Flint Ridge Campground			•		•					•	•	
Coastal Drive			•		•	•	•			•	•	

KLAMATH OVERLOOK: *2.5 mi. W. of Hwy. 101 on Requa Rd./Patrick Murphy Memorial Dr.* Also known as Requa Overlook, this picnic spot on a high bluff offers an elevated view of the mouth of the Klamath River. Restrooms are available. This is the trailhead for the Coastal Trail segment leading north to Lagoon Creek, a distance of four miles. Whales can be seen offshore during migration.

REQUA: *Hwy. 101, 18.5 mi. S. of Crescent City.* This hamlet overlooking the mouth of the Klamath River has been the site of habitation for thousands of years. A large Yurok Indian village named Re'kwoi was located here until the residents were driven out by white settlers who came to the area in the 1850s. Commercial fish canneries were established in the late 19th century, and a ferry service across the river began in 1896, using a 1,700-foot-long cable. By 1902 the town had a store, two saloons, a dance hall, a livery stable, and a hotel, but it went into decline beginning in 1926 when the Douglas Memorial Bridge was built upstream, linking Del Norte County by road with the rest of the state. Requa was bypassed, and today it is a low-key sport fishing and tourist area.

Klamath River mouth

KLAMATH RIVER: *Hwy. 101, 19 mi. S. of Crescent City.* After the Sacramento, the Klamath is California's second largest river in volume and is well known for its year-round salmon and steelhead fishing. Anglers on the river and beaches also take trout and red tail perch. The area's fishing character is saluted in the annual Salmon Festival held in August in the town of Klamath. The festival is sponsored by the Yuroks, California's largest Native American tribe, and one of the largest in the United States still living on or near their ancestral lands. Salmon barbecues are accompanied by games, Native American dances, demonstrations, and a craft fair. Yurok reservation lands extend from the mouth of the Klamath River upstream for 43 miles, stretching one mile on each side.

The Klamath River originates in the Cascade Mountains of Oregon and flows generally southwestward for 263 miles on its way to the sea, joined along the way by the Salmon, Scott, Shasta, and Trinity Rivers. Although the salmon are central to the local Native American culture, as well as the recreational and commercial economy of coastal communities, the number of fish in the Klamath River has been dramatically reduced by dams that block them from reaching some 350 miles of their historic range. Logging and other damage to the watershed have contributed to the decline of the anadromous fish population. Restoration efforts continue to be discussed. Other fish in the river and estuary include starry flounder, Pacific lamprey, American shad, and striped bass; Dungeness crab are found near the river mouth. The mouth of the river is a top spot to view wildlife in numbers, including seals, sea lions, and birds. A small pod of gray whales is often present in the ocean near the mouth of the Klamath, even outside of the whale's fall-to-spring migration season.

Access to the ocean is available on both sides of the Klamath River. On the north side, the Requa Rd. takes visitors to the Requa Overlook with its elevated view of the river mouth. On the south side, Klamath Beach Rd. leads along the river delta to the coast, where it joins Coastal Drive. Intrepid surfers have been known to ride the waves at the mouth of the river, despite treacherous currents, pounding surf, and the shark-attracting presence of plentiful marine mammals.

Kayaking on the Klamath River

TOWN OF KLAMATH: *E. of Hwy. 101, 19 mi. S. of Crescent City.* The original settlement known as Klamath City was founded on the south bank of the Klamath River in 1851, during the brief period of gold fever. Later another town known as Klamath grew up on the north bank of the river, but it was completely washed away in the disastrous flood of December 1964, which was caused by torrential winter rains and a sudden warm spell that rapidly melted snow in the Siskiyou Mountains. The 1926 Douglas Memorial Bridge on Hwy. 101 was also swept away, although its southern terminus remains on Klamath Beach Drive, flanked by statues of golden bears. The "new town" of Klamath was established on higher ground east of Hwy. 101 and today is a small residential community and fishing resort, as well as home of the Yurok Tribal Headquarters. Numerous campgrounds, trailer parks, and resorts catering to sport fishermen are located in the vicinity of Klamath. North of town on Hwy. 101 is a long time roadside attraction, Trees of Mystery, marked by a giant statue of Paul Bunyan with Babe the Blue Ox.

FLINT RIDGE CAMPGROUND: *1 mi. S. of Klamath River mouth on Coastal Dr.* From Hwy. 101 on the south bank of the Klamath River, follow Klamath Beach Rd. west to Coastal Dr. and turn south one mile to reach this primitive campground, located one-quarter mile inland from the parking area. In a forest setting there are ten well-spaced campsites, each with a picnic table, barbecue grill, and bear locker. Restrooms available; no running water. From the campground, hike west on the Coastal Trail and then north to a wide sandy ocean beach south of the Klamath River mouth. The broad beach, scattered with driftwood, is backed by a steep bluff crowned by dense forest.

The Flint Ridge Trail, which is a link in the Coastal Trail, connects the Flint Ridge Campground with the Klamath River near the old Douglas Memorial bridge site on Klamath Beach Rd. at the intersection of Alder Camp Road. The 4.5-mile-long trail crosses over steep hills, passing through meadows and redwood and alder forests. Although strenuous, with many switchbacks, the trail provides an alternate hike-in route to the campground or a day-hike option for visitors traveling on Hwy. 101.

COASTAL DRIVE: *S. of Klamath River mouth to Newton B. Drury Parkway.* An eight-mile scenic drive winds along the steep bluffs between the Klamath River and Prairie Creek Redwoods State Park. The road is perched high above the sea, and the views of open water and undeveloped land are dramatic. The road is narrow, with limited pull-out opportunities; it is also unpaved in parts and not recommended for trailers. One pull-out located one-half mile south of the Flint Ridge Trailhead overlooks an abandoned World War II–era radar station that was disguised as a farmhouse. One mile farther south of the old radar station is the High Bluff Overlook, which has picnic tables, barbecue grills, and restrooms, as well as a grand view of the mountainous coast. A pull-out located one mile from Newton B. Drury Parkway provides trail access to Carruthers Cove.

Along Highway 101

Rugged northern California coast (Greenwood Cove, Mendocino County) ©2004, Tom Killion

Rivers and Streams

THE LARGEST coastal rivers in California are found in the north, where 60 percent of the state's annual rainfall occurs. River runoff, the amount of water discharged through surface streams, is determined by a combination of factors, including local geology, topography, drainage area, and rainfall patterns. As they flow down from their headwaters toward the coast, rivers carve steep, narrow canyons through the mountains. As they approach the coast, they lose speed, depositing sediment to build broad floodplains with rich, deep soils, such as the Eel River delta. Coastal rivers also carry sediments to beaches, playing a crucial role in replenishing sand.

On their way to the ocean, California's coastal streams and rivers flow through the canyons and valleys of coastal mountains, linking forest, grassland, and marsh. Riparian woodlands develop along streambanks and floodplains, providing important habitat for wildlife and cooling water temperatures for anadromous fish species. Coastal wetlands and estuaries form where the rivers enter the sea, creating important rearing habitat for the young of many marine and anadromous organisms. Rivers transport nutrients, sediments, and oxygen through the watershed, and life flourishes in their path.

King (or Chinook) salmon (*Oncorhynchus tshawytscha*) are the longest-lived and largest salmon species, averaging 36 inches in length and weighing approximately 22 pounds. All Pacific salmon are members of the genus *Oncorhynchus*, meaning "bent snout"; you can see from this photo that this name aptly describes this genus. Salmon spawn in large streams or large tributaries of streams. The spawning season varies considerably from one stream to the next, depending on such conditions as distance to the sea from the spawning grounds.

King salmon

The name of the **coastal cutthroat trout** (*Oncorhynchus clarki clarki*) is derived from the brilliant slash of orange or red that usually marks the lower jaw line. The coastal cutthroat trout is unlike most of the other salmon species in that it has the ability to spawn more than one time. In addition, coastal cutthroat have evolved a unique strategy for survival: some fish spend their entire lives in small tributary streams, while others become anadromous, migrating to the ocean and returning to spawn, similar to salmon. Resident cutthroat living in streams may be only a few inches long as adults, while sea-run cutthroats may reach a length of 20 inches or more. These trout occur in coastal streams and rivers from the Eel River in California north to Alaska.

Coastal cutthroat trout

American dipper

The **American dipper** (*Cinclus mexicanus)* can be best described as a water lover. The dipper's nest is always located by a stream, sometimes even behind a waterfall. This curious little bird actually has the capability of walking on the bottom of the stream with swift running water over its head. The dipper uses its wings to help keep its balance and to "fly" through the water, while foraging for aquatic insects under stones in the stream. The name "dipper" comes from the bird's habit of rapidly bobbing by bending its legs, causing a quick up-and-down action. American dippers have a lovely melodious song that can be described as bubbling and wren-like. Fortunately, the song is loud enough to be heard over rushing water. Their singing is not restricted to courting time, and you might be able to hear a dipper singing on your next trip to a rushing coastal river.

Common mergansers

Common mergansers (*Mergus merganser*) are easily recognized by their long pointy scarlet bills. In addition, both sexes sport a ragged crest of feathers on the back of their heads. Usually found on streams, rivers, and lakes, they are divers, swimming easily underwater in search of prey. They eat a wide variety of fish depending on the season. The merganser's bill is serrated and hooked, allowing the bird to grasp and hold on to its slippery meal.

Coastal giant salamander

The **coastal giant salamander** (*Dicamptodon tenebrosus)* is the largest terrestrial salamander that occurs in the United States, reaching lengths of up to 14 inches from nose to tail tip. The salamanders are generally dark brown with a lighter tan or gray marbling of spots. This species can be found in old-growth or well-established second-growth Douglas-fir and broadleaf maple forests. By day, salamanders use damp mosses, downed logs, and other debris near cold streams and mountain lakes to hide under. They have hardened toenails that help them to dig for their prey of insects and snails, and to burrow in the damp soft forest floor for protection. This is the only salamander known to have a true voice, and when startled it may give a little bark.

The **Pacific aquatic garter snake** *(Thamnophis atratus)* is the only snake known to engage in a behavior called "tongue fishing." The snake sits on the bank of a small stream in an ambush position. If it sees a fish approaching, it sticks its tongue on the water surface and wiggles the tip as though to mimic an insect falling into the water. Adult aquatic garter snakes are active foragers who feed chiefly on aquatic giant salamanders in streambed substrates. Juvenile snakes primarily use the ambush tactics to capture juvenile salmonids along stream margins.

Pacific aquatic garter snake

The **river otter** (*Lutra canadensis*) is almost impervious to cold because of an outer coat of coarse guard hairs, plus a dense, thick undercoat that helps to "waterproof" the animal. The long and slender body, the small eyes and ears, and the webbed feet are all adaptations to make the river otter more efficient in the water. The otters' small ears and nostrils can close tightly when they are in the water, and they are excellent swimmers and divers. Otters are well known for their intelligence and their playful antics.

River otter

Sandbar willow (*Salix exigua*) grows in sandy soil along streams and rivers of the northern California coast. These plants can form a dense impenetrable thicket with numerous small diameter stems. This species spreads by underground root suckers that take advantage of the moist environment. The leaves are narrow in form, and yellow to gray-green in color. Native Americans chewed the inner bark of willow for relief from headaches (the bark contains salicin), and the young foliage is rich in vitamin C. The slender, flexible stems were also used to weave baskets.

Sandbar willow

Western azalea

Western azalea (*Rhododendron occidentale*) hosts buttery white to pink flowers with golden markings on the upper petal, and is a treat for the senses. When in bloom, a bower of azaleas is perhaps the single most fragrant flowering shrub in California, inducing smiles and a zombie-like backtracking to the plants you just passed. The peak bloom is usually sometime in May and lingers into June. In fall, the leaves turn all shades of flaming red, yellow, and orange. In general, western azaleas can be found along streambanks and creeks.

Big-leaf maple

Big-leaf maple (*Acer macrophyllum*) is well named, as its leaves are enormous. It is a beautiful, deciduous tree found growing along creeks and streams. The large maple leaves turn wonderful shades of yellow or gold in autumn before they drop. The big-leaf maple contributes to the diversity of riparian forests and provides cover for many species of small mammals and birds. Native Americans used the bark for making rope, and the wood for carved bowls, utensils, and canoe paddles. Traditionally, the sap was boiled into a maple syrup.

Page opposite: Patricks Point State Park, Humboldt County

Humboldt County

DEL NORTE
Prairie Creek Redwoods SP
101
Redwood Cr
Orick
Humboldt Lagoons SP
Redwood NP
Big Lagoon
Patrick's Pt.
Patrick's Point SP
Trinidad Head
Trinidad SB
Trinidad
Little River SB
Little R
Clam Beach County Park
Mad R
McKinleyville
Six Rivers NF
Klamath
River
96
Trinity
River
Redwood
Cr.
299
255
Arcata
Arcata Bay
Eureka
Humboldt Bay
Fort Humboldt SHP
Six Rivers NF
Humboldt Bay NWR
HUMBOLDT
Eel River
Fernbridge
Fortuna
211
Ferndale
MATTOLE RD
Bear R
36
Eel
River
PACIFIC OCEAN
Petrolia
Mattole R
Humboldt Redwoods SP
MATTOLE RD
AVENUE OF THE GIANTS
WILDER RIDGE RD
Lost Coast
King Range NCA
ETTERSBURG
HONEYDEW RD
BRICELAND THORN RD
Redway
Garberville
Benbow
Benbow Lake SRA
Richardson Grove SP
SHELTER COVE RD
Pt. Delgada
Shelter Cove
MENDOCINO
2200
1800
1400
1000
600
200
N
0
10
20 Miles
0
10
20 Kilometers
Depth Contours Shown in Meters

Humboldt County

ON THE PACIFIC COAST of the continental U.S., Humboldt County projects farther west than any other land area. Eureka is almost as far west as it is north of San Diego; Eureka is farther west than Seattle and Portland, too. Humboldt County's coastal climate exhibits a strong marine influence; average high temperatures range between 55 and 65 degrees, winter and summer. A few miles inland, summer afternoon temperatures are considerably warmer. Rainfall ranges between 40 and 100 inches annually.

The coastline is varied, including sheltered lagoon and bay waters, miles of sand dunes and remote beaches, and forests of redwood, Douglas-fir, and Sitka spruce. California's largest coastal roadless area, known as the "Lost Coast," lies in the southern part of the county. The Humboldt Bay area is the main urban center on the North Coast, but the riches of the outdoors are never far away.

The Greater Eureka Chamber of Commerce maintains a visitor center at 2112 Broadway, Eureka; call: 707-442-3738. The Humboldt County Convention and Visitors Bureau is located at 1034 Second St., Eureka; call: 707-443-5097.

Orick Chamber of Commerce; call: 707-488-2885.

Trinidad Chamber of Commerce; call: 707-677-1610.

McKinleyville Chamber of Commerce, 1640 Central Ave.; call: 707-839-2449.

Arcata Chamber of Commerce, 1635 Heindon Rd.; call: 707-822-3619.

Ferndale Chamber of Commerce; call: 707-786-4477.

Fortuna Chamber of Commerce, 735 14th St.; call 707-725-3959.

Narrated tours of Humboldt Bay, offered May through September:
Humboldt Bay Harbor Cruise, 707-445-1910.

Seasonal fishing trips for salmon, albacore, and rockfish:
King Salmon Charters, 707-441-9075.

Salmon and steelhead fishing trips on the Eel, Mattole, Van Duzen, and Mad Rivers, as well as the Smith, Klamath, and Trinity Rivers farther north:
Brice Dusi's Guide Service, 707-444-2189.
Mad River Outfitters, 707-826-7201.

Rafting and kayak trips on area rivers:
Redwoods & Rivers, 1-800-429-0090.

Kayak, canoe, and paddle boat rentals; full-moon kayak tours; and tours of Humboldt Bay with a naturalist:
Hum Boats at Woodley Island Marina, Dock F, Eureka, 707-444-3048.

Recreational equipment sales and rentals:
Northern Mountain Supply, Eureka, 707-445-1711.
Adventure's Edge, Arcata, 707-822-4673.

Kayak tours and instruction:
North Coast Adventures, Trinidad, 707-677-3124.

Bicycle rentals in Arcata:
Revolution Bicycle Repairs, 707-822-2562
Life Cycle Bike Shop, 707-822-7755.

Surfing and fishing equipment:
Salty's Surf & Tackle, Trinidad, 707-677-0300, and Eureka, 707-445-0200.

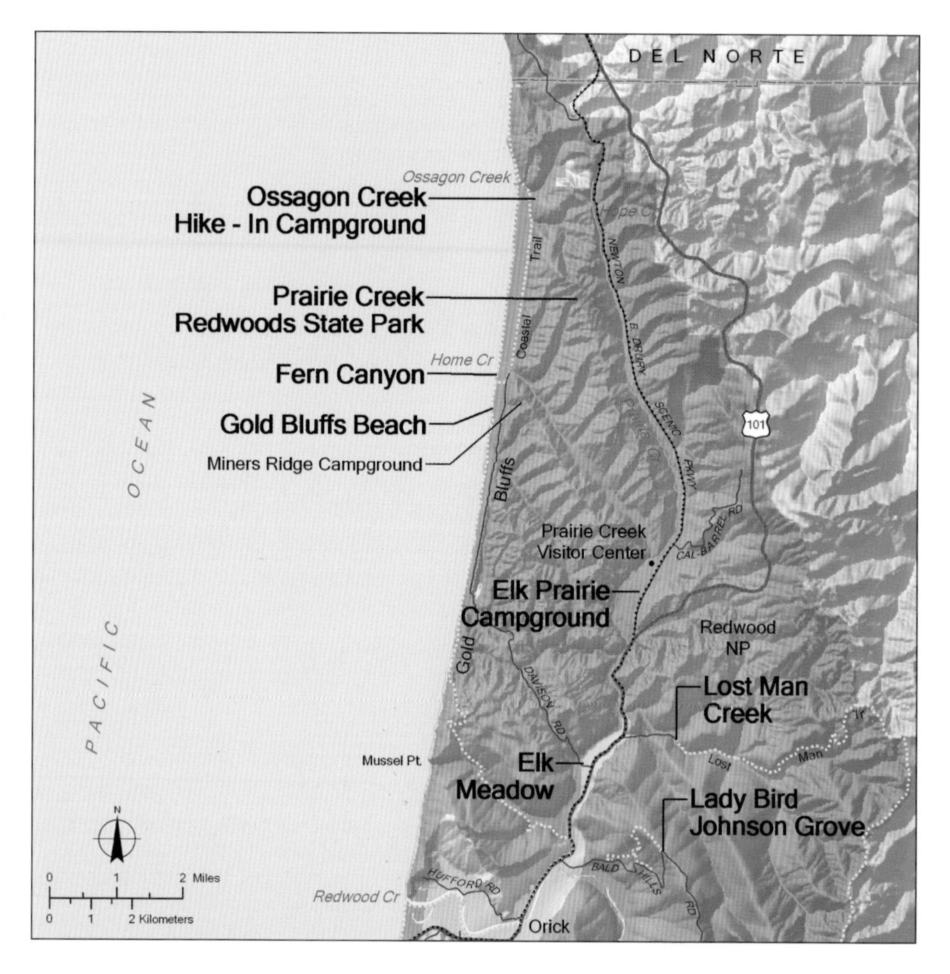

Honey mushrooms (*Armillaria* spp.), Prairie Creek Redwoods State Park

Northern Humboldt County

	Sandy Beach	Rocky Shore	Trail	Visitor Center	Campground	Wildlife Viewing	Historic Building	Fishing or Boating	Facilities for Disabled	Restrooms	Parking	Fee
Ossagon Creek Hike-in Campground			•		•	•				•		•
Prairie Creek Redwoods State Park	•	•	•	•	•	•		•	•	•	•	•
Fern Canyon			•			•			•	•	•	•
Gold Bluffs Beach	•		•		•	•		•	•	•	•	•
Elk Prairie Campground			•	•	•	•			•	•	•	•
Lost Man Creek			•						•	•	•	
Elk Meadow			•			•			•	•	•	
Lady Bird Johnson Grove			•						•	•	•	

OSSAGON CREEK HIKE-IN CAMPGROUND: *On the Coastal Trail, W. of Newton B. Drury Scenic Parkway.* Three primitive campsites with picnic tables, pit toilets, and bear lockers are located in a meadow on a roadless stretch of coast north of Fern Canyon. Campers must park at the Elk Prairie Campground, register, and hike seven miles on the Miners Ridge Trail to the Ossagon Creek campsites. Open all year, but not recommended in winter due to possible flooding; three-day camping limit. Day visitors to the Ossagon Creek area may hike or bicycle on the Coastal Trail from the Fern Canyon parking lot at the end of Davison Rd. or hike a 1.8-mile-long trail that climbs steeply over the coastal ridge from the Newton B. Drury Scenic Parkway; watch for the pull-out at Hope Creek at milepost 132.83.

PRAIRIE CREEK REDWOODS STATE PARK: *W. of Hwy. 101, 5.5 mi. N. of Orick.* Prairie Creek Redwoods State Park, a unit of Redwood National and State Parks, includes a full spectrum of visitor facilities. At the southern end of the park on Newton B. Drury Scenic Parkway is the park's visitor center, with a nature museum, bookstore, and a display about the Yurok Indians, whose ancient homeland encompassed the modern park. Over 70 miles of hiking, bicycling, and nature trails crisscross the park. Trails that begin at or near the visitor center include a 4.5-mile-long trail to the ocean and the quarter-mile-long self-guided Revelation Trail designed for the blind, with guide ropes and Braille signs. Developed campgrounds are available at Elk Prairie and Gold Bluffs Beach,

Prairie Creek Redwoods State Park

Black bear

and there are two hike-in campgrounds at Ossagon Creek and on the Miners Ridge Trail.

Prairie Creek Redwoods State Park includes over 14,000 acres of forest, prairies, and a ten-mile-long beach. Most of the forest is dominated by old-growth coast redwood, with western hemlock, Douglas-fir, and, near the coast, Sitka spruce and red alder. The forest understory is very dense, due to high precipitation. Ferns, mosses, and lichens thrive here. The Rhododendron Trail, located near the Prairie Creek Visitor Center, offers a showy display of western azalea and rhododendron blooms in May and June. Roosevelt elk may be seen browsing in Elk Prairie or at Gold Bluffs Beach, and other park wildlife includes the black bear, gray fox, bobcat, and flying squirrel.

Hwy. 101 skirts around the park, while the Newton B. Drury Scenic Parkway (old Hwy. 101) offers a scenic alternative route, paralleling Prairie Creek. Along the parkway, several pull-outs provide access to forest trails, some leading to the ocean. Other scenic drives include three-mile-long unpaved Cal-Barrel Rd., which winds through old-growth redwood forest (RVs and trailers prohibited), and eight-mile-long unpaved Davison Rd., which leads to Gold Bluffs Beach and Fern Canyon (vehicles limited to eight feet wide and 24 feet long; trailers prohibited). For park information, call: 707-464-6101 ext. 5301.

On a drizzly day, walk into the redwoods where you are sheltered by the canopy of branches hundreds of feet overhead and the brick-red forest duff is soft underfoot. The colors of bark and leaves are intensified by rain, and every surface glistens. The forests, streams, and meadows of Prairie Creek Redwoods State Park and Patrick's Point State Park lured filmmaker Stephen Spielberg to shoot scenes for his 1997 film *The Lost World: Jurassic Park* here. The redwood forest with its emerald carpet of ferns is unmistakable in the film, although if you look closely, you will see a few non-native palm trees, imported to create a more tropical appearance. Other films made in Humboldt County include parts of the Star Wars movie *Return of the Jedi*, filmed by George Lucas at the Tall Trees Grove in Redwood National and State Parks.

FERN CANYON: *End of Davison Rd., 8 mi. W. of Hwy. 101.* Fern Canyon is a narrow slot cut by Home Creek through Gold Bluffs. The canyon has 50-foot sheer walls covered with eight species of ferns; five-finger ferns are most abundant. Rainbow trout are found in the creek, and the western red-legged frog and tailed frog inhabit the canyon, along with winter wrens and dippers. Facilities include a gravel parking lot and a flat quarter-mile-long trail into the canyon. Davison Rd. is unpaved and bumpy, and it dips through shallow creeks. Fern Canyon can also be reached by trails, including a 4.5-mile-long trail from Elk Prairie.

GOLD BLUFFS BEACH: *On Davison Rd., 6 mi. W. of Hwy. 101.* Gold Bluffs Beach is broad and uninterrupted by rocks or seastacks. The campground is exposed to sea breezes but backed by forest and meadows where Roosevelt elk graze. There are 25 tent or RV campsites among low sandy dunes; RVs up to 24 feet long and 8 feet wide can be accommodated; no trailers allowed on gravel Davison Road. There are restrooms, firepits, showers, and wheelchair-accessible campsites; campfire programs are offered. Dogs restricted to the campground; due to snowy plover habitat, seasonal closure of certain beach areas may be in effect. Fees for camping and day use; no reservations.

On a low rise overlooking Gold Bluffs Beach are three hike or bike campsites on Miners Ridge Trail. Running water, pit toilet, firepits, and bear lockers; no reservations. Campers must park at the Elk Prairie Campground, register, and hike on a moderate trail 4.5 miles to the Miners Ridge campsites. Three-day camping limit. Three environmental campsites in a forest setting are also available; check in at the entry kiosk on Davison Rd. for the lock combination.

ELK PRAIRIE CAMPGROUND: *Off Newton B. Drury Scenic Parkway, 5.5 mi. N. of Orick.* Elk Prairie Campground, located in an old-growth redwood forest with views of nearby Elk Prairie, has 75 tent or RV sites, accommodating trailers up to 24 feet and RVs up to 27 feet in length; dump station available. Hike or bike sites also available. Campsites have picnic tables, bear lockers, and firepits; four sites are wheelchair accessible. There are showers, ranger-led walks, junior ranger programs, and campfire programs. The Elk Prairie Campground also includes a picnic area; the 1.4-mile-long Elk Prairie Trail offers views of Roosevelt elk. Fees for camping and day use. For camping reservations, call: 1-800-444-7275.

LOST MAN CREEK: *1 mi. E. of Hwy. 101.* A quiet picnic area in the redwood forest, next to a creek, is located east of Hwy. 101; the turnoff is .5 mile north of Davison Road.

ELK MEADOW: *Davison Rd., W. of Hwy. 101.* A picnic area in Redwood National and State Parks overlooks a meadow where Roosevelt elk graze. Trails from the parking area include the 2.5-mile-long Trillium Falls Trail, the 3.4-mile-long Davison Trail, and the 2.8 mile-long Streelow Creek Trail; the latter two trails allow bicycle use.

LADY BIRD JOHNSON GROVE: *2.5 mi. E. of Hwy. 101, N. of Orick.* One mile north of Orick on Hwy. 101, take Bald Hills Rd. east to a picnic area with a one-mile-long loop trail through a ridge-top redwood and Douglas-fir forest.

After major gold discoveries on the American River in 1848 and Trinity River in 1849, gold fever struck the coast of Humboldt and Del Norte Counties. In 1850, gold in fine particles mixed with layers of black sand was discovered in the cliffs along what is now known as Gold Bluffs Beach. The difficulty of separating the valuable metal from the surrounding material, as well as the rough terrain and the lack of harbors, made large-scale mining impractical. Gold seekers persevered, nevertheless, and in the peak year of 1888 one million dollars in gold and platinum was taken from the beach.

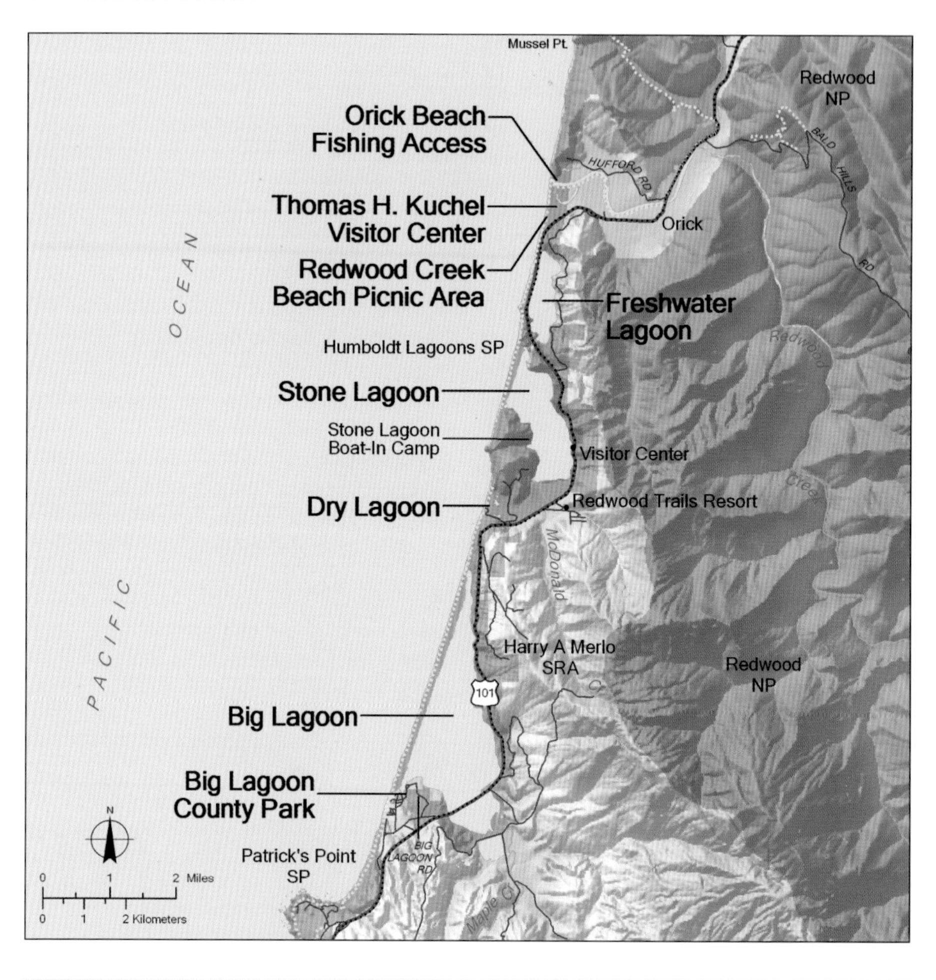

Freshwater Lagoon

Redwood Creek to Big Lagoon

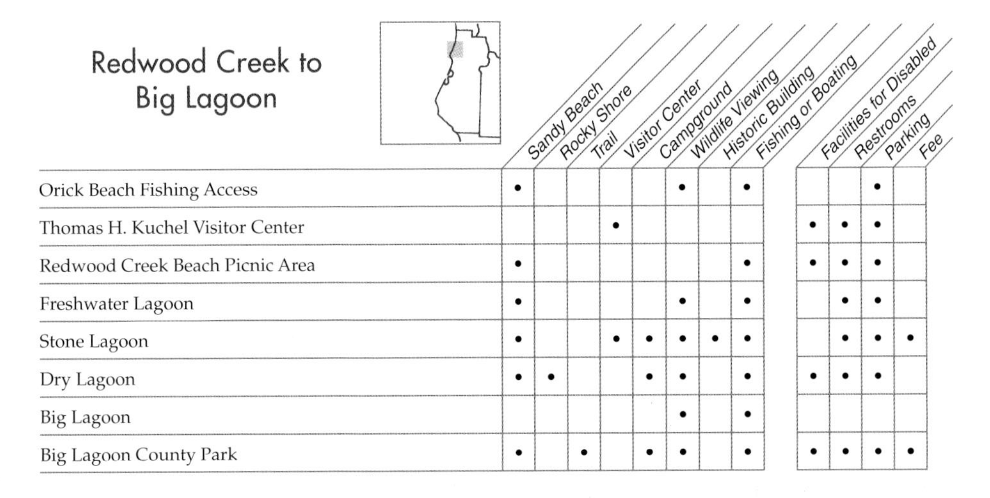

	Sandy Beach	Rocky Shore	Trail	Visitor Center	Campground	Wildlife Viewing	Historic Building	Fishing or Boating	Facilities for Disabled	Restrooms	Parking	Fee
Orick Beach Fishing Access	•					•		•			•	
Thomas H. Kuchel Visitor Center				•					•	•	•	
Redwood Creek Beach Picnic Area	•							•	•	•	•	
Freshwater Lagoon	•					•		•		•	•	
Stone Lagoon	•			•	•	•	•	•		•	•	•
Dry Lagoon	•	•			•	•		•	•	•	•	
Big Lagoon						•		•				
Big Lagoon County Park	•		•		•	•		•	•	•	•	•

ORICK BEACH FISHING ACCESS: *End of Hufford Rd. off Hwy. 101, 1.5 mi. N.W. of Orick.* On Hufford Rd., take the left fork to reach the north side of the Redwood Creek mouth. Campers and trailers are not recommended on the rough and narrow road, which ends on a dike at the creek mouth. Fishing for salmon and steelhead trout, and access to the beach north of the creek. No facilities.

THOMAS H. KUCHEL VISITOR CENTER: *W. of Hwy. 101, 2 mi. S.W. of Orick.* The visitor center at the southern approach to Redwood National and State Parks offers information about park activities, exhibits on the redwood forest, gift shop, and picnic area. A boardwalk offers an easy stroll through the Redwood Creek estuary, where herons and shorebirds may be seen. Bald eagles may also be spotted along the creek. West of the visitor center is a broad sandy beach scattered with driftwood. Coast walks and ranger presentations take place during the summer; the visitor center is open year-round, 9 AM to 5 PM. For information, call: 707-464-6101 ext. 5265.

REDWOOD CREEK BEACH PICNIC AREA: *W. of Hwy. 101, 2 mi. S.W. of Orick.* The picnic area shares a highway turn-off with the Thomas H. Kuchel Visitor Center. There are picnic tables, barbecue grills, and restrooms in an open setting separated from the sandy beach by low dunes. Nearshore rocks stand in a row, like sentinels. Long beach walks are possible here; to the south, the sand stretches for more than a mile, on a spit between Hwy. 101 and Freshwater Lagoon. To the north of the picnic area, when the summer sandbar is present at the mouth of Redwood Creek, one can walk in a more remote setting, away from the highway, for more than two miles along a narrow beach backed by steep, forested bluffs, until reaching the massive rock at Mussel Point. Trapper and explorer Jedediah Smith found his party's overland progress blocked by the same promontory in 1828 when seeking harbors for potential use in the fur trade.

FRESHWATER LAGOON: *E. of Hwy. 101, 3 mi. S. of Orick.* Freshwater Lagoon was sealed off from the ocean in the early 1950s by the construction of Hwy. 101, which runs along the lagoon's barrier beach. The mile-long lagoon is used by waterfowl, mink, and river otters, and is stocked with trout. There is a parking area and boat ramp at the north end of the lagoon, east of Hwy. 101. On the west side of the highway are picnic tables, dunes, and a long sandy beach, for day use only. The National Park Service plans facility improvements, including a paved link in the California Coastal Trail that would parallel the shore on the west side of Hwy. 101. Large but variable waves and undertow make this a challenging location for surfing.

STONE LAGOON: *W. of Hwy. 101, 5 mi. S. of Orick.* Stone Lagoon is part of Humboldt Lagoons State Park. A visitor center on the east side of the lagoon is operated by the

Stone Lagoon

North Coast Redwood Interpretive Association. Adjacent to the visitor center is a picnic area, and a second picnic location is at the north end of the lagoon, adjacent to the ocean beach, which offers six miles of beachcombing, whale watching, and agate hunting. There are no facilities at the ocean picnic area. The lagoon's sand spit is breached by winter storms, allowing anadromous fish to enter the lagoon and its tributary, McDonald Creek. The lagoon is a good place to fish for salmon and trout and to look for birds such as grebes, loons, rails, and wintertime migratory ducks and geese.

Six boat-in campsites are available on the southwest side of Stone Lagoon. No reservations; register at the visitor center, where a boat can be launched. The campsites have fire rings, food lockers, and chemical toilets. Call: 707-488-2041.

The Redwood Trails Resort located south of Stone Lagoon on the east side of Hwy. 101 offers 58 RV/trailer campsites and 30 tent sites, as well as horseback riding, trout-fishing lake, store, showers, laundry, and waste dump station. Call: 707-488-3895. A 19th century red schoolhouse, now a museum, is open to visitors; Roosevelt elk frequently browse in the nearby meadow.

DRY LAGOON: *W. of Hwy. 101, 6.5 mi. S. of Orick.* Turn west off Hwy. 101 at the sign for Dry Lagoon State Park (now part of Humboldt Lagoons State Park) to reach a day-use area next to an ocean beach, scattered with driftwood and backed by forested slopes. There are also six environmental campsites available; no reservations. Register in person at Patrick's Point State Park to receive the padlock combination to the campground. Campsites have firepits and chemical toilets; no running water. Dogs not allowed. Call: 707-488-2041.

Dry Lagoon is actually a freshwater marsh, which provides habitat for ducks, bitterns, and herons. Gold strikes in 1849 on the Klamath and Trinity Rivers brought prospectors who mined the barrier beaches along the Humboldt Lagoons for gold, with limited success.

BIG LAGOON: *W. of Hwy. 101, 8 mi. N. of Trinidad.* The largest of the Humboldt coastal lagoons, Big Lagoon is breached several times each winter by ocean waves. The lagoon's brackish waters support marshes of tules, sedges, and saltgrass, and the lagoon forms an important resting and feeding site for thousands of birds during winter migrations. Some 30 species of fish inhabit the lagoon, including salmon, trout, and starry flounder. On the slopes to the east of the highway is a forest of old-growth redwood and Sitka spruce, inhabited by black bears; a Roosevelt elk herd lives in the marshland. Although Hwy. 101 passes close by, there are no developed facilities in the lands bordering the lagoon, known as the Harry A. Merlo State Recreation Area. Highway pullouts provide places to access the lagoon for fishing or wildlife viewing.

BIG LAGOON COUNTY PARK: *Off Hwy. 101, 8 mi. N. of Trinidad.* Turn west off Hwy. 101 on Big Lagoon Rd. to reach a county park at the south end of Big Lagoon, overlooked by a tiny residential community with the same name. There are day-use facilities next to a sandy beach, and camping facilities nearby, in a sheltered setting; fees for both. Thirty campsites, one with electrical hookup, accommodate RVs and trailers. Boat ramp available. Dogs allowed on leash; no horses. Call: 707-445-7652.

Big Lagoon County Park

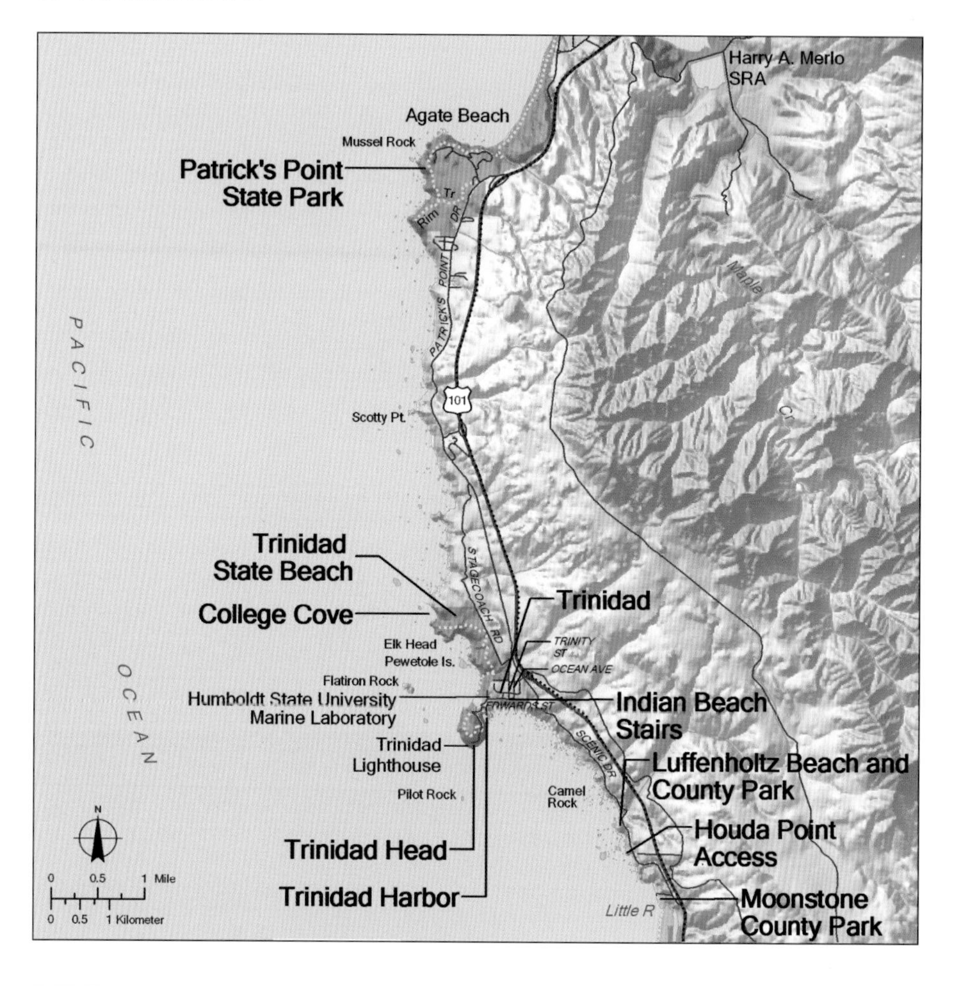

Patrick's Point State Park

Patrick's Point / Trinidad

	Sandy Beach	Rocky Shore	Trail	Visitor Center	Campground	Wildlife Viewing	Historic Building	Fishing or Boating	Facilities for Disabled	Restrooms	Parking	Fee
Patrick's Point State Park	•	•	•	•	•	•	•	•	•	•	•	•
Trinidad State Beach	•	•	•			•		•	•	•	•	
College Cove	•	•	•			•		•		•	•	
Town of Trinidad	•	•	•			•	•	•	•	•	•	
Trinidad Head			•			•	•				•	
Trinidad Harbor	•	•	•			•		•		•	•	
Indian Beach Stairs	•	•	•				•				•	
Luffenholtz Beach and County Park	•		•			•		•		•	•	
Houda Point Access	•		•			•		•			•	
Moonstone County Park	•					•		•		•	•	

PATRICK'S POINT STATE PARK: *Patrick's Point Dr. off Hwy. 101, 5.3 mi. N. of Trinidad.* Patrick's Point State Park offers one square mile of high ocean cliffs, driftwood-scattered beaches, and dense forests of Douglas-fir, pine, and red alder. This well-maintained park contains an unusually broad array of coastal habitats, and offers varied opportunities for recreation and learning.

There are 125 campsites with tables, firepits, restrooms, and wheelchair-accessible showers, along with group and hike or bike sites. For day use there are several picnic areas and six miles of hiking trails. The Rim Trail follows a route along the ocean bluff originally used by Native Americans who lived in the area. From the trail, visitors can see whales, sea lions, and harbor seals, as well as the coastline from Big Lagoon in the north to Trinidad Head in the south. An old-growth Sitka spruce forest is traversed by the Octopus Tree Trail, and two wheelchair-accessible trails lead to ocean overlooks. Agate Beach at the north end of the park is scattered with small pieces of agate and jade and has tidepools with sea anemones, sea stars, and rock crabs.

A visitor center offers activities such as agate jewelry workshops and basket weaving classes, and campfire programs are available. A reconstructed Yurok village known as Sumêg, with traditional-style family houses, a sweat house, changing houses, and a dance house, is located within the state park. Local Yuroks continue to use the village to pass on cultural traditions to their youths and to share them with the public. A nearby Native American garden contains plants used by the Yuroks for medicines, baskets, and other purposes.

Offshore rocks near Mussel Rock provide breeding habitat for seabirds such as pigeon guillemots, black oystercatchers, and cormorants. The lush undergrowth in the forest includes salal, bracken fern, and azaleas. Dogs are allowed on leash within the state park, except on beaches and trails. For information, call: 707-677-3570.

TRINIDAD STATE BEACH: *Off Trinity St., N. of Main St., Trinidad.* This day-use state park includes a long, narrow sandy beach facing forested Pewetole Island, picnic facilities, and trails linking to other beaches both north and south. There is a natural arched rock near the north end of the beach. Restrooms have running water, and there is an outdoor shower. The trail to the beach is one-half mile long, descending over 100 feet.

Flatiron Rock, which lies less than a mile offshore from Trinidad Beach, supports a breeding colony in late spring/early summer of tens of thousands of common murres. Other seabirds that nest on offshore rocks include pigeon guillemots, black oystercatchers, and cormorants. Large numbers of storm-petrels build nests on the rocks as well, but the birds' nocturnal habits make them all but invisible to visitors.

COLLEGE COVE: *Stagecoach Rd., 1 mi. N. of Trinidad State Beach main entrance.* A smooth sandy beach lies at the base of forested slopes at College Cove, part of Trinidad State Beach. The beach is sheltered by Elk Head, the promontory to the north that overlooks College Cove Beach. From the College Cove parking area, hike north along a trail through the forest to Elk Head. Other trails run south to the main part of Trinidad State Beach. Horses, but no dogs or motorbikes, allowed on trails.

TOWN OF TRINIDAD: *Off Hwy. 101, 14 mi. N. of Arcata.* Located on a marine terrace above a small harbor, Trinidad is the former site of a Yurok Indian village, Tsurai. The largest and southernmost permanent settlement of the Yuroks, Tsurai was established 5,000 years ago, and was occupied until 1916. The Yuroks used redwood canoes to fish and hunt seals in Trinidad Bay. The village site is now marked by a plaque at the south end of Ocean Street.

The Spanish explorers Heceta and Bodega sailed into the bay in 1775 and erected a cross on Trinidad Head on Trinity Sunday, thereby giving the bay its name. Trinidad Bay was forgotten until gold strikes on the Klamath and Trinity Rivers beginning in 1849 created the need for a convenient port to supply the mines. The oldest town on the northern California coast, Trinidad was founded in 1850 and had a population of 300 prospectors within three months, but as the economy shifted from mining to logging, the larger and more protected Humboldt Bay soon became the county's main port.

Trinidad is the site of the Humboldt State University Marine Laboratory, which has a wheelchair-accessible visitors area with an aquarium containing ocean fish from the Trinidad area and a whale skull and exhibit on the history of whaling in Trinidad. The laboratory is located near the west end of Edwards St., on Ewing Street. Open 9 AM–5 PM Monday–Friday, 10 AM–5 PM weekends, except in summer; call: 707-826-3671.

Trinidad Harbor

Stand on the shore at Patrick's Point, facing west. You live on a flat, circular expanse of landscape, surrounded by ocean. The earth floats gently over the ocean, riding in a rhythm too subtle to be detected. The waves roll in at your feet, caused by the rising and falling of the dome of the sky against the plane of the sea, far out at the edge of the world. Beyond the sky, there are regions no human has visited. Heroes sometimes paddle far out to the edge of the sea and, with perfectly timed motions, slip under the edge of the sky dome as it rises and falls. You are in Yurok country, which extends from the mouth of Little River in Humboldt County to the mouth of Damnation Creek in Del Norte County, as well as inland along the Klamath River to Bluff Creek. The center of your world is a place called Qe'nek, on the Klamath River.

Every islet and offshore rock before you is named, and there are many stories involving the beings who lived here before humans came to the land. The lives of the Yurok people are intricately entwined with the land, the rivers, and the forests that surround them. Before the 19th century invasion of their lands by settlers and miners, the homes of the Yuroks were typically near water, above either the banks of the Klamath or the shore of the ocean. Trails linking villages were well established, including a coastal trail that paralleled the shore the length of their territory. Those who lived inland traveled along the river in dugout canoes made of redwood. Fishing was important; the word for salmon is *nepu*, which means simply *that which is eaten*. Those Yurok who lived along the coast and south of the Klamath River, the Ner-er-ner, used heavier canoes to launch at specific locations into the ocean where they hunted sea lions and collected mussels from offshore rocks. The Ner-er-ner also ate clams and ocean-growing plants, and they caught surf fish and smelt for consumption and trade. If a whale came ashore, there was a feast.

Among many Yurok stories is one told by those who lived upriver, about two seastacks—and a hero named No'ots. A girl who lived inland from Yurok country on the Klamath River one day digs a strange bulb while gathering food. She has been warned against the double-stalked bulb by her grandmother, but she persists. The girl pulls up the root, which grows into a baby boy. He calls her mother, but she rejects him, and No'ots is raised by the girl's grandmother.

Soon he is a man, and he earns his stepmother's affection by shooting a white deer in the sky. But instead of remaining at home with her, he makes himself a boat and paddles down the Klamath River, singing a song to protect himself, until he reaches the sea. His stepmother realizes that No'ots will leave her, and she chases after him. When she sees him striking out across the ocean for Pulekuk beyond the sky, she breaks her pestle in two and flings a piece at him. It strikes his headdress and feathers go flying. Still No'ots continues across the sea, and she throws the other half of her pestle at him. Now she knows he is lost to her. The first chunk of pestle lies where it landed, five miles west of the mouth of Redwood Creek, known as Sekwona, or Redding Rock. The other half of her pestle is Pekwtso, the rock on which stands the St. George Reef lighthouse, northwest of Crescent City. And the feathers that flew from the headdress that No'ots wore? Now there are seabirds, everywhere on the ocean.

Trinidad Head

Trinidad Head is an enormous block of altered igneous rock known as metagabbro. This rock is a part of the Franciscan Complex, a chaotic mixture of sediments from mainland North America and rocks scraped off the ocean floor as the Pacific plate dived beneath the North American plate 60 to 120 million years ago. The metagabbro making up Trinidad Head is an isolated block of ocean floor incorporated into this mélange—akin to a raisin in a pudding. The metagabbro stands high above the waves because of its hardness relative to the rocks formerly surrounding it. Waves have deposited sand in the lee of the head, building up a spit connecting the head with the mainland. Such a coastal landform is known as a tombolo, and Trinidad Head is a classic example.

TRINIDAD HEAD: *S.W. of town of Trinidad.* Trinidad Head juts out to sea creating a sheltered bay below the little town of Trinidad. A 1.5-mile-long loop trail leads from the parking area at the west end of Edwards St. around the 362-foot-high, sheer-sided rock, in places through a virtual tunnel of dense vegetation including Sitka spruce, blue blossom ceanothus, and ferns. The trail is steep in places, but otherwise easy walking, and benches spaced along the path offer dramatic vistas in all directions. Perched halfway up Trinidad Head and not accessible from the trail is a lighthouse which has been in continuous operation since 1871; the light was automated in the 1970s. Seabirds including pelagic cormorants and pigeon guillemots nest on the rock faces of the headland, as well as on rocks seaward of the head.

TRINIDAD HARBOR: *W. end of Edwards St.* A pier, restaurant, and gift shop are located at Trinidad Harbor. Inquire at the gift shop about ocean fishing trips. A marine railway, used for launching small boats, is open May–October; fee for launching. For information, call: 707-677-3625. The small sandy beach is a good place to put in a kayak. From the harbor, the Galindo Trail leads up the bluff to the town of Trinidad.

INDIAN BEACH STAIRS: *Foot of Trinity St. at Edwards St.* In 1949, the Trinidad Civic Club built a replica of the Trinidad Head lighthouse to serve as a memorial to fishermen lost at sea. Adjacent to the memorial lighthouse is a small parking area and benches overlooking the picturesque harbor. A long and steep wooden staircase, without railings, leads down the cliff to Indian Beach, which can also be reached by trails from Edwards St., Wagner St., and the end of Parker Creek Road.

LUFFENHOLTZ BEACH AND COUNTY PARK: *Scenic Dr., 2 mi. S. of Trinidad.* At a promontory above the beach is a parking area with picnic tables, firepit, and restrooms. From the parking area, a short trail leads out onto the headland, overlooking Luffenholtz Beach; access to the beach is along a short, steep trail with wooden stairs that starts on Scenic Drive 200 yards north of the main parking area. Camel Rock, which is located offshore, holds major breeding colonies of Leach's storm-petrels and other seabirds such as cormorants and black oystercatchers. Tidepools are located among the rocks at the south end. Luffenholtz Beach is somewhat wind- and wave-sheltered by Trinidad Head and offshore rocks. For information, call: 707-445-7651.

HOUDA POINT ACCESS: *Scenic Dr., 2.3 mi. S. of Trinidad.* A roadside parking area with benches serves several short but steep and rocky trails that lead down the bluff to two pocket beaches, a popular surfing area. Another short trail leads out onto a point with a view overlooking the beach to north and south. The flat sandy beach is backed by cliffs and sea caves.

MOONSTONE COUNTY PARK: *Scenic Dr., 3 mi. S. of Trinidad.* A parking lot at sea level provides access to a wide sandy beach located at the mouth of the Little River, with ample room for beach games. Surfing, boogie boarding, and wading are popular here, as are clamming and fishing. Dogs are allowed on the beach.

On New Year's Eve 1914, enormous storm waves swept the Trinidad coast. As the Trinidad Lighthouse keeper watched in awe, waves washed over 103-foot Pilot Rock. Then a wave washed over the lighthouse itself, located nearly 200 feet above sea level, breaking windows and damaging the light mechanism. Accounts differ regarding the consequences; a plaque at Trinidad Head reads that three keepers were killed by the wave, while other accounts mention no loss of life.

Houda Point Access

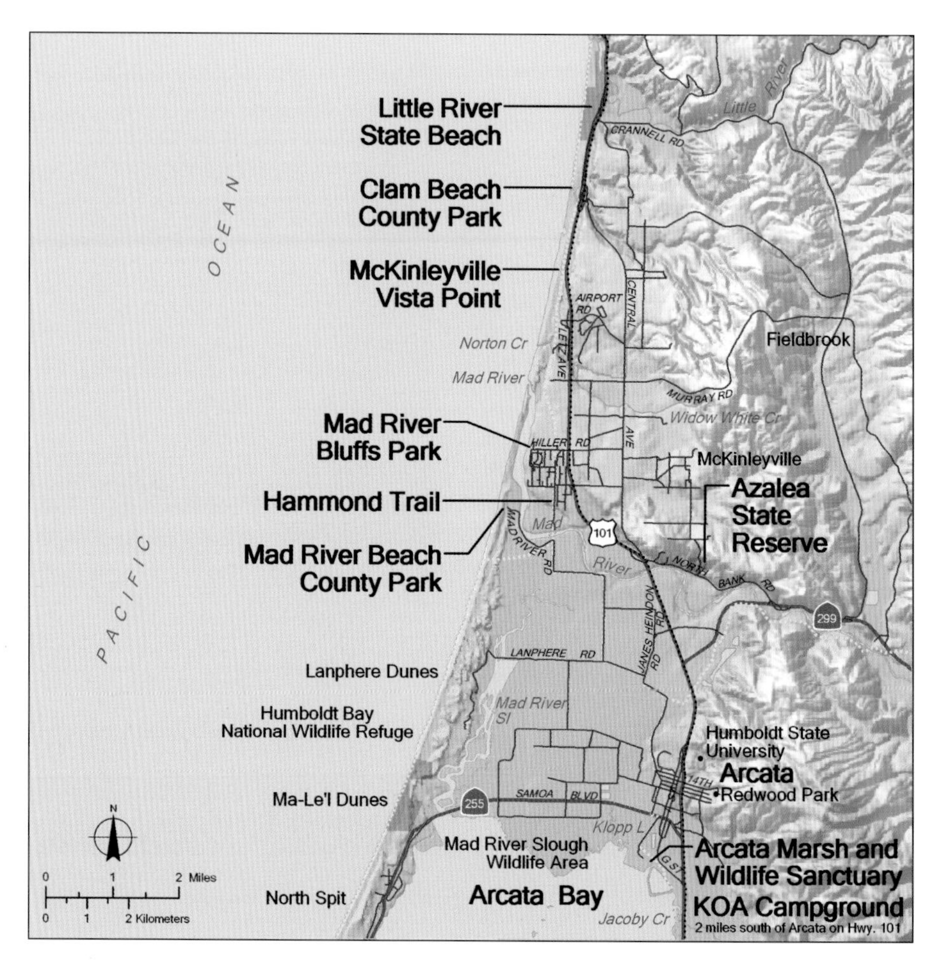

Clam Beach County Park

Arcata Area

	Sandy Beach	Rocky Shore	Trail	Visitor Center	Campground	Wildlife Viewing	Historic Building	Fishing or Boating	Facilities for Disabled	Restrooms	Parking	Fee
Little River State Beach	•					•		•		•	•	
Clam Beach County Park	•				•			•		•	•	•
McKinleyville Vista Point											•	
Mad River Bluffs Park			•								•	
Hammond Trail			•								•	
Mad River Beach County Park	•					•		•		•	•	
Azalea State Reserve			•								•	
Arcata							•	•	•	•	•	
Arcata Marsh and Wildlife Sanctuary			•	•		•		•	•	•	•	
Arcata Bay			•			•		•	•	•	•	
KOA Campground					•				•	•	•	•

LITTLE RIVER STATE BEACH: *W. of Hwy. 101 at Crannell Rd. exit, 8.5 mi. N. of Arcata.* Little River State Beach comprises the northern portion of a nearly two-mile-long sandy beach, extending from the mouth of the Little River; Clam Beach County Park makes up the southern part of the beach. Little River State Beach is undeveloped, used for strolling, fishing, and picnicking. The beach is backed by dunes, part of the huge dune system that runs south for miles to the mouth of Humboldt Bay and beyond. Dune vegetation includes yellow bush lupine, beach strawberry, and European beach grass. Good views of Trinidad Head to the north. Call: 707-488-2041.

CLAM BEACH COUNTY PARK: *W. of Hwy. 101, 7.5 mi. N. of Arcata.* This broad, dune-backed beach is popular for camping, as well as day-use beachcombing, clamming, fishing, and picnicking. There are parking and restroom facilities, cold running water, firepits, and both tent and RV campsites, located close to Highway 101. Camp host available; fee required for camping, with a three-day limit. No reservations taken. This beach is a potential snowy plover nesting area; watch for seasonal restrictions on use. For information, call: 707-445-7651.

MCKINLEYVILLE VISTA POINT: *Hwy. 101, 6.5 mi. S. of Trinidad, southbound only.* Parking area provides panoramic view of Clam Beach and the headlands to the north. During the 1990s, the mouth of the Mad River migrated north to a location below the vista point; it has since moved south again.

MAD RIVER BLUFFS PARK: *End of Hiller Rd., W. of Hwy. 101.* This 74-acre park includes trails through beach pine and spruce forest and grassland habitat to and along the coastal bluff, with scenic views overlooking the lower Mad River and the ocean. Dogs allowed on leash or under voice control. Access to the beach from the bluffs is planned.

HAMMOND TRAIL: *W. of Hwy. 101, from the Mad River to McKinleyville.* The Hammond Trail is a hiking, bicycling, and equestrian trail that runs through open fields, woods, and residential areas near the coast. To reach the southern trail terminus, exit Hwy. 101 for Mad River Beach County Park (see below) and park at the former railroad bridge across the Mad River. From the bridge, a 2.2-mile-long trail segment extends north to the end of Murray Rd., west of Hwy. 101, where a small parking area offers another trailhead, as well as an access to the bank of the Mad River. A second section of the Hammond Trail extends from the north end of Letz Ave. (take Airport Rd. west from Hwy. 101) to Clam Beach County Park. Plans are underway to connect the two trail segments, via a bridge over Widow White Creek.

MAD RIVER BEACH COUNTY PARK: *End of Mad River Rd., W. of Hwy. 101.* The park is 4.2 miles from the Janes Rd. exit off Hwy. 101; turn right on Heindon Rd., left on Iverson Rd., and right on Mad River Rd. This long and narrow park on a sandspit that separates the Mad River estuary from the ocean offers day use on the sandy ocean beach and a good place to launch a small boat in the Mad River. There are two separate parking lots, one with a boat ramp, picnic tables, and non-wheelchair-accessible restroom adjacent to the river, and the other next to the sandy ocean beach and high dunes. The Mad River is a major spawning ground and nursery for salmon and steelhead; also found here is an anadromous fish called eulachon, or candlefish, which reaches its southern limit here. Dogs allowed on the beach, but do not trespass or allow dogs onto the adjacent private farm lands. Call: 707-445-7651.

The Mad River was named in 1849 by the exploring party of Josiah Gregg after a dispute between Gregg and the rest of the party. The river, which flows northwest from the Coast Ranges for 110 miles before entering the ocean west of McKinleyville, once flowed into Humboldt Bay; Mad River Slough at the north end of Arcata Bay is part of its former course. In recent years, the river mouth has taken the opposite course and gradually moved northward, until the state's Department of Transportation was forced to take measures to keep the river from undermining Hwy. 101.

Great Kinetic Sculpture Race, Arcata

AZALEA STATE RESERVE: *E. of Hwy. 101 on North Bank Rd.* The major draw of this 30-acre reserve is the profusion of native western azaleas in bloom during April and May. A short self-guided nature trail, mostly level, leads through fragrant woods inhabited by songbirds. There are picnic tables near the parking lot. Call: 707-488-2041.

ARCATA: *Hwy. 101, 7 mi. N. of Eureka.* Home to Humboldt State University, Arcata offers typical college-town bookstores and cafes, and some more unusual attractions. The town has its own 600-acre community redwood forest, which is managed for sustainable harvesting of timber as well as for habitat protection and recreation; hiking trails can be reached from Redwood Park at the east end of 14th Street. The town's innovative wastewater treatment system utilizes natural processes in a series of ponds that treat effluent and also benefit wildlife.

Annual events include the Great Kinetic Sculpture Race, held over a three-day period on Memorial Day weekend. Entrants create people-powered vehicles that must both float and roll, in order to complete the course, which runs from Arcata along the North Spit, across Humboldt Bay, and through Eureka to the finish line in Ferndale. For information, call: 707-845-1717. Arcata also hosts Godwit Days, an annual spring festival celebrating the marbled godwit and other birds of the redwoods, bays, marshes, and mudflats. The festival offers field trips, workshops, and boat excursions to destinations such as Indian Island in Humboldt Bay, Arcata Marsh, and the open ocean for views of pelagic birds and marine mammals. There are also opportunities for visitors to see rare bird species such as the marbled murrelet, spotted owl, and snowy plover. Call: 1-800-908-9464.

ARCATA MARSH AND WILDLIFE SANCTUARY: *Off G St., Arcata.* The Arcata Marsh Interpretive Center and trailheads are located on G St., one-half mile south of Samoa Boulevard. The Arcata Marsh and Wildlife Sanctuary, supported by agencies such as the

State Coastal Conservancy and winner of a Ford Foundation "Innovations in Government" award, has evolved over more than 20 years to make use of natural processes to dispose of the town's wastewater, while simultaneously providing recreational opportunities and wildlife habitat. Today, the marsh is a pleasant setting for strolling on more than a mile of trails, fishing for trout in Klopp Lake, learning about wetlands, and birding. In spring, look for black-bellied plovers, swallows, and purple martins; in summer: herons and marsh wrens; in fall: black-crowned night herons and sandpipers. Marsh tours are offered every Saturday at 2 PM, starting at the visitor center. Audubon Society members lead Saturday morning bird walks at 8:30 AM, rain or shine, starting at the Klopp Lake parking lot at the south end of "I" Street. For information on the marsh, call: 707-826-2359. No horses; dogs must be leashed.

ARCATA BAY: *Northern part of Humboldt Bay.* One of the few public access points to the northern portion of Humboldt Bay is a concrete boat ramp at the southern end of "I" St., near Arcata Marsh. This an access point for sport boating, duck hunting, and sport shellfish harvesting. Arcata Bay is shallow, and boat access is available only at tides higher than +3.0 feet above mean lower low water. Restrooms, picnic tables, and interpretive displays.

Farther west along the shore of Arcata Bay is the Mad River Slough Wildlife Area, reached off Samoa Boulevard. The Fay Slough Wildlife Area is on the southeast edge of Arcata Bay; turn off northbound Hwy. 101 at Harper Motors, then turn left. Both areas offer hunting and wildlife observation, with day use only, managed by the California Department of Fish and Game; call: 707-445-6493.

KOA CAMPGROUND: *4050 N. Hwy. 101, 2 mi. S. of Arcata.* This privately operated campground offers 165 sites, including 29 for tent campers. Facilities include picnic tables, barbecue grills, swimming pool, and hot tub. There is also a laundry, store, snack bar, and recreation hall, and bicycles are available for rent. Two of the restrooms are wheelchair accessible. Call: 707-822-4243 or 1-800-562-3136.

Egret in Arcata Marsh

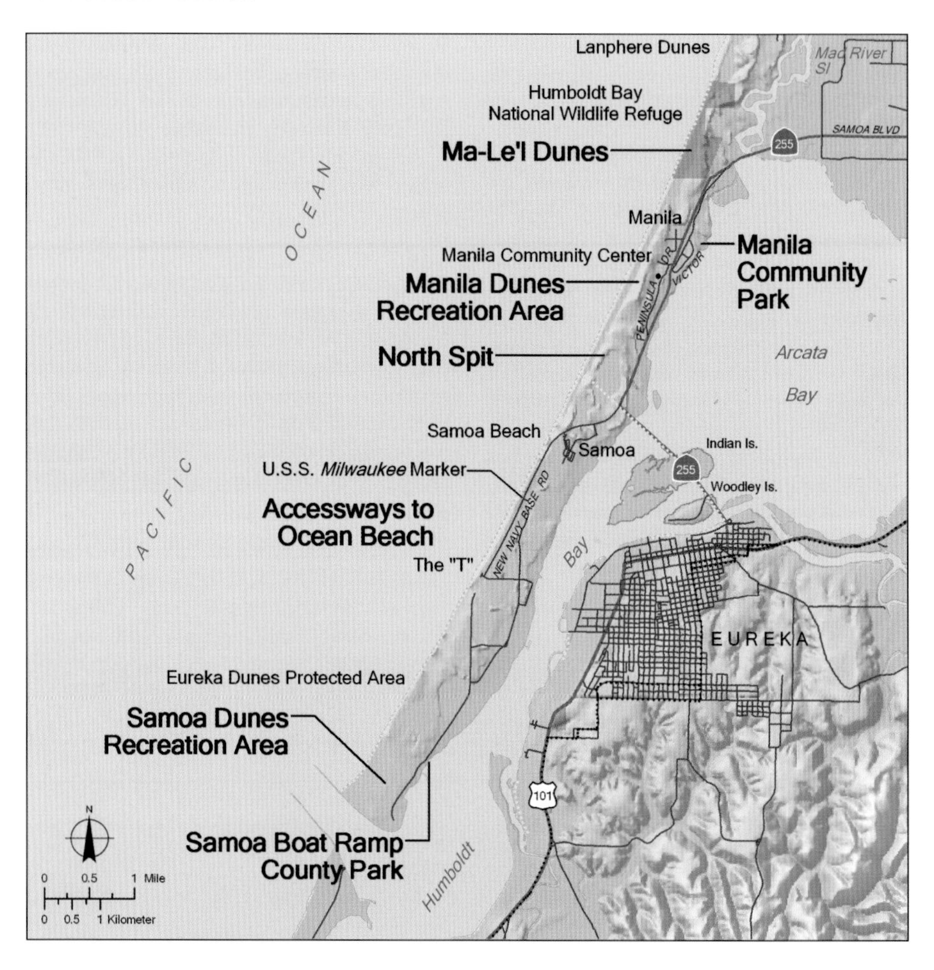

Sanderlings, North Spit

North Spit / Humboldt Bay

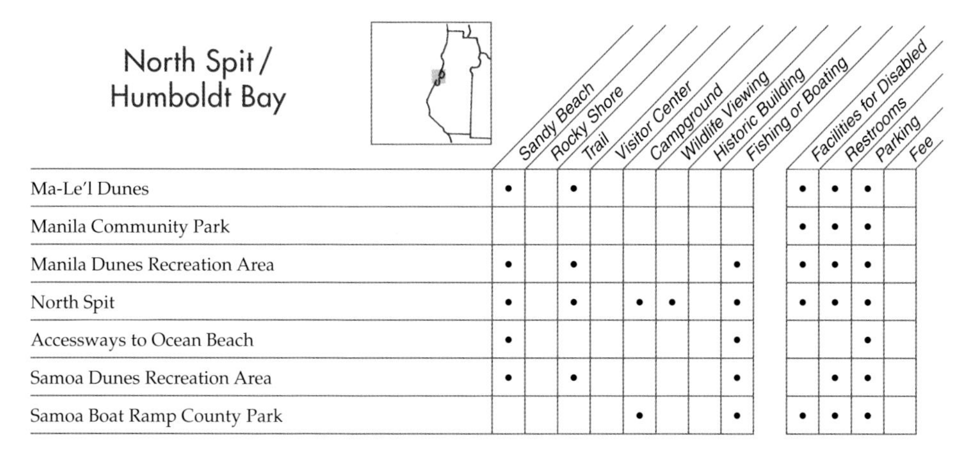

	Sandy Beach	Rocky Shore	Trail	Visitor Center	Campground	Wildlife Viewing	Historic Building	Fishing or Boating	Facilities for Disabled	Restrooms	Parking	Fee
Ma-Le'l Dunes	•		•						•	•	•	
Manila Community Park									•	•	•	
Manila Dunes Recreation Area	•		•					•	•	•	•	
North Spit	•		•		•	•		•	•	•	•	
Accessways to Ocean Beach	•							•			•	
Samoa Dunes Recreation Area	•		•					•		•	•	
Samoa Boat Ramp County Park					•			•	•	•	•	

MA-LE'L DUNES: *W. side of Hwy. 255, Manila.* This newly expanded recreation area is co-operatively managed by the U.S. Fish and Wildlife Service and the Bureau of Land Management. The northern part of Ma-le'l Dunes, a unit of the Humboldt Bay National Wildlife Refuge, is scheduled to open in 2005 and will offer wildlife observation, photography, and canoe or kayak landing/launching. Facilities will include picnic areas, restrooms, and trails through the dune forest and onto the beach; a wheelchair-accessible trail is proposed. Gate to be closed at sundown; dogs, horses, and off-highway vehicle use not allowed. Call: 707-743-5406.

The southern end of the Ma-le'l Dunes, managed by the Bureau of Land Management, offers hiking and equestrian access to the ocean, as well as wildlife observation. Dogs are allowed on leash or under voice control. Planned improvements include parking to accommodate passenger vehicles and equestrian trailers, picnic tables, restrooms, and interpretive kiosks. Open sunrise to sundown. Camping and off-highway vehicle use not allowed. Call: 707-825-2300.

The Humboldt Bay National Wildlife Refuge also includes the Lanphere Dunes unit located farther north, to the west of Arcata. The Lanphere Dunes is managed primarily for the protection of environmentally sensitive plants and lichens, as well as bird and animal life. Public access is allowed only through special permit or guided walks; for information, call: 707-444-1397.

MANILA COMMUNITY PARK: *E. end of Victor Ave., Manila.* A park with play equipment and restrooms is located on the bay side of the North Spit. Walk along the creek to reach the shoreline.

MANILA DUNES RECREATION AREA: *1611 Peninsula Dr., W. of Hwy. 255, Manila.* Over 100 acres of beach and dune habitat are accessible through the Manila Community Center. Loose sand trails lead through a dune forest of beach pine and Sitka spruce and a dune mat plant community, including wildflowers such as the endangered Humboldt Bay wallflower and beach layia. Dogs allowed on leash or under voice control; horses allowed on designated trails. For recreation area information, call: 707-445-3309. Friends of the Dunes offer guided walks on the second and fourth Saturday of every month at 10 AM; call: 707-444-1397. The Friends also offer a brochure and map to recreation sites all around Humboldt Bay.

NORTH SPIT: *Off Hwy. 255, S.W. of Arcata.* The North Spit separates Arcata Bay from the ocean and forms part of a 34-mile-long dune system, the largest in northern California. The North Spit offers beaches, dune forest, wetlands, and recreational opportunities including hiking, picnicking, fishing, and off-highway vehicle use. Stay on trails to protect native plants and wildlife. The North Spit is a habitat area for the western snowy plover, a small shorebird that nests in open sandy areas along the beach between March and September. During the snowy

Manila Dunes Recreation Area

plover nesting season, keep to the wet sand, maintain dogs on leash, and watch for posted rules or closures.

ACCESSWAYS TO OCEAN BEACH: *New Navy Base Rd., S. of Hwy. 255 intersection.* Where New Navy Base Rd. runs parallel and close to the ocean, there are three pedestrian accessways to the beach: Samoa Beach, the U.S.S. *Milwaukee* Marker, and the "T." At each location, parking at pull-outs; no facilities. The rough-sand beach with rolling surf invites surf fishing and beachcombing; from the entrance to Humboldt Bay, the beach extends uninterrupted about 12 miles north. There is day use only, and vehicles are not allowed on this part of the beach.

SAMOA DUNES RECREATION AREA: *S. end of New Navy Base Rd., off Hwy. 255, North Spit.* A 300-acre day-use recreation area is located adjacent to the Coast Guard station and immediately north of the mouth of Humboldt Bay. There are trails for off-road vehicles and motorcycles, as well as picnic tables, fire rings, and restrooms. No water available. Beach access by trail only to Eureka Dunes Protected Area, located north of Samoa Dunes Recreation Area. Open from one hour before sunrise until one hour after sunset. Call: 707-825-2300.

Fishing from the jetty is popular, but use caution as waves sometimes break over it. A popular but challenging surfing spot is found near the jetty and harbor mouth; waves up to 15 to 18 feet in height along with strong currents demand experience.

SAMOA BOAT RAMP COUNTY PARK: *New Navy Base Rd., 4.4 mi. S. of Hwy. 255 intersection.* This park on the bay (east) side of the North Spit offers the only camping on the peninsula. The shoreline is armored here, and the campground offers little shelter. There are 24 sites for RVs, plus hike or bike-in sites; picnic tables, firepits, and restrooms. Pets allowed on leash. A boat ramp will accommodate boats up to 40 feet long, 10 tons maximum.

USS *Milwaukee*, January 13, 1917

Treacherous currents and sandbars make navigation at the entrance to Humboldt Bay a challenge. In the winter of 1916–17, two naval vessels came to grief in the surf of the Samoa peninsula, one of them, ironically, in a vain effort to free the other. The 150-foot-long submarine *H-3* was headed for Humboldt Bay in December 1916 when she suffered engine trouble during a heavy fog and stranded on the Samoa peninsula. Authorities decided to tow the submarine back out to sea, using the 426-foot-long, 9,700-ton cruiser *Milwaukee*. A heavy cable was attached to the *H-3*, and on a high tide on January 13, 1917, in the middle of the night, the *Milwaukee* throttled up its 22,000 horsepower engines and tugged. But the *H-3* was stuck fast, and in the swirling fog and darkness, the *Milwaukee* gradually swung parallel to the beach, instead of pulling out to sea, and soon both ships were hopelessly mired. The 438 crew members were taken off the *Milwaukee* and sent to the Samoa Cookhouse for a hearty meal, and the *Milwaukee* soon began to break up in the surf. After salvage operations, the remainder of the hulk was left to be gradually enveloped in the sand, where nothing is visible today except sometimes at low tide. A monument to the *Milwaukee* is located on New Navy Base Road, near the shipwreck site. As for the *H-3*, she was later pulled across the Samoa peninsula, like a giant redwood log, and relaunched in Arcata Bay.

Bays and Estuaries

BAYS AND ESTUARIES form where fresh and salt water mingle at the mouths of rivers. These tidally influenced areas support salt marsh, mudflats, and sometimes eelgrass beds. They are biologically diverse and provide important rearing habitat for many ocean creatures. Plants and animals that occur in estuaries must be able to cope with fluctuations in water depth and salinity; indeed, many of the organisms that occur there have remarkable adaptations. If you take a trip to the northern California coast, there are many beautiful and ecologically important bays and sloughs to visit. Take the time to look closely at the insects, birds, and plants that are readily observed. You may notice one of the many remarkable ways that members of this biological community have evolved to deal with these challenging conditions.

Pickleweed

Pickleweed (*Salicornia virginica*) is a halophyte, meaning that it is saltwater-tolerant. The plant sends salt to the tips of its pickle-like branches, and in the fall, the branches turn red and the tips wither and drop off, taking the extra salt with them. There are joints in the branches that allow the tip to break off. In this way pickleweed rids itself of excess salt. In the spring, new, green shoots sprout from the old rootstalks. The scientific name *Salicornia* means "salt horns" and describes the fact that the top sections of the branches are filled with salt.

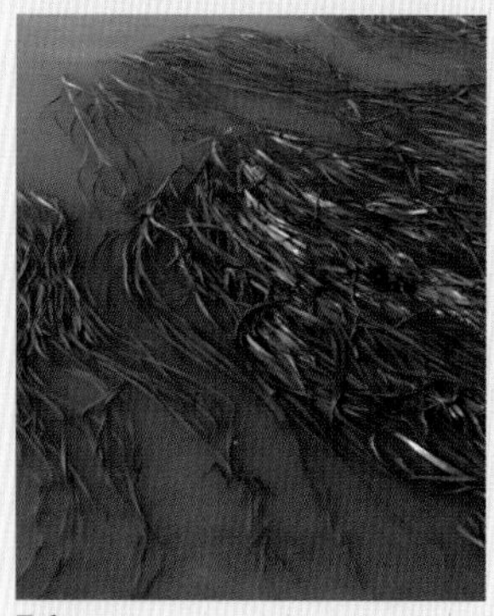

Eelgrass

Eelgrass (*Zostera marina*) is not seaweed (algae), but a perennial flowering plant that grows submerged or partially floating in quiet bay waters. Eelgrass beds have long been recognized as important estuarine ecosystems. They provide critical nursery areas for flatfish and a variety of marine and terrestrial invertebrates. In addition, winter waterfowl such as Pacific brant (*Branta bernicla*) and American wigeon (*Anas americana*) forage on eelgrass. Unfortunately, eelgrass beds are declining worldwide due to habitat loss; run-off and sediment deposition from agriculture and forestry; oyster culturing; and shellfish harvesting. Humboldt Bay has the largest concentration of eelgrass between Willapa Bay in Washington and Mexico.

The **Olympia oyster** (*Ostrea lurida*) is a marine bivalve with an irregularly shaped, fluted shell. Formerly abundant, the Olympia oyster was an important food source for many coastal Native American tribes. Oysters usually inhabit low tidelands or estuaries that remain inundated with water during low tide. Although the Olympia oyster is small, seldom reaching more than 1.5 inches in diameter, it provides a high quality food source, and early settlers quickly took advantage of oyster stocks. However, over-harvest and water pollution have had profound effects on oyster populations. As filter-feeders, oysters take in large quantities of seawater (about 20 to 30 quarts per hour) to extract phytoplankton. Any pollutants or pathogens that are present are also extracted and quickly become concentrated in the oysters' tissues.

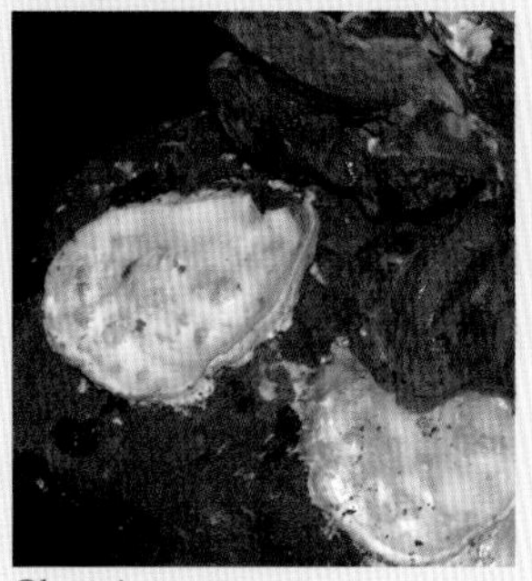
Olympia oyster

The **heart cockle** (*Clinocardium nuttallii*) is a medium-sized clam that has a heart-shaped profile when viewed from the side. Like oysters, cockles are filter-feeders, feeding by capturing particles suspended in the water column. They use short siphons on either side of the shell to filter water for plankton. The heart cockle prefers quiet bays in which the substrate consists of fine, muddy sand. They are particularly fond of eelgrass beds. The heart cockle is hermaphroditic; this means that both the eggs and sperm are released simultaneously from one individual during spawning.

Heart cockle

Lewis's moon snail (*Polinices lewisii*) is a large gastropod typically found partially buried in sand and mud beaches. Gastropod means "stomach-foot." The moon snail preys mainly on clams, which it reaches by digging with its large foot. The snail then uses its sandpaper-like tongue, called a radula, to drill a small hole near the hinge of the clam. Digestive juices are released into the hole, allowing the moon snail to suck the clam up through its proboscis straw. If you find a clam shell with a perfectly round hole drilled in it, it was likely eaten by a moon snail. The best times to look for moon snails are during low tides in spring and summer.

Lewis's moon snail

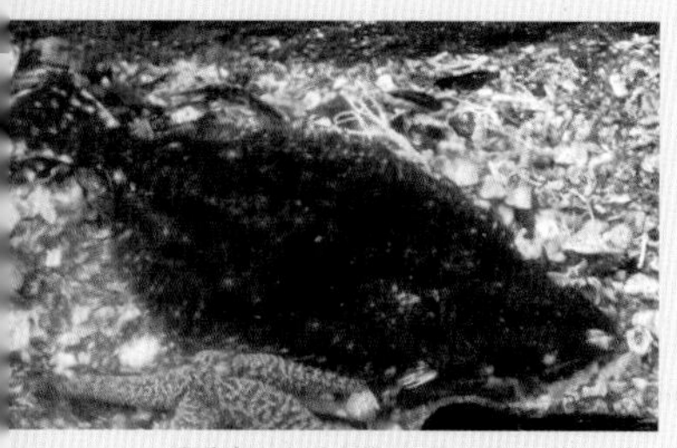
California halibut

The **California halibut** (*Paralichthys californicus*) swims on its side and is known as a "flatfish." Flatfish have both eyes on one side of their body. They begin life as normal-looking fish, with an eye on each side of the head. But in a few weeks, one eye starts migrating around the head to join the other eye. When the change is complete, the halibut is still less than one inch long, but ready to live life sideways. Halibut are able to match their skin color with the sandy or pebble bottom upon which they rest. A halibut hides itself by burying its body up to its eyes in the sandy sea floor.

Black-necked stilt

The **black-necked stilt** (*Himantopus mexicanus*) is a tall, spindly, dark-backed shorebird with a long neck and a thin, straight black bill. Its extraordinarily long pink legs are one of its most distinguishing features. Stilts are well adapted for foraging for food in mudflats. They use their long slender beaks to probe for aquatic invertebrates, and they have partial webbing between their toes, allowing them to maneuver easily through the mud. During the nesting season stilts can become very aggressive, flying low over intruders and vocalizing loudly.

Pacific brant

The **Pacific brant** (*Branta bernicla*) is the smallest of the wild geese, measuring about two feet in length with a wingspan of about four feet. The brant has a distinctive and striking white band on its neck. These beautiful geese nest in tundra along the Yukon delta in Alaska. They travel 3,000 miles during migration to spend the winter along the Baja California peninsula. Brant do not migrate in V's like Canada geese or straight lines like snow geese. They fly in unorganized groups. The brant's dietary staple is eelgrass, and threats to eelgrass put this species at risk. During migration, brant can be seen on rocky beaches along the shore, picking through the tidepools for a meal or foraging in eelgrass beds. Humboldt Bay is a critical stopping point for brant along their spring migration route.

The **bufflehead** (*Bucephala albeola*) is our smallest diving duck. These little ducks will suddenly disappear from the water's surface to search for aquatic insects and small fish, sometimes popping up far from their former location. A breeding male bufflehead has a black head with a white triangular patch and a black back and wings. Wait for the sun to catch the male's head just right so that the brilliant green and purple sheen is reflected; the colors are absolutely stunning. Adult females and first-year males are brownish rather than black, and the white marking on the face is duller and smaller in size than in adult males. The bufflehead is commonly observed between October and April in bays and estuaries along coastal California.

Bufflehead

The **snowy egret** (*Egretta thula*) can be distinguished from the great egret by its smaller size, black bill, and yellow feet. In the breeding season, long lacy plumes called aigrettes grow from the snowy egret's head. In the latter part of the 19th century, it became fashionable to adorn women's hats with aigrettes. The result was that egrets were hunted almost to extinction. These regal birds have several feeding strategies. You may see them stalking their prey in shallow water, shuffling their feet to stir up the mud, or running in the shallows to scare up movement of aquatic organisms. However, sometimes you may see them standing poised, still as a statue, using the sit-and-wait method to ambush their prey.

Snowy egret

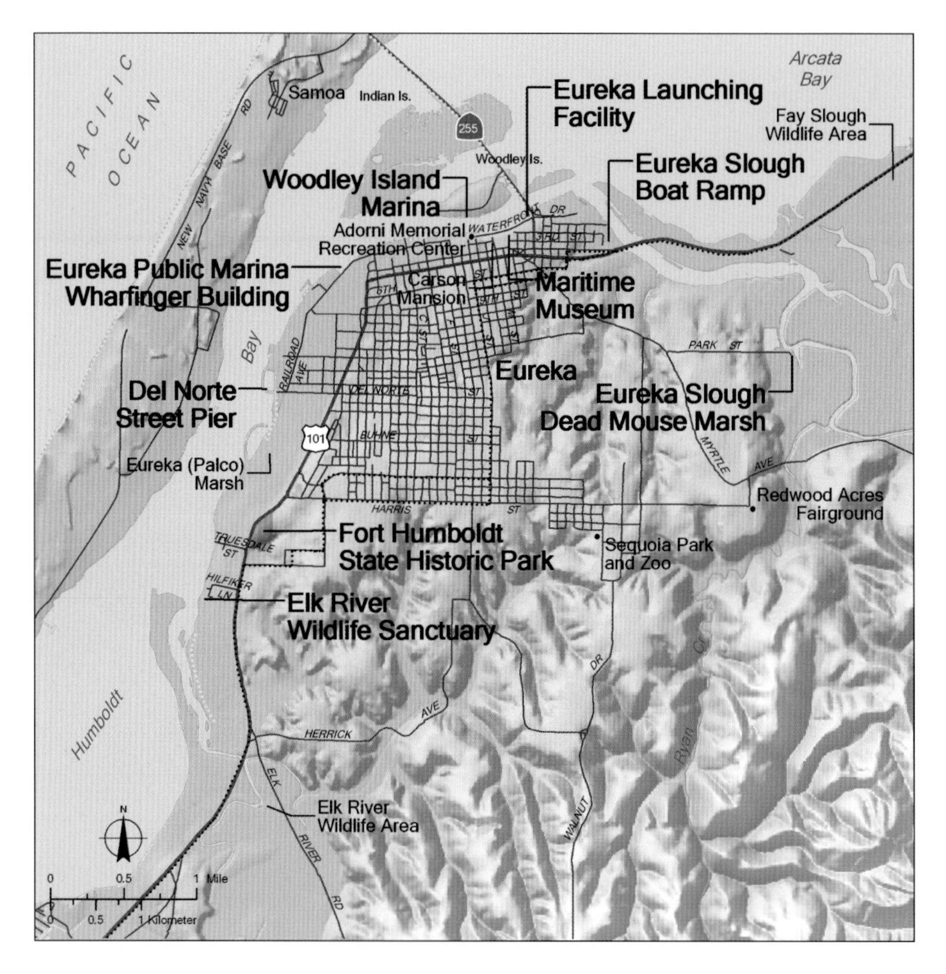

Channel between Eureka waterfront and Indian and Woodley Islands

Eureka

	Sandy Beach	Rocky Shore	Trail	Visitor Center	Campground	Wildlife Viewing	Historic Building	Fishing or Boating	Facilities for Disabled	Restrooms	Parking	Fee
Eureka Slough Boat Ramp						•		•			•	
Eureka Launching Facility			•					•		•	•	
Woodley Island Marina						•	•	•		•	•	
Eureka				•		•	•	•	•	•	•	
Maritime Museum							•			•	•	
Eureka Slough / Dead Mouse Marsh			•			•					•	
Eureka Public Marina and Wharfinger Building								•	•	•	•	
Del Norte Street Pier			•			•		•			•	
Fort Humboldt State Historic Park				•			•		•	•	•	
Elk River Wildlife Sanctuary			•			•					•	

EUREKA SLOUGH BOAT RAMP: *E. end of 3rd St. off Hwy. 101.* Behind the Target store, there is a small concrete public ramp for hand-launching of boats into Eureka Slough, near the northern portion of Humboldt Bay. Check tide tables, because accumulated sediments in the slough make access to the ramp difficult at low tide. Parking available; no other facilities.

EUREKA LAUNCHING FACILITY: *Waterfront Dr., under the Samoa Bridge, Eureka.* A public launch ramp for trailered or small boats is located beneath the Hwy. 255 bridge. A waterfront trail extends west of the ramp through a waterfront park, where the annual summer Blues by the Bay music festival is held, to the Adorni Memorial Recreational Center with its associated waterfront boardwalk. Views of Humboldt Bay and the fishing fleet at Woodley Island Marina.

WOODLEY ISLAND MARINA: *Off Hwy. 255, Eureka.* Woodley Island is part nature preserve, not open to public access, and part marina, which is shared between recreational and commercial boats. Overlooking the harbor is a statue of a fisherman, a memorial to those lost at sea. The tower of the old Table Bluff Lighthouse, once located on the ocean shore south of Humboldt Bay, is also located in the marina. There are 237 berths, and facilities include one and two-ton hoists, sewage and bilge pump-out, utilities, fire protection, and 24-hour security. Hot pay showers and laundry for tenants only. Coffee shop and bar open to the public. There is no boat ramp available, but canoes and kayaks can be launched here. For information on services available at the marina, call: 707-443-0801. Sailing lessons and boat rentals are available at HumBoats at F Dock; call: 707-443-5157. For charter fishing boat information; call Woodley Island Ship Shop: 707-442-7115.

Eureka waterfront

The Wiyot people have lived around *Wigi* (Humboldt Bay) since time immemorial. Traditionally, the Wiyot people would hunt the plentiful wildlife, fish for salmon, dig clams, and gather berries and plants for medicine, food, and basketry. An effort to renew ancient traditions is now focused on Indian Island in the bay where a tragic episode took place nearly 150 years ago. Prior to 1860, the Wiyot people maintained a village known as Tuluwat where an annual World Renewal Ceremony was held. Early on February 26 of that year, as the Indians slept, a group of settlers armed with hatchets and knives sneaked onto the island and massacred men, women, and children. The attack was denounced by some, including writer Bret Harte who worked for a local newspaper, but the murderers were never identified or brought to trial. Army troops removed the remaining Wiyots from their villages to Fort Humboldt and later to reservations outside their historic homeland.

Indian Island is the center of the Wiyot people's world. In 2000, the Wiyot Tribe purchased 1.5 acres at the northeastern end of the island, which had been lost to them since 1860. In 2004, the Wiyot Tribe took possession of 67 additional acres, which were gifted by the City of Eureka in a ceremony of peace.

The Tribe has been working to restore saltmarsh habitat and to clean up pollutants left from a long-time ship repair facility that operated on the island. Additional restoration efforts are proposed for the ancient village of Tuluwat, and the Wiyot Tribe plans to revive the tradition of the World Renewal Ceremony, with dancing, celebration, and the asking of their creator's blessings for all people and for the land. Public access to Indian Island is not available.

Proposed pier, canoe landing area, and Circle of the Tribes Gathering Area on Indian Island

Woodley Island Marina

EUREKA: *Hwy. 101, on Humboldt Bay.* The town's name is Greek for "I have found it," a motto adopted by the fledging state of California in 1850, when it proved popular with miners and traders. Eureka became prominent as a commercial center and shipping point for fish and lumber, and its 19th century origins are visible in the Victorian architecture that lines many of the streets in historic Old Town, a shopping district that extends from C to M St., between the bay and 3rd St. The most-photographed of all the historic structures is probably the Carson Mansion at 2nd and M Sts.; the massive redwood building features ironwork, stained glass, turrets, and out-sized ornamentation in a variety of styles. Now a private club, the mansion is not open to the public.

Eureka is moving to expand public access to its waterfront, although the industrial character of much of the shoreline persists. Still, there are numerous points at which visitors can launch a small boat, cast a fishing line, walk along the waterfront, or view the fishing fleet on the bay. One of the city's initiatives is the boardwalk at the foot of F St., where a pedestrian esplanade extends

Other attractions in Eureka include:

The City of Eureka's Adorni Memorial Recreational Center, including a public accessway along the bay, foot of J St.

Free Thursday night concert series on summer evenings at 6 PM, F St. waterfront plaza.

The Children's Discovery Museum, 3rd and F Sts. in Old Town, 707-443-9694.

The Clarke Historical Museum, with collections of baskets and implements from the Wiyot, Yurok, Karuk, Hupa, and other California Native American tribes, 240 E St., 707-443-1947.

The Morris Graves Museum of Art, 636 F St., 707-442-0278.

The Sequoia Park and Zoo, a pleasant small zoo in a forest of redwood trees, with a visitor center, gift shop, paved paths, and picnic tables, 3414 W St., 707-442-6552.

Camping for RVs only, Redwood Acres Fairground, 3750 Harris St., 707-445-3037.

four blocks along the waterfront and offers views of Humboldt Bay, wildlife, and sometimes the Humboldt State University crew rowing in the channel. Plans are proceeding for a fisherman's terminal with public pedestrian access at the foot of C St., where the old Lazio's pier and fish restaurant once stood and where harbor cruises are available seasonally on the *Madaket*; call: 707-445-1910. By the end of 2005, construction of a new pedestrian walkway along C St., from First St. to the bay, is planned. Additional new development, including visitor-serving commercial facilities, is proposed adjacent to the boardwalk between D and F Sts.

MARITIME MUSEUM: *423 1st St., Eureka.* The Humboldt Bay Maritime Museum in Old Town Eureka has collections of nautical photos, models, and other materials related to the maritime history of northern California; call: 707-444-9440.

EUREKA SLOUGH/DEAD MOUSE MARSH: *E. end of Park St., Eureka.* Trails lead along dikes through an area of restored salt marsh and freshwater marsh habitat; good birding. No facilities.

EUREKA PUBLIC MARINA AND WHARFINGER BUILDING: *Waterfront Dr., Eureka.* Waterfront Dr., along with 2nd St. and Railroad Ave., extends along the northwest edge of Eureka's bayfront from Old Town to the Del Norte St. Pier. In this reach of shoreline, west of Commercial St., is a public marina with paved paths along the water, a small boat harbor with a concrete boat ramp, fuel docks, pumpout station, and berths. Overlooking the marina is the Wharfinger Building, a modern facility used for community events and conferences. Picnic tables and the Humboldt Yacht Club are located in the harbor area. There is also fishing access to the bay at the end of Commercial St., when vessels are not berthed there.

DEL NORTE STREET PIER: *Foot of Del Norte St., off Hwy. 101.* A public fishing pier is located at the end of Del Norte St. There are picnic tables and views of the industrial waterfront of Eureka and shorebirds on the mudflats

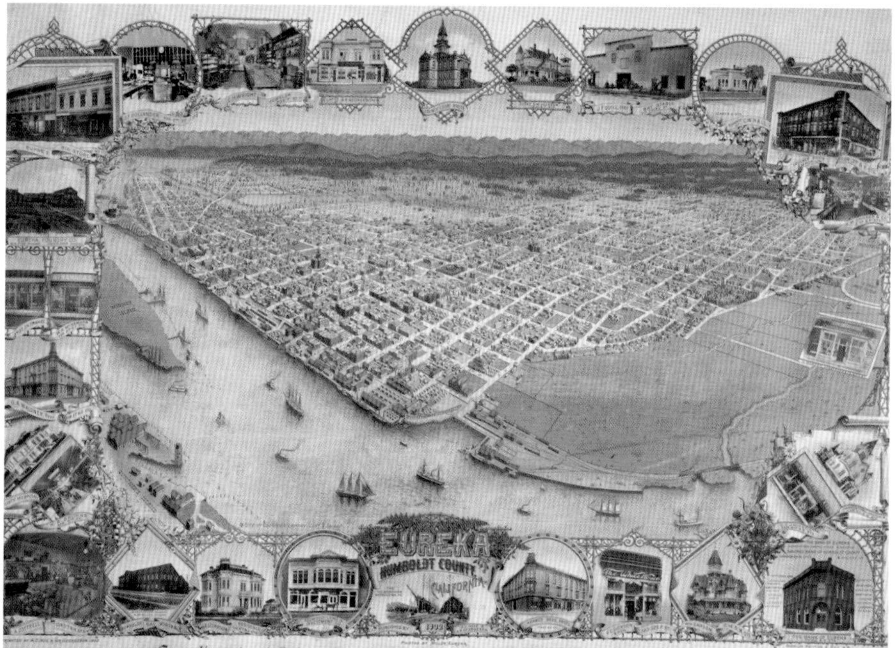

Bird's-eye view of Eureka ca.1902

at low tide. A stretch of Eureka's waterfront trail extends south from the Del Norte Street Pier through the restored Eureka Marsh, also known as the Palco Marsh for the Pacific Lumber Company facility that once occupied the site, as far as the Bayshore Mall.

FORT HUMBOLDT STATE HISTORIC PARK: *3431 Fort Ave., off Hwy. 101, Eureka.* Troops arrived at Eureka in 1853, sent there with the purpose of pacifying the Native Americans whose lands were suddenly overrun with miners and settlers. The generally peaceful Indians of the coast were drawn into conflict with the newcomers, and a number of skirmishes ensued. Despite the establishment of Fort Humboldt, frequent battles continued until 1865, when remaining Indians were forced to move onto reservations. Among those posted to the fort was the 32-year-old Ulysses S. Grant, who held the rank of captain while he was there in 1854. The park maintains an interpretive trail and visitor center with exhibits on the logging and mining history of the area, as well as the fort's military origins; call ahead for wheelchair-accessible tours. There are picnic areas with a view of Humboldt Bay. Open 8:30 AM–4 PM daily; call: 707-445-6567.

ELK RIVER WILDLIFE SANCTUARY: *W. end of Hilfiker Ln., W. of Hwy. 101.* The City of Eureka incorporates natural wetland processes into its wastewater disposal system, much as does the City of Arcata at the better-known Arcata Marsh. From the bayside paths there are views of Humboldt Bay and the North Spit, located across the channel. River otters may be spotted here, along with pelicans, cormorants, and shorebirds. There is also access to the bay at the end of Truesdale St.; no facilities. The Department of Fish and Game maintains the Elk River Wildlife Area upstream on the Elk River, on the inland side of Hwy. 101, for fishing and birding. Take Elk River Rd. east of Hwy. 101; call: 707-445-6493.

Elk River Wildlife Sanctuary

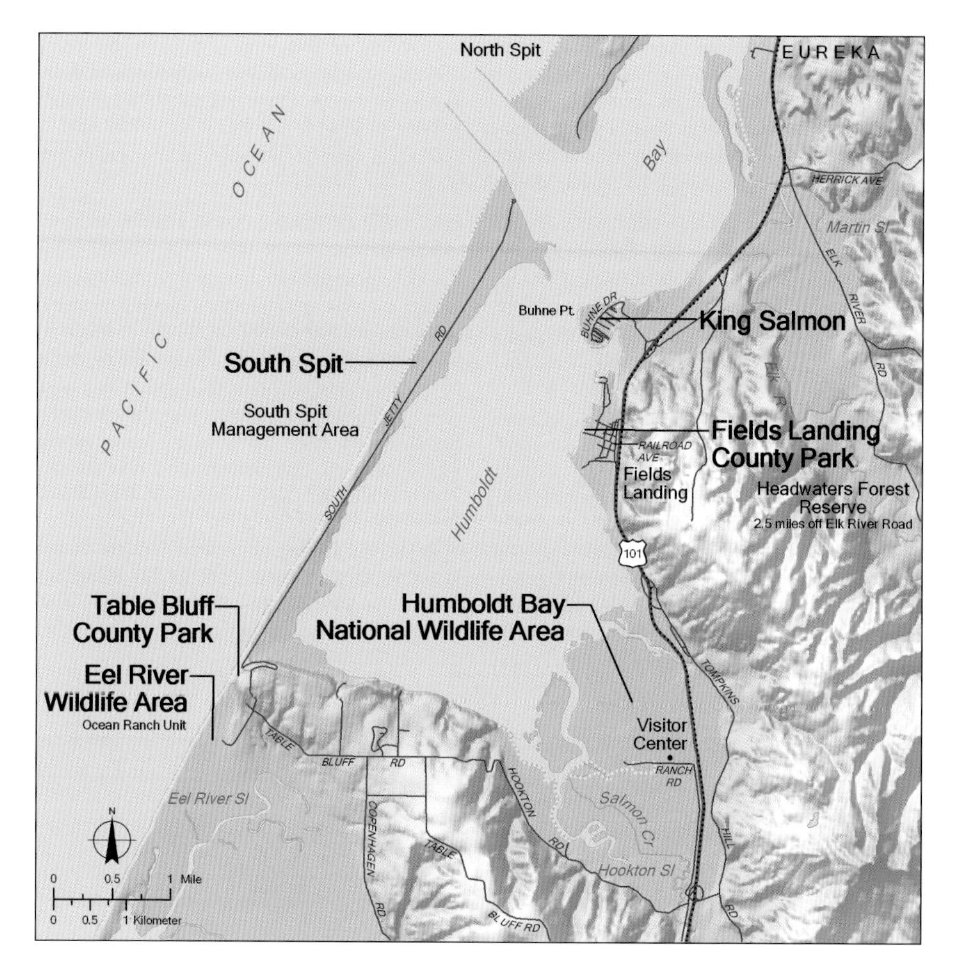

Yellow sand verbena

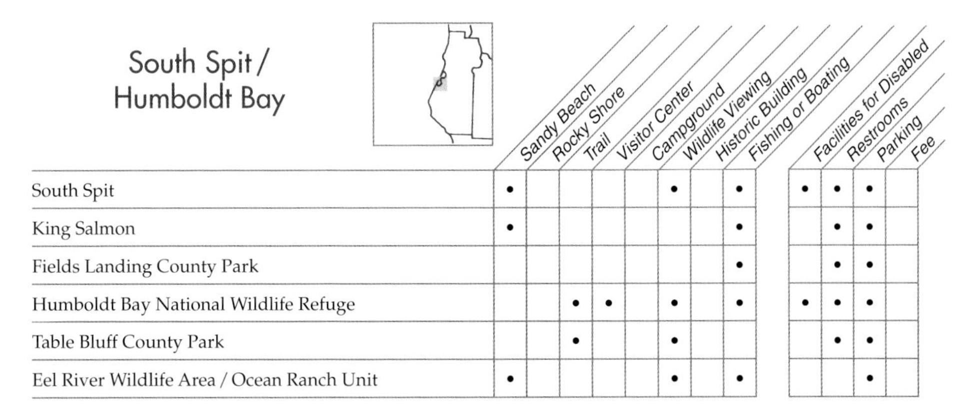

South Spit / Humboldt Bay

	Sandy Beach	Rocky Shore	Trail	Visitor Center	Campground	Wildlife Viewing	Historic Building	Fishing or Boating	Facilities for Disabled	Restrooms	Parking	Fee
South Spit	•					•		•	•	•	•	
King Salmon	•							•		•	•	
Fields Landing County Park								•		•	•	
Humboldt Bay National Wildlife Refuge			•	•		•		•	•	•	•	
Table Bluff County Park			•			•				•	•	
Eel River Wildlife Area / Ocean Ranch Unit	•					•		•			•	

SOUTH SPIT: *W. of Hwy. 101, S. of Eureka.* Use the Hookton Rd. exit off Hwy. 101. The South Spit is a four-mile-long strip of low dunes and fine-sand beach, separating the ocean from the southern part of Humboldt Bay. Long a remote and sometimes rowdy recreational spot, the spit has come under closer management in recent years. The South Spit Wildlife Area, cooperatively managed by the Department of Fish and Game and the Bureau of Land Management, offers picnic tables, fire rings, and restrooms at both ends of the spit. Public fishing access at the jetty, but use caution; waves can be dangerously large. Off-road vehicle access restricted to the wave slope; speed limit: 15 mph. In November, the South Spit's east side is popular for hunting Pacific brant.

A sparse coastal strand community of yellow bush lupine, European beach grass, beach strawberry, and yellow sand verbena occupies the low dunes on either side of South Jetty Rd., which runs the length of the peninsula. Songbirds and rodents on the spit are preyed upon by kestrels, northern harriers, owls, and foxes; terns, gulls, and shorebirds frequent the sandy beach. Snowy plovers nest here; watch for seasonal restrictions on vehicle and beach use. Open daily from one-half hour before sunrise to one-half hour after sunset. Pets allowed only if leashed. Call: 707-825-2300.

KING SALMON: *Off Buhne Dr., W. of Hwy. 101, 1 mi. N. of Fields Landing.* Take the King Salmon Dr. exit off Hwy. 101. Sport fishing charter boats, RV and trailer camping, and fishing and boating equipment are available at two privately operated facilities, Johnny's Marina and RV Park: 707-442-2284, and E-Z Landing RV Park and Marina: 707-442-1118. A picnic area with restrooms is adjacent to the PG&E power plant on King Salmon Drive.

FIELDS LANDING COUNTY PARK: *Foot of Railroad Ave., Fields Landing.* Use the Fields Landing exit from Hwy. 101 to reach this park overlooking Humboldt Bay. Facilities include a public concrete boat ramp, picnic tables, and restrooms. Small boats or kayaks can also be launched here.

HUMBOLDT BAY NATIONAL WILDLIFE REFUGE: *Humboldt South Bay, W. of Hwy. 101.* Use the Hookton Rd. exit off Hwy. 101; on the west side of the highway, take Ranch Rd. north to the Wildlife Refuge visitor center. Humboldt Bay is a major feeding and resting site for many kinds of migratory water birds. The Wildlife Refuge is one of the most important areas in the U.S. south of Alaska for Pacific brant. These geese feed in the large eelgrass beds of Humboldt Bay during winter and spring prior to flying north to arctic nesting grounds for the summer. Shorebirds such as willets, marbled godwits, and dunlin also use the bay's mudflats for feeding, and the bay serves as an important spawning and nursery area for fish, crabs, and other marine life. In winter, 100,000 birds may be present, feeding or resting in the bay and surrounding wetlands.

The Shorebird Loop Trail, less than two miles long, starts at the visitor center and takes in some of the refuge's best shorebird

If you visit the jetties at Humboldt Bay or the waterfront at Crescent City, you will notice concrete units that look like enormous sculptures. The shape of these units is an "H," where one side of the H has been rotated by 90 degrees. These concrete units, called dolosse (singular dolos), are named for their resemblance to a South African game piece made from the anklebone of a sheep. Dolosse have been designed for use in high wave energy environments, and the first time they were used in the United States was at Humboldt Bay.

Waves have been a regular threat to navigation along the North Coast; winter storm waves can reach 40 feet at the entrance to Humboldt Bay and 35 feet at Crescent City. Engineers have worked to improve navigational safety in these areas, building the Humboldt jetties in the late 1800s and the breakwater at Crescent City in the 1930s. While improving shipping safety, these structures were often damaged during storms. The Corps of Engineers determined that individual stones would need to be 80 to 100 tons each to be able to withstand the wave forces, and such large stones are nearly impossible to obtain.

Dolos units are designed specifically to dissipate wave forces in high-energy areas and to replace these very large stones. The twisted H shape helps these units nest together, and 42-ton dolosse provide the same stability as 80 to 100-ton quarry rocks. In 1971, 5,000 dolosse were added to the ends of the Humboldt Jetties. These dolosse measure 15′ x 15′ x 15′ and contain over 19 cubic yards of concrete and almost 1,500 pounds of reinforcing steel. Two years later, dolosse were added to the breakwater at Crescent City. These dolosse had no reinforcing steel, and within 12 years of initial placement approximately 75 percent of the units had broken. Most of the dolosse that can be seen today at Crescent City are 42-ton reinforced concrete units that were installed in 1986. No significant breakage has been observed in the units with reinforcing steel, and the breakwater has needed only minor maintenance since these dolosse were installed.

Dolosse

viewing areas, including seasonal wetlands and a permanent freshwater pond. The refuge's wildlife resources are described on interpretive panels. Look for deer, songbirds, tree frogs, and river otters, in addition to waterfowl and shorebirds. The best times to observe wildlife here are from November to April, an hour or two before and after high tide. Waterfowl hunting on the Salmon Creek unit takes place on Tuesdays and Saturdays by permit only. The Salmon Creek unit and visitor center are open daily from 8 AM to 5 PM. No pets allowed. Call: 707-733-5406.

The Hookton Slough unit of the refuge includes trails that offer fishing or wildlife viewing; take Hookton Rd. west from Hwy. 101. The Hookton Slough Trail follows a tidal slough past grasslands, freshwater marsh, mudflats, and open water along the edge of Humboldt Bay. An out-and-back walk is three miles long. Look for shorebirds and waterfowl from fall through spring, and herons, snowy egrets, osprey, and harbor seals anytime. Fishing is allowed from shore along the Hookton Slough Trail for sand dabs, English sole, leopard sharks, or young Dungeness crab. Fishing and wildlife viewing from the bay itself are also possible; a dock for launching small non-motorized boats is located at the Hookton Slough Trail parking area. The Hookton Slough unit is open daily from sunrise to sunset; no pets allowed.

TABLE BLUFF COUNTY PARK: *Table Bluff Rd. off Hookton Rd., 5 mi. W. of Hwy. 101, Loleta.* Humboldt Bay is separated from the Eel River to the south by a ridge that terminates above the ocean in a 165-foot-high flat bluff. The Table Bluff Reservation of the Wiyot Tribe is located here. Table Bluff County Park has views of Humboldt Bay and ocean, short paths, and restrooms. There is access to the Table Bluff unit of Humboldt Bay National Wildlife Refuge. Open one hour before sunrise until one hour after sunset.

EEL RIVER WILDLIFE AREA/OCEAN RANCH UNIT: *End of Table Bluff Rd. W. of Hwy. 101.* Wildlife viewing and hunting, with limited access due to conservation needs. Fox, deer, river otter, and marine mammals are found here. Vehicles are restricted to the wave slope (wet sand) and the back dune road. Dogs must be leashed between March 1 and September 15, due to snowy plover habitat. Parking available. Call: 707-445-6493.

The Headwaters Forest Reserve, consisting of over 7,000 acres of timberland formerly owned by Pacific Lumber Co., is managed jointly by the Bureau of Land Management and the State of California. The purpose of the reserve is to protect the upper watersheds of Elk River and Salmon Creek, which provide habitat for the marbled murrelet, king and coho salmon, steelhead, and salamanders.

Public access into the reserve is available from the northwest at the end of Elk River Rd., where there is parking and a kiosk with hiking information. A 5.5-mile trail follows an abandoned logging road to the old-growth forest; day use only.

Limited public access to the southern part of the reserve is available May through November 15, on guided hikes on weekends. On Tuesdays and Wednesdays, the trail is reserved for school groups. For reservations on guided hikes, call: 707-825-2300.

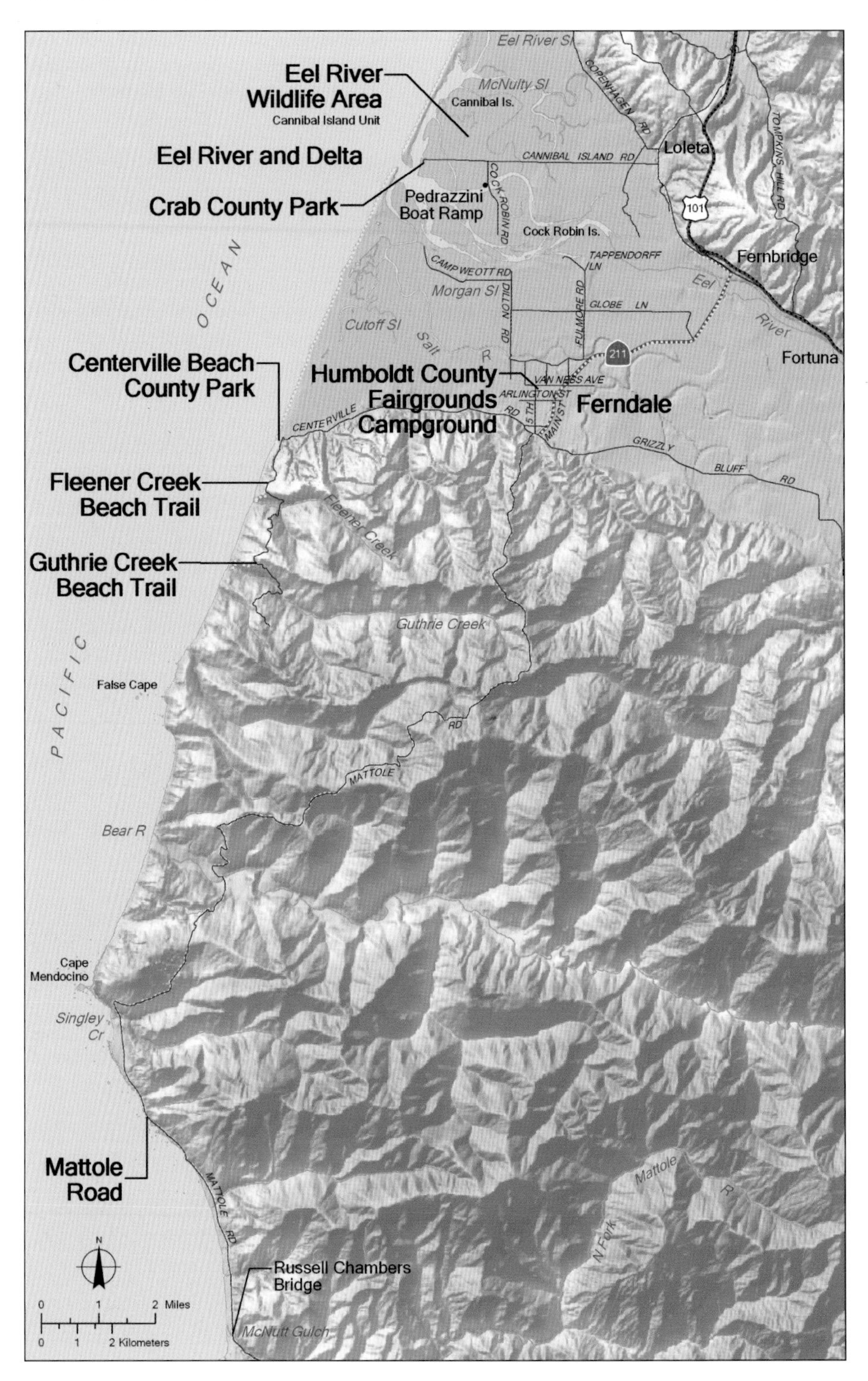
Eel River Wildlife Area
Cannibal Island Unit
Eel River and Delta
Crab County Park
Centerville Beach County Park
Fleener Creek Beach Trail
Guthrie Creek Beach Trail
Mattole Road
Humboldt County Fairgrounds Campground
Ferndale
Pedrazzini Boat Ramp
Cannibal Is.
Cock Robin Is.
Loleta
Fernbridge
Fortuna
Russell Chambers Bridge
False Cape
Cape Mendocino
PACIFIC OCEAN
Eel River Sl
McNulty Sl
Morgan Sl
Cutoff Sl
Salt R
Eel River
Fleener Creek
Guthrie Creek
Bear R
Singley Cr
McNutt Gulch
Mattole R
N Fork
COPENHAGEN RD
TOMPKINS HILL RD
CANNIBAL ISLAND RD
COCK ROBIN RD
CAMP WEOTT RD
TAPPENDORFF LN
DILLON RD
FULMORE RD
GLOBE LN
VAN NESS AVE
ARLINGTON ST
5TH
MAIN ST
CENTERVILLE RD
GRIZZLY BLUFF RD
MATTOLE RD
101
211
N
0 1 2 Miles
0 1 2 Kilometers

Eel River Valley South

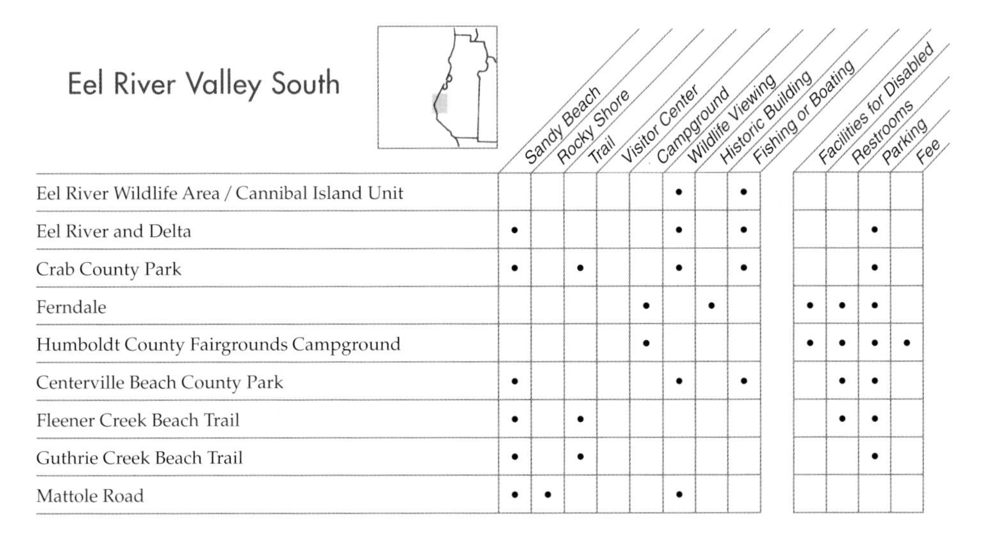

	Sandy Beach	Rocky Shore	Trail	Visitor Center	Campground	Wildlife Viewing	Historic Building	Fishing or Boating	Facilities for Disabled	Restrooms	Parking	Fee
Eel River Wildlife Area / Cannibal Island Unit						•		•				
Eel River and Delta	•					•		•			•	
Crab County Park	•		•			•		•			•	
Ferndale					•		•		•	•	•	
Humboldt County Fairgrounds Campground					•				•	•	•	•
Centerville Beach County Park	•					•		•		•	•	
Fleener Creek Beach Trail	•		•							•	•	
Guthrie Creek Beach Trail	•		•								•	
Mattole Road	•	•				•						

EEL RIVER WILDLIFE AREA / CANNIBAL ISLAND UNIT: *W. end of Cannibal Island Rd., 4 mi. from Loleta.* The Eel River Wildlife Area includes wetlands and a large riparian forest, habitat for wildlife that includes foxes, deer, river otters, and many species of birds. The Cannibal Island unit of the wildlife area borders McNulty Slough and offers hunting and wildlife viewing; no facilities. For information, call: 707-445-6493.

EEL RIVER AND DELTA: *W. of Hwy. 101 at Fernbridge.* The Eel River emerges from narrow river canyons south of Fortuna into a broad, marshy estuary, much of it devoted to agriculture. The Eel is California's third largest river, carrying ten percent of the state's annual runoff, and its delta provides essential habitat for migratory birds using the Pacific Flyway. The delta's raptor population, which includes ospreys, red-tailed hawks, kestrels, and endangered peregrine falcons, is one of the largest along the coast. The estuary also serves as a spawning and nursery area for ocean fish such as the starry flounder and Pacific herring.

The lower delta, with its maze of river channels and sloughs, is well suited for kayaking, fishing, hunting, or wildlife observation. There are numerous points of access to the water's edge, although signs and facilities are scarce, and private farmlands are all around. The lower river delta is accessible both from the north side, off Cannibal Island Rd., and from the south side, off Dillon and Camp Weott Roads.

Pedrazzini Boat Ramp, on Cock Robin Island Rd. off Cannibal Island Rd., consists of a parking area and a ramp for launching small boats or kayaks into the Eel River. From here one could boat upriver toward Fernbridge, or west toward the river mouth and then north along Eel River Slough.

On the south side of the Eel River, there are public access points at the ends of Camp Weott and Dillon Roads. Parking at both locations is very limited, and surrounding land is private. At the Dillon Rd. access point, a flat, grassy, one-eighth-mile trail leads to the river bank, providing good views of waterfowl, great blue herons, and harbor seals hauled out on the sand. Tappendorf Ln., Singley Rd. near Fernbridge, and Sandy Prairie Rd. near Fortuna also provide access to the Eel River.

CRAB COUNTY PARK: *W. end of Cannibal Island Rd., 4 mi. from Loleta.* The park is located near the Eel River mouth, which is visible from the small sandy parking lot. Grass-covered dunes border the river, and paths lead north and south along the water for fishing and wildlife viewing. Look across the channel for harbor seals hauled out on the spit. Dogs, horses, and beach fires allowed.

Ferndale

Sometimes turbulent water where the river approaches the sea may make this a less favorable site for kayak launching than Pedrazzini Boat Ramp, a mile upstream. For information, call: 707-445-7651.

FERNDALE: *Hwy. 211, 5 mi. S.W. of Hwy. 101.* Well known for its Victorian architecture, this small town developed as a prosperous farm community and dairying center for the Eel River Delta. Floods in the 1950s and 1960s and a decline in the dairy business threatened to turn Ferndale into a ghost town. But the low cost of buildings attracted artists who helped breathe life back into Ferndale, and today the flourishing community is home to antique stores, gift shops, art galleries, and eating establishments along Main Street. The Ferndale Museum at 3rd and Shaw Streets contains artifacts from the town's history; call: 707-786-4477.

HUMBOLDT COUNTY FAIRGROUNDS CAMPGROUND: *5th St. between Van Ness and Arlington Sts., Ferndale.* The campground is located in a flat, grassy area outside the county fairgrounds, a block west of Main Street. Facilities include 94 campsites, electrical hookups, water at each site, and restrooms with showers; no picnic tables. Fee for camping. Reservations required during the fair in August; call: 707-786-9511.

CENTERVILLE BEACH COUNTY PARK: *Centerville Rd., 5 mi. W. of Ferndale.* This park marks the edge of the low-lying Eel River Delta and the beginning of mountainous ridges that extend south into the King Range. A large parking area is located next to a long, sandy beach on an undeveloped stretch of coastline. Restrooms and firepits available; dogs and horses allowed. Along the eastern edge of the beach are low sand dunes covered with grass and wildflowers, showing some damage from off-road vehicles that are allowed on the beach. Fishing for surfperch or smelt; the steep beach and riptides make swimming inadvisable. Day use only; for information, call: 707-445-7651.

FLEENER CREEK BEACH TRAIL: *Centerville Rd., 6 mi. S.W. of Ferndale.* A trailhead with small gravel parking area, picnic tables, information kiosk, and pit toilet is planned for this property newly acquired by the Bureau of Land Management. A half-mile trail will lead from the blufftop overlooking Fleener Creek across former grazing lands to a narrow sandy beach. Facilities are anticipated to be open by 2005; call: 707-825-2300.

GUTHRIE CREEK BEACH TRAIL: *Centerville Rd., 8 mi. S.W. of Ferndale.* A parking area, restroom, and hiking trail to the beach are located on the former Lost Coast Ranch, acquired for the public in 2001 by the Bureau of Land Management, with assistance by the State Coastal Conservancy and The Conservation Fund. The mile-long trail leads down to the mouth of Guthrie Creek on a narrow sandy beach backed by high wave-eroded cliffs. Call: 707-825-2300.

MATTOLE ROAD: *Singley Creek to McNutt Gulch.* An eight-mile stretch of road borders the coast at ocean level. Sandy beaches, dunes, and rocky areas are accessible at four locations: mile marker 25.02; mile marker 24.02 near Devils Gate; 500 feet south of mile marker 23.07; and on the north side of the Russell Chambers Bridge across McNutt Gulch, just before the road turns inland, where there is a path to the beach through a gap in the fence. The northern paths are 50 feet wide, and the path at McNutt Gulch is 20 feet wide. Private property is adjacent; do not trespass. Limited shoulder parking.

Cape Mendocino, the grass-covered headland jutting into the ocean north of Singley Creek, is the westernmost point in the continental United States. This prominent feature was used for navigation by famous explorers including Francis Drake in 1579, Sebastián Rodriguez Cermeño in 1595, Sebastián Vizcaíno during 1602–3, and George Vancouver, who set out in 1792 to survey on behalf of England the extent of Spanish holdings on the Pacific coast.

No access to the cape, but there is a good view on the south side from the Mattole Road. The Cape Mendocino Lighthouse was located here from 1868 until 1999, when it was moved to Mal Coombs Park in Shelter Cove.

Mattole Road

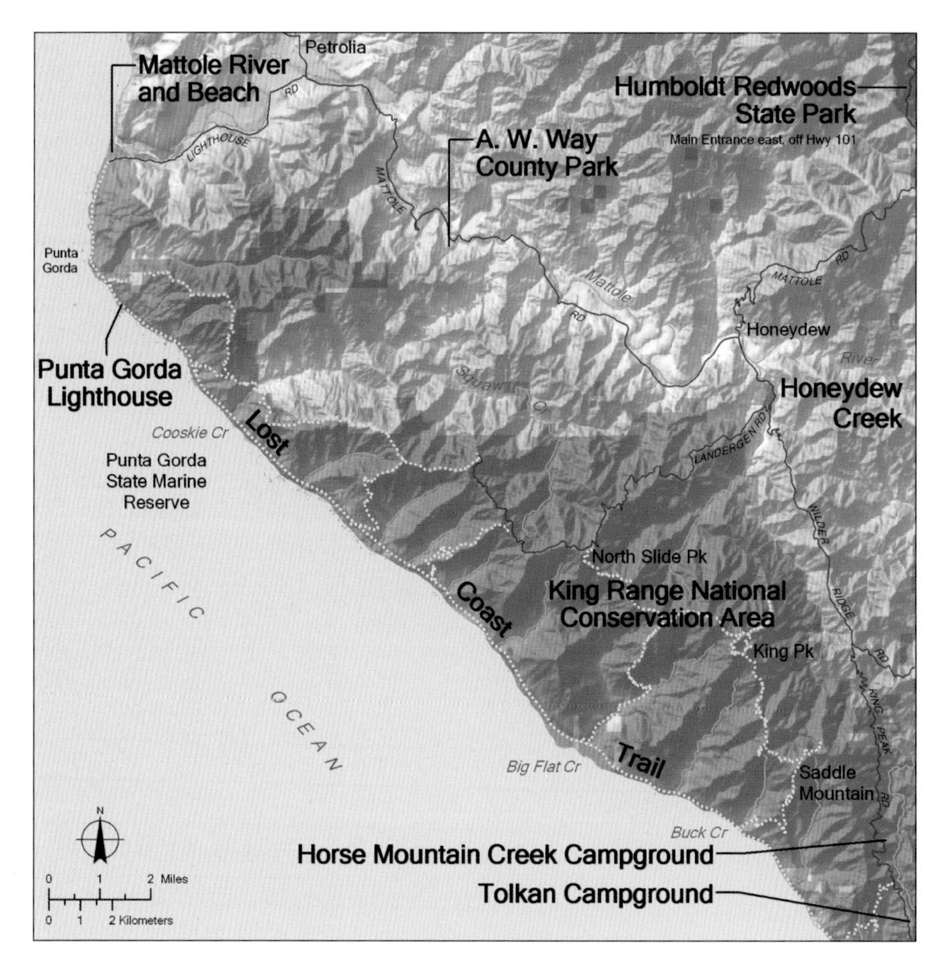

Mattole River camp

Mattole River / King Range

	Sandy Beach	Rocky Shore	Trail	Visitor Center	Campground	Wildlife Viewing	Historic Building	Fishing or Boating	Facilities for Disabled	Restrooms	Parking	Fee
Mattole River and Beach	•		•		•	•			•	•	•	•
Punta Gorda Lighthouse			•				•					
A. W. Way County Park					•			•		•	•	•
Humboldt Redwoods State Park			•	•	•			•	•	•	•	•
King Range National Conservation Area	•	•	•	•	•	•		•		•	•	•
Lost Coast Trail	•	•	•		•	•		•		•	•	
Honeydew Creek					•					•	•	
Horse Mountain Creek and Tolkan Campgrounds					•					•	•	

MATTOLE RIVER AND BEACH: *W. end of Lighthouse Rd., 5 mi. W. of Petrolia.* Day-use area and campground are located at the end of a paved road at the Mattole River mouth, at the northern terminus of the Lost Coast Trail. The Mattole River estuary attracts black turnstones and other shorebirds, and in fall and winter common loons, western grebes, and other diving ducks may be seen. Harbor seals and sea lions gather on offshore rocks, migrating gray whales pass close by, and there are tidepools to explore. Nestled in sand dunes, there are 14 campsites with picnic tables, barbecue grills, running water, and restrooms. Fee for camping.

PUNTA GORDA LIGHTHOUSE: *3 mi. S. of Mattole River Beach on Lost Coast Trail.* From the Mattole Beach, a 2.4-mile segment of the Lost Coast Trail leads to Punta Gorda and 1.4 miles beyond the point to the Punta Gorda Lighthouse, offering either the beginning of a multi-day hike through the King Range National Conservation Area or a moderately challenging afternoon out-and-back along the wilderness coast. The lighthouse was closed in 1951 because of the high cost of maintenance. Parts of the Lost Coast Trail cannot be negotiated at high tide; check a tide table before setting out. Dress for changeable weather any time of year, and carry water. Offshore is the Punta Gorda State Marine Reserve, where the taking of marine plants and animals is prohibited; for information, call: 707-445-6493.

A.W. WAY COUNTY PARK: *Mattole Road, 6 mi. S.E. of Petrolia.* Thirty campsites in a grassy meadow by the Mattole River; picnic tables, firepits, restrooms, and running water. Fee for camping and day use. "Mattole" means clear water in native language, and the river lives up to its name, snaking to the ocean through towering ridges of the King Range. In summer, the valley of the Mattole enjoys sunshine, sheltered from coastal fog.

HUMBOLDT REDWOODS STATE PARK: *Hwy. 101, 20 mi. N. of Garberville.* The park includes the largest remaining contiguous old-growth coast redwood forest in the world. Camping and day use; fishing on the Eel River in fall and winter. The visitor center is one mile south of Weott on Avenue of the Giants, at Burlington Campground. A scenic loop drive can be taken from Hwy. 101 via Bull Creek Rd., which becomes Mattole Rd., leading west to the coast and then north to Ferndale near Hwy. 101. The road is paved, although steep and winding; allow at least a half day to drive the 70-mile loop.

KING RANGE NATIONAL CONSERVATION AREA: *Mattole River Beach to S. of Shelter Cove.* Most of California's 34 million residents live near the coast, and freeways and urban development line much of the state's shoreline. The King Range presents a strong contrast: a 35-mile stretch of California coastline without even a rural highway. Shoreline road-building along the steep and unstable 4,000-foot slopes of the King Range was simply not feasible, and Hwy. 101 instead skirts the east side of the mountains. The result is what is known as the Lost Coast, much of which lies in the 60,000-acre King Range Conservation Area. The Bureau of Land Management (BLM) emphasizes the wilderness values of this unique area, and developed tourist facilities are few. Visitors should plan accordingly.

The steep slopes of the King Range are the result of geological activity, as three giant chunks of the earth's crust press together at an offshore point known as the Triple Junction. The mountains have risen as much as 66 feet in the last 6,000 years, including a three-foot rise within seconds during the 1992 Petrolia earthquake. The shoreline is lined by narrow beaches, intersected by ridges and intervening streams. Douglas-fir-covered slopes are mixed with chaparral and grassland. The King Range receives high rainfall, up to 100 inches or more per year, but summers are generally rainless. Near the shoreline, summer fogs alternate with dry winds from the interior.

Over 70 miles of hiking trails are available, from peaks such as King, North Slide, and Saddle Mountain, down to ocean level along the Lost Coast Trail. Inland trails are often steep, and most back-country roads are not suitable for trailers; some unimproved roads are best negotiated with four-wheel drive vehicles or mountain bikes. In addition to the Lost Coast Trail, which runs the length of the shoreline, the King Crest Trail runs along an inland ridge, while the Horse Mountain Creek Trail, the Buck Creek Trail, and others connect to the coast. Roads and trails are not well marked, and it is essential to have a map in hand when exploring the back country; visit the King Range visitor center at 768 Shelter Cove Rd., nine miles east of Shelter Cove, call: 707-986-5400, or see www.ca.blm.gov/arcata/king_range.html.

The King Range has seven public campgrounds; dogs allowed on leash. No reservations, except for the group camp at Nadelos, southeast of Shelter Cove. Camping on BLM public lands is also permitted, except at locations specifically closed, such as at Shelter Cove; check signs or BLM maps. Outside developed campgrounds, permits are required for campfires or stoves, and campfires are prohibited during the fire season.

Trail just north of Big Flat, Lost Coast

Big Flat Creek with King Peak in background, Lost Coast

Water from coastal streams must be purified before drinking, and water sources on ridge trails are limited.

LOST COAST TRAIL: *From Mattole River Beach to Black Sands Beach, Shelter Cove.* The 24.8-mile trail offers a true wilderness experience along the beach, between surf on one side and steep slopes on the other. Everything seems big here: high cliffs, giant ochre stars in tidepools, the vast Pacific. In spring and summer, wildflowers bloom near the beach or along hanging creeks that cascade onto the sand; look for beach aster, farewell-to-spring, and monkeyflower. In early morning, look in the sand for tracks of raccoons, bears, or shorebirds. Rattlesnakes are also present.

If planning to hike the whole trail, start at the Mattole River Beach end in order to have prevailing winds at your back. Hiking on sloping sand and slippery rocks can be tough going; wear sturdy footgear, not beach sandals. Punta Gorda and two trail segments of several miles each are impassable at high tide; check trailhead signs and consult a tide table in advance. Campers bring tents or use driftwood shelters; be prepared for changeable weather. Bears visit the coast too; food and scented items must be stored overnight in animal-resistant canisters, which can be rented at BLM offices and the Petrolia Store. No toilets are available on the trail. A few private cabins lie along the route; do not trespass. Special recreation use permits are required for all organized commercial and non-commercial trips. Call: 707-986-5400.

The Lost Coast Trail Transport Service offers a shuttle between all trailheads in the King Range National Conservation Area and in Sinkyone Wilderness State Park. Hikers can park at one trailhead, get a shuttle ride to another, and then hike back to their vehicles. Call: 707-986-9909. The Shelter Cove Campground and Deli also offers shuttle information; call: 707-986-7474.

HONEYDEW CREEK: *Wilder Ridge Rd., 1.1 mi. S. of Honeydew.* A campground by a creek has picnic tables, firepits, restrooms, and water. Trailers not recommended on Wilder Ridge Rd. south of the campground.

HORSE MOUNTAIN CREEK AND TOLKAN CAMPGROUNDS: *King Peak Rd., 3.6 and 6.5 mi. N. of Shelter Cove Road.* Two campgrounds on a ridge above the Lost Coast; picnic tables, firepits, restrooms, and water.

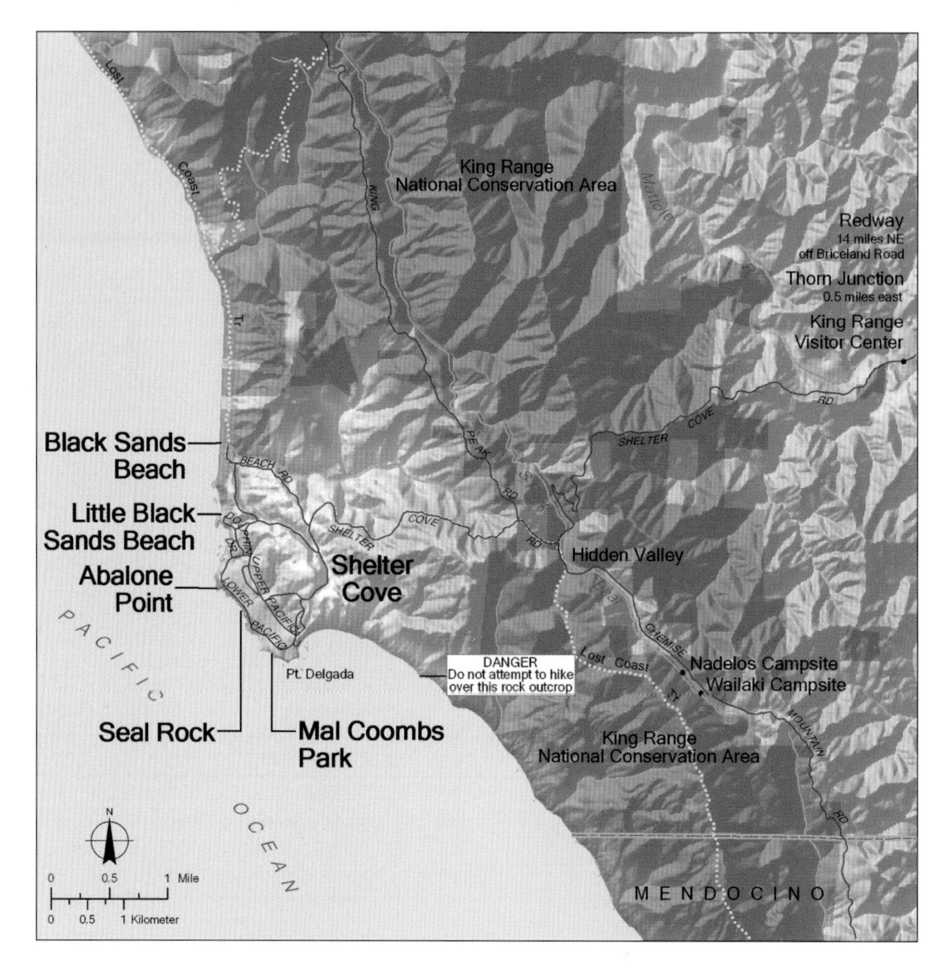

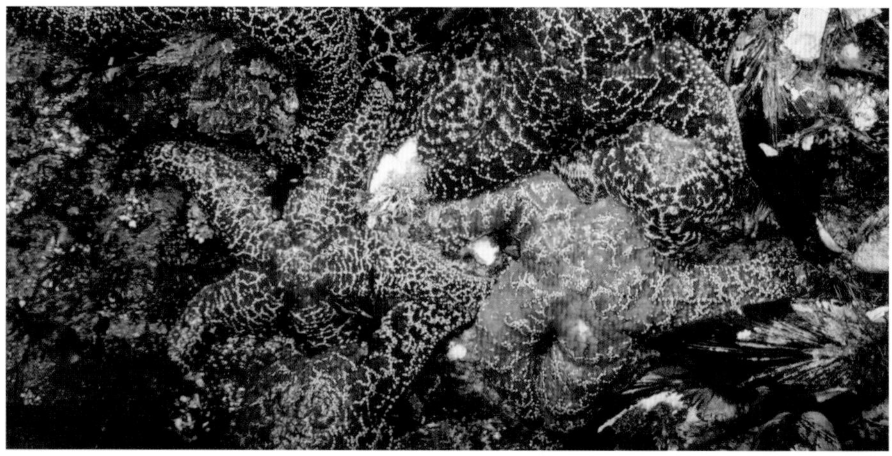

Ochre sea stars

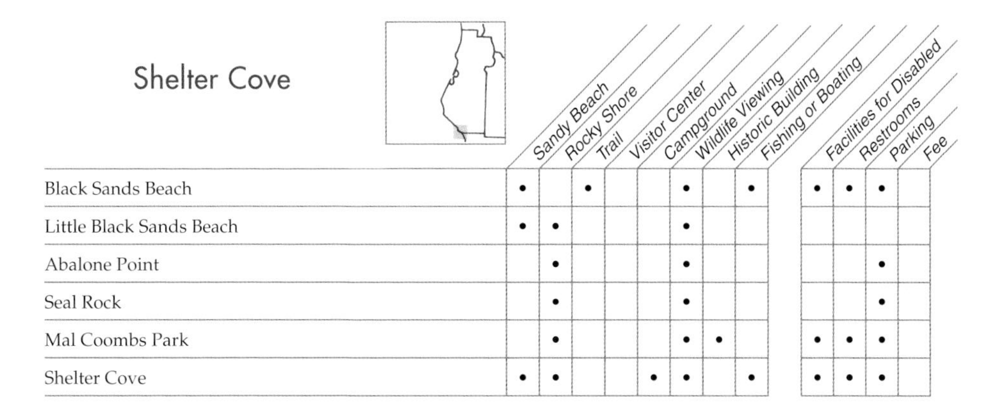

Shelter Cove

	Sandy Beach	Rocky Shore	Trail	Visitor Center	Campground	Wildlife Viewing	Historic Building	Fishing or Boating	Facilities for Disabled	Restrooms	Parking	Fee
Black Sands Beach	•		•			•		•	•	•	•	
Little Black Sands Beach	•	•				•						
Abalone Point		•				•					•	
Seal Rock		•				•					•	
Mal Coombs Park		•				•	•		•	•	•	
Shelter Cove	•	•			•	•		•	•	•	•	

BLACK SANDS BEACH: *N. end of Beach Rd., Shelter Cove.* At the north end of Shelter Cove, at the end of Beach Rd., is a dark gray beach backed by high forested slopes. This is the southern terminus of the Lost Coast Trail, which leads north to Mattole River Beach. About four miles north of the Black Sands Beach parking area, the trail is not passable at high tide; plan your hike accordingly. A bulletin board at the trailhead describes the Lost Coast Trail segments where high water limits access and gives tide times for upcoming weeks. There are also interpretive panels describing the active geology of the area. Erosion of the cliffs along the coastline contributes sand to the beach. The sand's dark color reflects the dark color of the sandstone making up the cliffs. This type of sandstone is called graywacke, a well-cemented mixture of sand-sized quartz, feldspar, and rock grains in a matrix of mud. Graywacke is common in the Franciscan Complex and was formed from sediments swept into deep water by underwater mud flows.

There are picnic tables and two parking areas, including a separate wheelchair-accessible parking area part-way down the bluff; restrooms are near the upper parking area. Black Sands Beach is intersected by several small streams that flow down the steep slopes, and there is a verdant growth of wildflowers on the bluffs. Visible to the north is the King Range, which rises

Black Sands Beach

Cape Mendocino Lighthouse relocated to Shelter Cove

abruptly from sea level to a height of 4,000 feet in less than three miles. Golden eagles, bald eagles, and peregrine falcons, all endangered species, can be found in the mountains. For information on Black Sands Beach, the King Range, or the Lost Coast Trail, call: 707-986-5400 or 707-825-2300.

LITTLE BLACK SANDS BEACH: *W. end of Dolphin Dr., Shelter Cove.* Dolphin Dr., off Lower Pacific Dr., ends abruptly at the bottom of a hill, where the ocean continues to erode the bluffs. There are no facilities, but there is an overlook at the edge of a black sand beach offering views of the sea, shorebirds, and cliff swallows.

ABALONE POINT: *Lower Pacific Dr., Shelter Cove.* This picnic area one-quarter mile west of Upper Pacific Dr. offers views north toward the Lost Coast; interpretive panels describe the geology of the area. Wildflowers including sea thrift, butter and eggs, and lupine grow on the low coastal bluff at Abalone Point, where there is access to abundant tidepools. California mussels, purple sea stars, and Pacific goose barnacles inhabit the rocky shore. The Sinkyone Indians once spent summers here, fishing for salmon, hunting seals, and collecting clamshells for trade with the Northern Pomo Indians. The Sinkyone people were driven out when ranchers and loggers settled grasslands near the sea in the 1850s. By 1886, a wharf had been built at the cove, and milk, wool, and tanbark were exported.

SEAL ROCK: *Lower Pacific Dr., Shelter Cove.* A small blufftop park one-quarter mile south of Abalone Point has parking, three picnic tables, and interpretive panels with information about marine mammals, including gray whales that can be seen migrating past Shelter Cove between November and late

Seal Rock

May. Harbor seals haul out on the rocks here. No beach access.

MAL COOMBS PARK: *Lower Pacific Dr., Shelter Cove.* The park is located one-half mile south of Seal Rock, near Point Delgada. An open grassy area overlooking the sea has picnic tables and barbecue grills. The old Cape Mendocino Lighthouse, built in 1868, has been moved here from its original location some 35 miles to the north, after being declared surplus property in 1994. From the low bluff, three dozen steps lead down to the shore, where anemones, hermit crabs, sea cucumbers, and oysters inhabit the rocks and tidepools. Birds such as black oystercatchers and double-crested cormorants may be seen in the area.

SHELTER COVE: *W. end of Shelter Cove Road.* Take Briceland-Thorn Rd. 20 miles west from Hwy. 101 (Business Route) in Redway, north of Garberville, to Shelter Cove Road. The King Range National Conservation Area office of the Bureau of Land Management is located on Shelter Cove Rd., near Thorn Junction; open 8 AM–4:30 PM during summer and more limited hours in winter; call: 707-986-5400. A large color map and guide to the King Range is for sale at the office as well as at local businesses in Shelter Cove. Shelter Cove is a privately owned residential community at Point Delgada within the King Range National Conservation Area. There are motels, a private campground and RV park, a daylight airstrip surrounded by a nine-hole golf course, and limited services.

At the south end of the loop formed by upper and lower Pacific Drives is the main cove, where there is a small fine-sand beach and boat launch managed by the Humboldt County Harbor District. Surfers ride several breaks near the cove, although the shoreline has rocky hazards. Mario's Marina complex on the 100-foot-high bluff above the cove includes boat launching and dry storage facilities, as well as a restaurant and motel; call: 707-986-1145. There is also a bait and tackle shop, where charter fishing boat referrals are available; call: 707-986-1234. Salmon fishing is popular here, along with bottom fishing, crabbing, surf fishing, and abalone hunting. The Shelter Cove Campground and Deli offers information on shuttle services for backpackers; call: 707-986-7474.

The sheer cliffs adjoining the cove protect it from northwest winds and waves. Point No Pass south of the cove is a rock promontory hundreds of feet high that blocks access by foot along the beach; the Coastal Trail leading into Mendocino County is routed well inland, east of the mountain ridge.

Fish cleaning station at Shelter Cove

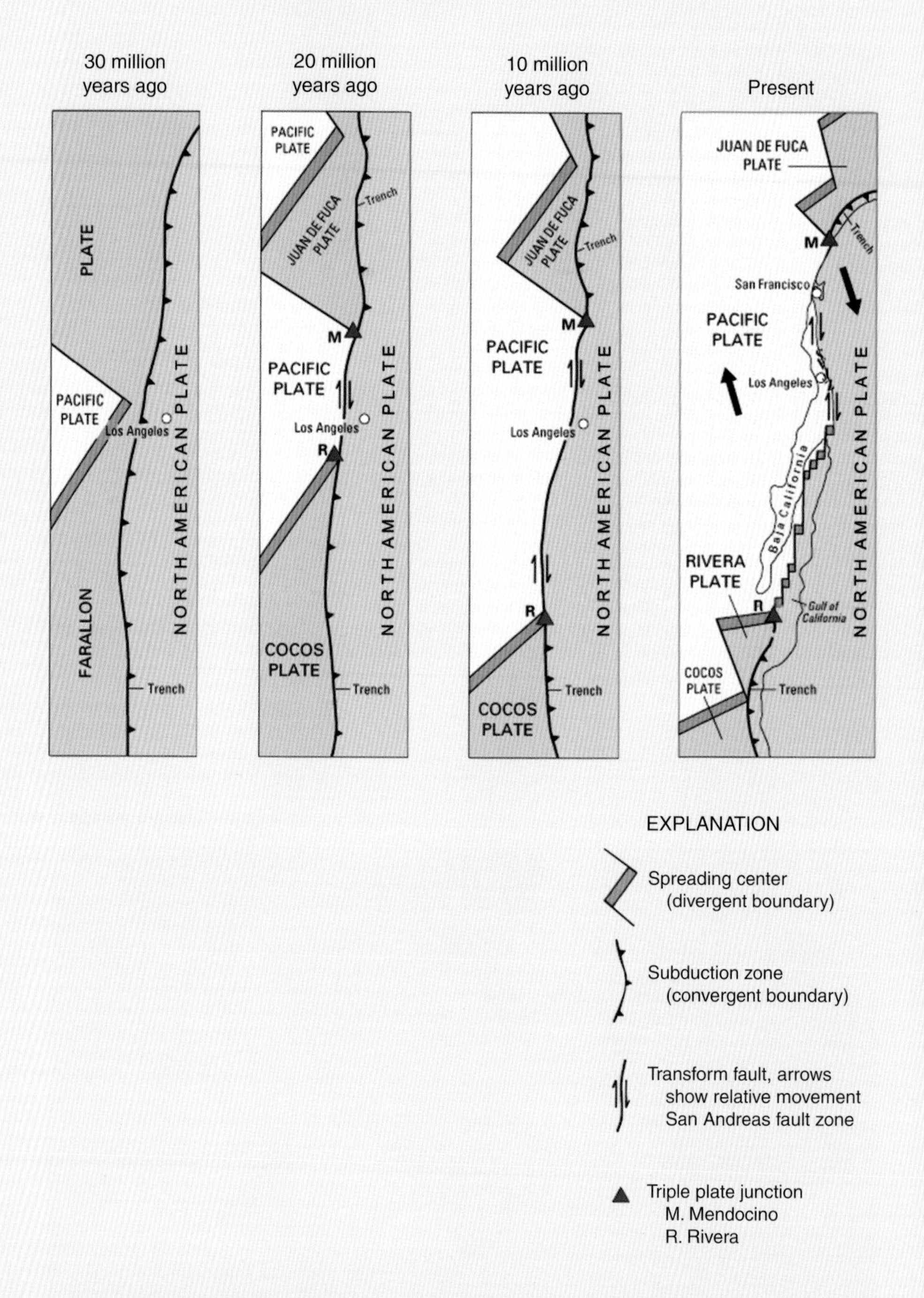

Adapted from U.S. Geological Survey

Geology of the Coast

THE CALIFORNIA COAST owes its rugged beauty to the fact that it is a geologically active place. Like the rest of the west coast of North America, California lies in a zone of collision between the shifting plates that make up the Earth's crust. As the nature of these plate interactions has changed, the shape of the California coast has changed as well.

Prior to about 28 million years ago, the entire California coast was a zone of *subduction*, along which the Farallon plate, made up of ocean crust, dived beneath the westward-advancing North American plate. The descending oceanic plate brought water into the Earth's mantle, lowering its melting point and generating molten rock, or magma. The magma migrated upward, and upon reaching the Earth's surface, a chain of volcanoes was produced. The roots of these volcanoes can still be seen in the granitic and altered volcanic rocks of the Sierra Nevada. During subduction, marine sediments were scraped off the Farallon plate, forming a thick wedge of sedimentary rock along the coast. Altered by heat and pressure, and sheared by tectonic processes, these sediments were transformed into an assemblage of sandstones, siltstones, shales, and cherts known as the Franciscan Complex. Found from Del Norte to San Luis Obispo Counties, this assemblage of weak and sheared rocks is particularly susceptible to landslides.

While subduction was occurring at the eastern edge of the Farallon plate along the coast of North America, new crust was being formed at the plate's western boundary, a so-called "spreading center." Here new ocean crust was formed as magma rose from below through cracks formed by diverging plates on either side of the spreading center. Across the spreading center to the west lay the Pacific plate. Approximately 28 million years ago, the westward drift of the North American plate caused that plate to override the spreading center, and it entered the subduction zone along the California coast. A "triple junction" was born—a point where three plates (North American, Farallon, and Pacific) come together. As subduction continued, the Farallon plate was broken into smaller plates, and the triple junction migrated to the north along the coast. A similar triple junction migrated to the south. This migration was along a series of faults across which rocks moved horizontally, and which later evolved into the San Andreas fault system. The fragments of the Farallon plate north of the triple junction are now known as the Juan de Fuca and Gorda plates, and the triple junction is known as the Mendocino triple junction.

The Mendocino triple junction marks a major dividing line on the California coast. North of the triple junction, the coast is still a subduction zone, known as the Cascadia subduction zone, and volcanoes such as Mt. Lassen and Mt. Shasta are generated by the subsiding Gorda and Juan de Fuca plates. Slippage of these plates beneath North America can generate very large earthquakes; there is evidence that the Cascadia subduction zone can generate earthquakes of magnitude ~9 once every 400–500 years. The last such earthquake occurred at about 9 PM on January 26, 1700; the time of the event is known with such precision because the arrival of a resulting tsunami was recorded in Japan. Between earthquakes, the coast is gradually uplifted by the subducting plate, and sea level relative to the coast is actually falling in northernmost California, despite the fact that absolute sea level is on the rise globally. South of the triple junction, the coast has been sheared by the series of faults carrying slivers of southern coastal California and Mexico northward relative to the rest of the state. The most famous of these faults is the San Andreas fault, which runs through interior and coastal California from the Gulf of California to the Mendocino triple junction at Cape Mendocino. Slippage along this and related faults has caused earthquakes as large as magnitude 8, such as the 1906 San Francisco earthquake. The City of Los Angeles is located west of the San Andreas fault, while San Francisco is east of the fault. If movement along the fault continues in the same direction and at the same rate as in the geologic past, in about 10 million years the two cities will lie adjacent to each other. Bends along these faults cause the rocks on opposite sides of the fault to be compressed, resulting in uplift and mountain building—and the spectacular scenery of the California coast.

In addition to the Franciscan Complex, rocks found on the California coast include younger rocks composed of sediments uplifted from the ocean floor, and occasional fragments of the older rocks that lie beneath them. These so-called "basement rocks" include the granite found on Point Reyes.

California's coast is an *emergent* coast—its rugged cliffs are the result of uplift of the land. Between periods of uplift, waves cut a flat, submarine *shore platform* which dips gently offshore. Waves continually advance this platform landward, creating a sea cliff. Sand carried to the shore by rivers may cover the shore platform, and be piled against the sea cliff to form a beach. Ancient shore platforms, sea cliffs, and beaches are preserved as marine terraces and terrace deposits when uplift of the land brings them above the sea. Sea cliffs in California commonly are capped by such ancient marine terrace deposits, representing former beaches now lying high above the waves.

Red chert, rock of the Franciscan Complex, at the Marin headlands

Cliffs and Bluffs

CALIFORNIA'S coastal cliffs and bluffs are the seaward edges of marine terraces, shaped by ocean waves and currents and uplifted from the ocean floor. Even the hardest rocks are subject to erosion in the harsh shoreline environment, and the organisms that occur here are tenacious. Ledges, gullies, slopes, and cracks provide spaces where soil can collect and seeds germinate. Buffered from the driving wind on protected ledges, seabirds such as the common murre and double-crested cormorant rest and build nests.

The roots of **bluff lettuce** (*Dudleya farinosa*) penetrate rock crevices on cliffs and bluffs, holding tight to steep slopes and rocky outcrops. The plant has a rosette of succulent leaves with a chalky appearance. The beautiful lemon-yellow flowers sit in clusters atop the stem. This species is a member of the live-forever family (Crassulaceae), whose members are usually perennial and have fleshy leaves. The succulent leaves of bluff lettuce serve the important function of water storage. Look for this species in bloom from May to September. As the name implies, the leaves of this plant are edible.

Bluff lettuce

Sea thrift (*Armeria maritima*) is a salt-tolerant plant often seen growing on rocky bluffs along the shore. It forms compact cushions and has wonderful pink flowers that contrast sharply with its tuft of dark evergreen foliage. The base of the flowers is surrounded by a calyx that has a paper-like texture. Reach down and feel the unusual, onion-skin-like calyx. The tight ball of pink flowers distinguishes this species from other flowers that grow along the coast. The name "thrift" refers to the narrow grass-like leaves, which reduce surface area and are thought to slow water loss in the salty air.

Sea thrift

Beach strawberries (*Fragaria chiloensis*) grow tiny, edible, red fruit that are generally less than one-half inch across. They are native to California's coastal bluffs and sand dunes. The beach strawberry has bright white flowers and three-part rounded, furry leaves. Horizontal runners, called stolons, arise from the base of the plant and spread out in every direction. Where the sto-

Beach strawberries

lons touch the soil, a new plant develops. If you chance upon a beach strawberry with ripe fruit, you may consider conducting a taste test. This species is one of two forefathers to the delicious cultivated strawberry.

Common murre

The **common murre** (*Uria aalge*) is an elegant, black-and-white pelagic seabird that only comes on shore to nest. It spends the other eight to nine months of the year offshore, in favorable feeding waters. Common murres nest in colonies on cliffs or among boulders and rocks, without using any nesting material. They lay one egg directly on the rock, where it is incubated for about a month. The egg is pear-shaped, or pyriform; if the egg is bumped, it pivots around its pointed end without falling off the ledge. Murre chicks leave the nest before they are fully grown. They jump out of the nest into the sea or onto the beach 30 to 50 feet below. Even if a murre chick lands on rocks or pebbles on the shore, it rarely gets hurt by falling down, and it walks directly toward the water. Once the chick reaches the sea, the male parent teaches it how to dive and catch fish.

Double-crested cormorant

Double-crested cormorants (*Phalacrocorax auritus*) dive for fish and marine invertebrates. After catching a fish, the cormorant surfaces, flips the fish in the air, and swallows it head-first. It has a long, hooked bill and an orange throat pouch. The double-crested cormorant has poorly developed oil glands and therefore is not waterproofed very well. You may catch sight of one perching on a rock and drying its feathers by stretching its wings out in the sun. This can be an impressive sight, because the wingspan of the double-crested cormorant is about four feet. Double-crested cormorants can also be seen flying in long lines, passing low over nearshore waters. They nest colonially in trees or on cliffs or rocky islands.

Page opposite: Hearn Gulch, Mendocino County

Mendocino County

HUMBOLDT
Lost
Coast
Sinkyone
Wilderness SP
PACIFIC
OCEAN
101
Leggett
1
162
To Wil
Rockport
Laytonville
Westport - Union
Landing SB
BRANSCOMB
RD
Westport
MENDOCINO
Longvale
Ten Mile R
SHERWOOD
RD
MacKerricher
SP
FORT
BRAGG
SHERWOOD RD
101
Fort Bragg
OAK AVE
Noyo R
Willits
Jug Handle SR
Jackson SF
Caspar
Caspar Headlands SB & SR
Pt. Cabrillo Light Station
20
Mendocino
Woodlands SP
Russian Gulch SP
Mendocino Headlands SP
Mendocino
Big R
Van Damme SP
Albion R
Albion
20
Lake
Mendocino
Navarro R
Navarro River
Redwoods SP
Ukiah
128
GREENWOOD
RD
Greenwood SB
Elk
Elk Cr
1
Hendy Woods SP
253
101
Boonville
Manchester SP
RD
Manchester
Garcia R
Pt. Arena
MOUNTAIN
VIEW
Hopland
175
Point Arena
128
Mailliard
Redwoods SP
Schooner
Gulch SB
200
600
1000
1400
1800
2200
2600
Anchor Bay
Gualala
N
Gualala R
0
10
20 Miles
0
10
20 Kilometers
Depth Contours Shown in Meters
SONOMA

Mendocino County

MENDOCINO'S COAST draws visitors from around the world to its rocky shoreline, sandy pocket beaches, and dense forests, as well as its well-known inns, art galleries, and gift shops. The county also offers some lesser-known recreational destinations. State parks such as Jug Handle State Reserve, Russian Gulch State Park, and Van Damme State Park offer much more than the shoreline activities for which they are noted; each has forests and meadows with miles of trails located inland and out of sight of Highway One.

There are also some major new parks-in-the-making: the newly created Stornetta Public Lands property surrounding the Point Arena Lighthouse adds dramatically to the miles-long public shoreline already available at neighboring Manchester State Park. The recent expansion of Mendocino Headlands State Park into the watershed of the Big River increases the recreational potential of that popular area.

Community events along the Mendocino coast include annual whale festivals in both Mendocino and Fort Bragg, the former in early March and the latter a few weeks later, with whale-watching cruises and walks, children's events, and more. An annual kite festival takes place in June in Fort Bragg, and spring garden tours are offered at several locations. Annual food-oriented events highlight salmon and abalone. For more information on events, as well as on lodging and recreational opportunities, contact the Fort Bragg–Mendocino Coast Chamber of Commerce at 707-961-6300 and the Redwood Coast Chamber of Commerce at 707-884-1182. The Mendocino Coast Audubon Society conducts field trips, bird walks, and ocean trips in search of birds and whales; call: 707-962-9413 or 707-964-6362.

Noyo Harbor at Fort Bragg is a major departure point for ocean fishing and wildlife trips. Salmon fishing is best from late spring through summer; crabbing is a year-round activity. Ocean trips often combine fishing for salmon or rock cod with crabbing, while some boats go in search of albacore or other sport fish. Gray whale-watching is best between January and April; some boats offer blue whale–watching trips in August and September. See p. 136 for more on Noyo Harbor.

Horseback riding north of Fort Bragg on ocean bluffs or beaches:
Lost Coast Trail Rides, 707-961-0700.
Ricochet Ridge Ranch, 707-964-7669.

Horseback riding through forest or at Manchester State Park:
Ross Ranch, 707-877-1834.

Kayaks, diving and surfing equipment:
Subsurface Progression, Fort Bragg, 707-964-3793.

Ocean kayak tours starting from Van Damme beach:
Lost Coast Kayaking, 707-937-2434.

Ocean kayak trips through sea caves from Greenwood State Beach:
Force Ten Ocean White Water Tours, 707-877-3505.

Tennis courts and ocean-view nine-hole golf course:
Little River Inn, Little River, 707-937-5942.

Canoe, kayak, or bicycle rentals:
Fort Bragg Cyclery, Fort Bragg, 707-964-3509.
Catch-A-Canoe, Mendocino, 707-937-0273.
Adventure Rents on the Redwood Coast, Gualala, 888-881-4386.

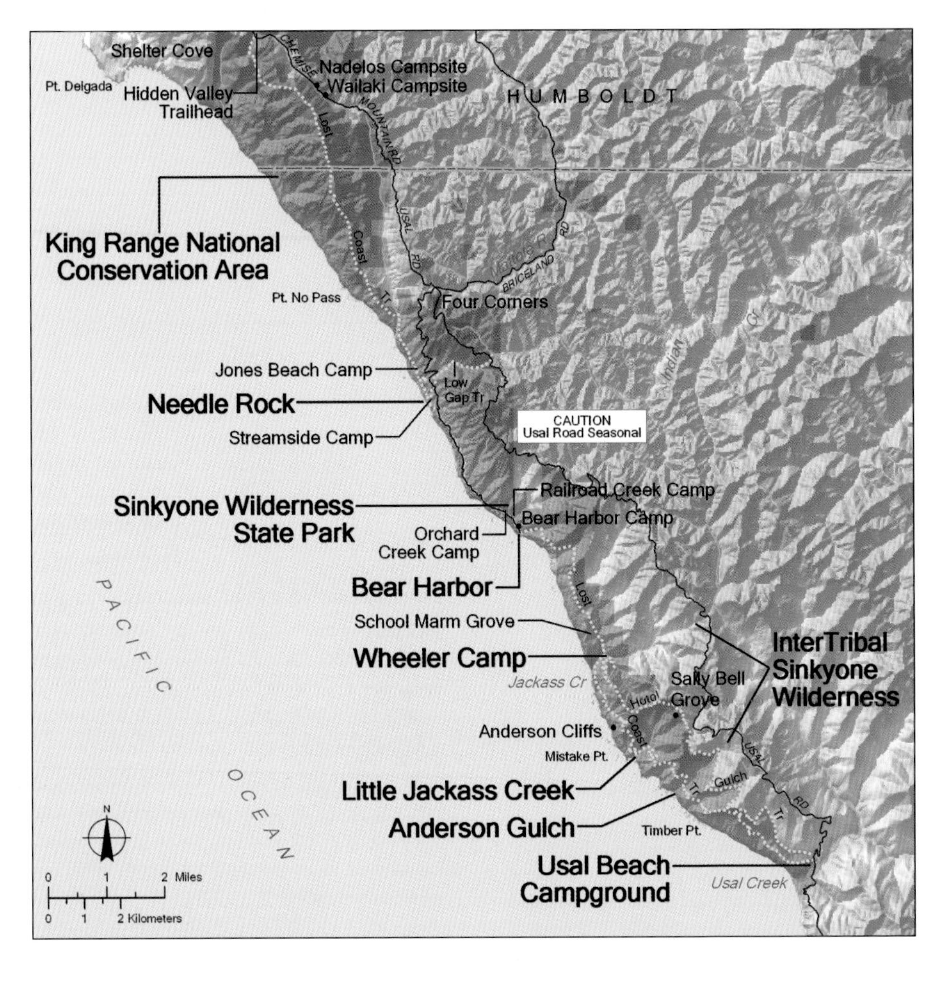

Sinkyone Wilderness State Park

Sinkyone Wilderness State Park

	Sandy Beach	Rocky Shore	Trail	Visitor Center	Campground	Wildlife Viewing	Historic Building	Fishing or Boating	Facilities for Disabled	Restrooms	Parking	Fee
King Range National Conservation Area			•		•						•	
Needle Rock		•	•	•	•	•	•			•	•	•
Sinkyone Wilderness State Park	•	•	•	•	•	•	•	•		•	•	•
Bear Harbor	•	•	•		•	•		•		•	•	•
Wheeler Camp			•		•					•		•
Little Jackass Creek			•		•					•		•
Anderson Gulch	•		•		•	•				•		•
Usal Beach Campground	•	•	•		•	•		•		•	•	•
InterTribal Sinkyone Wilderness			•		•							

KING RANGE NATIONAL CONSERVATION AREA: *Mattole River Beach (Humboldt Co.) to Four Corners (Mendocino Co.).* The southernmost portion of the King Range National Conservation Area is located in Mendocino County, south of Shelter Cove. Hikers headed south on the Lost Coast Trail toward Sinkyone Wilderness State Park may start at the Hidden Valley trailhead on Chamise Mountain Rd. in the National Conservation Area. The trail continues along a high ridge, past short spur trails to the Nadelos and Wailaki campsites, which lie east of the trail on Chamise Mountain Road. The Lost Coast Trail continues south into Sinkyone Wilderness State Park to Needle Rock, where there is a visitor center, campsites, and parking. Note that equestrians and mountain bikes are allowed in the King Range National Conservation Area but prohibited on hiking trails in the Sinkyone Wilderness State Park.

NEEDLE ROCK: *Briceland Rd./Mendocino Co. Route 435, 7 mi. S.W. of Four Corners.* A visitor center and picnic area are situated at this serene location in Sinkyone Wilderness State Park, in a meadow overlooking the sea. A historic ranch house serves as the visitor center, and the barn is open for overnight stays; two additional campsites are nearby. Swallows nest in the eaves of the old buildings. Restrooms, drinking water, and fire rings are available. Needle Rock is the only source of ready-to-drink water in the park; treat stream water elsewhere. The road to Needle Rock passes through coastal bluff scrub, forests of Sitka spruce, and deep alder-dominated canyons. Within a mile north of Needle Rock along the Coastal Trail are Jones Beach and Streamside Camps, each with three campsites.

SINKYONE WILDERNESS STATE PARK: *Between Four Corners and Usal Camp.* The park includes 7,367 acres, extending from the King Range National Conservation Area on the north to Usal Creek. There are undeveloped beaches and coastal mountains, redwood forest and alder groves, hiking opportunities and spectacular coastal views. Most of the park is roadless, and many of its attractions must be reached on foot. Vehicle access to the southern end of the park is via Usal Rd. (Mendocino Co. Route 431), which runs along a high ridge, to the east and mostly outside of the state park. Usal Rd. is unpaved but passable year-round from Hwy. One as far north as Usal Camp, and seasonally passable north of Usal Camp. At the north end of the park, Briceland Rd. (which becomes Route 435 in Mendocino Co.) provides the best vehicle access from Hwy. 101. Both Briceland and Usal Roads are steep, narrow, and unsuitable for RVs and trailers.

The rugged Lost Coast Trail parallels the coast through the entire length of the state park. Compared to the sea-level coastal trail in the King Range National Conservation Area, much of the Sinkyone trail passes through forests high above the ocean, with occasional access points between high cliffs down to cove beaches. From the north, the Lost Coast Trail enters the state park on a high ridge and leads south to Needle Rock, where there are campsites, a visitor center, and access to a beach. From Needle Rock, the Low Gap Trail heads east across the park, while a three-mile stretch of road continues south along a bluff above the rocky shore, ending one-half mile short of Bear Harbor. The Lost Coastal Trail, a link in the Coastal Trail, continues from that point to the southern boundary of the state park at Usal Camp, dipping in and out of steep gulches along the way. Plan on at least three days to walk the length of the park. Most of the park's campsites are located along the Lost Coast Trail. The Hotel Gulch Trail, which is being converted from an old road, runs roughly north and south along the eastern boundary of the state park at its southern end, connecting Usal Camp with Wheeler Camp. Hiking on the trail is permitted, but there are no campsites.

Sinkyone Wilderness State Park is open year-round for hiking and primitive camping; no reservations are taken. Dogs are not allowed on trails, but they are allowed on leash in certain designated areas; however, dogs are discouraged due to the presence of Roosevelt elk, which can be aggressive toward canines. Firewood is sold at the Needle Rock visitor center, and wood gathering is not allowed; fires are allowed only in steel rings. Mountain bikes are permitted on roads open to vehicles and prohibited on hiking trails. For recorded park information, including road conditions, call: 707-986-7711. Other information; call: 707-247-3318.

BEAR HARBOR: *Near S. end of Briceland Rd./Route 435.* Three groups of campsites are located near the end of the road, where the Lost Coast Trail continues southward. Orchard Creek Camp has three sites located less than 100 yards from the road, while Railroad Creek Camp's three sites are in a eucalyptus grove planted by early settlers. Three campsites at Bear Harbor Camp are located an easy, less-than-half-mile walk from the small cove beach, where fishing and beachcombing are possible. Pelicans can be seen on offshore rocks in summer.

WHEELER CAMP: *Lost Coast Trail, 12 mi. N. of Usal Camp, 4.7 mi. S. of Orchard Creek.* A logging town once was located here, in a valley that punctuates the high cliffs running north to Bear Harbor and south to Mistake Point.

Campsite at mouth of Jackass Creek

Scattered building foundations are still visible, where four campsites are scattered in the valley near the confluence of two creeks or in School Marm Grove at the old school site. At the mouth of Jackass Creek is a dark sand beach reachable from the Lost Coast Trail, which otherwise runs mostly high on the slopes, through the forest. Kingfishers, common yellowthroats, black phoebes, and other birds may be seen along the creek, and animal scat is evidence of the bears, mountain lions, and bobcats that inhabit the area.

LITTLE JACKASS CREEK: *Lost Coast Trail, 7.5 mi. N. of Usal Camp, 9.2 mi. S. of Orchard Creek.* While much of the Sinkyone Wilderness State Park has been logged in the past, old-growth redwoods remain in the upper Little Jackass Creek valley, in the Sally Bell Grove. The riparian vegetation along the lower stream also includes big-leaf maple trees. There are four campsites, two in a wooded area and two closer to the small cove beach sheltered by the high Anderson cliffs.

ANDERSON GULCH: *Lost Coast Trail, 5 mi. N. of Usal Camp.* This is the closest camp on the Lost Coast Trail to Usal Camp. A tiny pocket beach is located at the mouth of the narrow gulch, and two campsites are available on the slope above. Lush riparian vegetation lines the creek, and Douglas iris provide a springtime show. Listen for owls at dusk.

USAL BEACH CAMPGROUND: *Usal Rd., 6 mi. N. of Hwy. One turnoff at milepost 90.88.* There are 15 campsites accessible by two-wheel drive vehicles at Usal Beach Campground, located inland of the mouth of Usal Creek. This is the only car-camping location in the Sinkyone Wilderness State Park. Tables, fire rings, and pit toilets are available. The sandy beach, used for beachcombing, surf fishing, and abalone diving, extends for two miles. Look for osprey near the creek mouth.

INTERTRIBAL SINKYONE WILDERNESS: *E. of Sinkyone Wilderness State Park.* The InterTribal Sinkyone Wilderness Council, composed of 11 Northern California Indian tribes with ties to the Sinkyone region, manages some 3,800 acres of land east of the Sinkyone Wilderness State Park. The council has undertaken salmon habitat restoration programs and trained tribal members in heavy equipment operation to remove old logging roads. The council plans future construction of trails open to the public across the InterTribal Sinkyone Wilderness, linking with state park trails, as well as limited public camping facilities along the Usal Road. Contact: P.O. Box 1523, Ukiah 95482; 707-463-6745.

Sinkyone Wilderness State Park

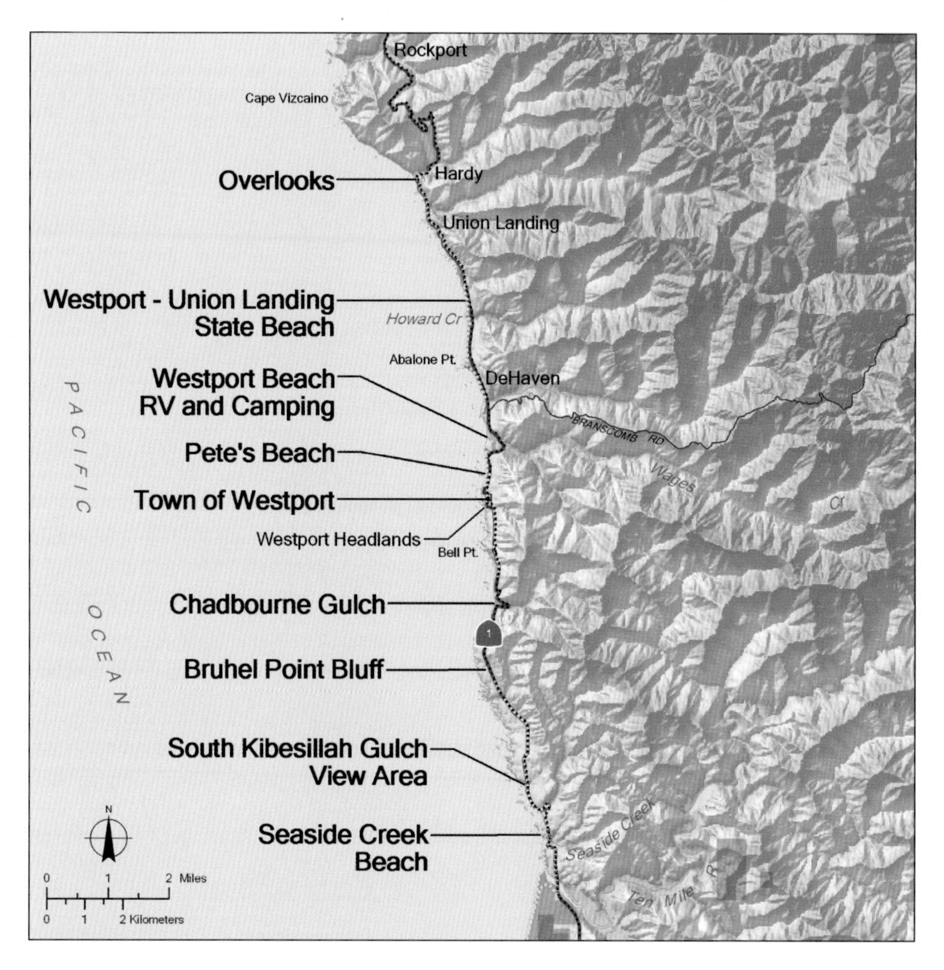

Westport-Union Landing State Beach

Westport Area

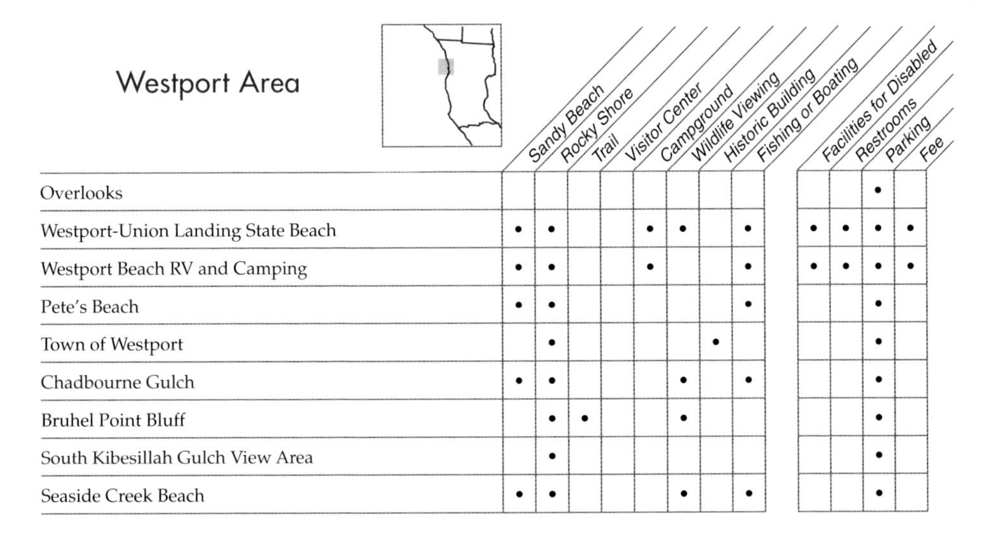

	Sandy Beach	Rocky Shore	Trail	Visitor Center	Campground	Wildlife Viewing	Historic Building	Fishing or Boating	Facilities for Disabled	Restrooms	Parking	Fee
Overlooks											•	
Westport-Union Landing State Beach	•	•			•	•		•	•	•	•	•
Westport Beach RV and Camping	•	•			•			•	•	•	•	•
Pete's Beach	•	•						•			•	
Town of Westport		•					•				•	
Chadbourne Gulch	•	•				•		•			•	
Bruhel Point Bluff		•	•			•					•	
South Kibesillah Gulch View Area		•									•	
Seaside Creek Beach	•	•				•		•			•	

OVERLOOKS: *Hwy. One at Hardy and Juan Creeks, 6.3 and 5.8 mi. N. of Westport.* Two roadside overlooks are located on the seaward side of Hwy. One at mileposts 83.5 and 82.96, just before the road turns inland. Views of mountainous coast and offshore rocks; no improved beach access or facilities.

WESTPORT-UNION LANDING STATE BEACH: *W. of Hwy. One, 3 mi. N. of Westport.* Day-use and camping facilities are spread along a narrow ocean bluff, between Hwy. One and the sea. The main day use area is at the mouth of Howard Creek, which bisects a sandy beach. The south end of the beach can be reached by a path down the bluff from the parking area south of Howard Creek. The north end of the beach is reached by a wooden stair from a second parking area north of the creek; a wooden boardwalk overlooking the beach and picnic area is wheelchair accessible. There are tidepools to explore, and anglers cast nets for smelt as the fish approach shore to spawn. Those willing to brave the cold and sometimes rough waters go abalone diving or spear fishing.

North and south of Howard Creek, 100 primitive campsites are spread along the open bluff, with views of rugged coast and mountains. Sunset vistas are outstanding. Restrooms, fire rings, and tables; no reservations taken. Fee charged for camping; no fee for day use. Call: 707-937-5804.

WESTPORT BEACH RV AND CAMPING: *Hwy. One, .5 mi. N. of Westport.* The sandy ocean beach at the mouth of Wages Creek is accessible from the privately run campground south of the creek. RV sites with hookups are adjacent to the beach; campground has playground, shuffleboard, and laundry. Surf nets for summer smelt fishing, poles, and other fishing equipment are available for rental, along with boogie boards. At low tides, abalone hunters clamber over the rocks at the north end of the beach in search of the giant mollusks; children paddle in the pool at the mouth of Wages Creek. Fee for day use or camping. Open all year; for information, call: 707-964-2964.

PETE'S BEACH: *Hwy. One, .3 mi. N. of Westport.* A quarter-mile-long sheltered beach of fine gray pebbles is located at the base of a bluff, north of Westport. An opening in the thicket at the south end of the unpaved pullout west of Hwy. One reveals a stairway to the shore. Indian paintbrush, lupine, salal, and succulents cover the slope. Several large seastacks near the shore are nesting grounds for cormorants and gulls, which may be seen sitting atop the rocks or clinging to the face of the rocks. Part of Westport-Union Landing State Beach; call: 707-937-5804.

TOWN OF WESTPORT: *Hwy. One, 7.5 mi. N. of Ten Mile River.* Westport consists of original simple wooden houses dating from the

Westport

19th-century logging period, a handful of bed-and-breakfast inns, and old roses gone rampant in vacant lots. A scattering of modern dwellings has not altered the timeless quality of the village. Established in the 1860s, the settlement was named Westport in the late 1870s by J. T. Rogers, who built a wharf and loading chute that could handle up to 150,000 board feet of lumber a day. Westport's major exports were shingles, railroad ties, and tanbark, which was the source of tannic acid used by San Francisco tanneries. The bluffs at Westport Headlands, across Hwy. One from the Westport store, offer views of the rocky coast and coastal mountains; funds for acquisition of the property by the Westport Village society were provided by the State Coastal Conservancy. A stairway to the beach and wheelchair-accessible access to the bluffs are expected to be completed in 2005, along with benches, a boat launch, parking, and signing.

CHADBOURNE GULCH: *Hwy. One, 5.8 mi. N. of Ten Mile River.* A wide sandy beach lies at the mouth of Chadbourne Gulch, where a small stream lined with thick riparian vegetation enters the ocean. The mile-long beach extends both north and south of the parking area. Popular for surf fishing, pokepoling, and abalone diving; known by surfers as Blues Beach. Offshore rocks provide nesting sites for black oystercatchers, gulls, cormorants, and pigeon guillemots; harbor seals are often sighted offshore.

BRUHEL POINT BLUFF: *W. of Hwy. One, 4.5 mi. N. of Ten Mile River.* Several trails lead less than one-quarter mile across the grassy blufftops to the shoreline. Access to trails is from an unnamed paved vista point south of a stand of Monterey pine trees at milepost 74.19. The trails lead to the rocky shoreline dotted with tidepools and punctuated by narrow gravel cove beaches. On the cliffs grow a golden-orange variety of Indian paintbrush. No facilities.

SOUTH KIBESILLAH GULCH VIEW AREA: *W. of Hwy. One, 2.1 mi. N. of Ten Mile River.* Blufftop overlook and parking area with a good view of offshore rocks and the Ten Mile Dunes extending south toward Fort Bragg. No beach access.

SEASIDE CREEK BEACH: *W. of Hwy. One, 1 mi. N. of Ten Mile River.* Small, undeveloped ocean beach directly off Hwy. One; shoulder parking. The beach is about one-half mile long, extending to the mouth of the Ten Mile River. Popular for beachcombing among driftwood, surf fishing, and sometimes for surfing. Shorebirds, including sanderlings, killdeer, and willets, may be seen here.

During the first 60 years of commercial redwood logging, beginning in the 1850s, ships provided the only means to move lumber products from the Mendocino coast to markets. The coastal mountain range deterred efforts to build a railroad into California's interior until 1912, when the Union Lumber Company's California Western Railroad finally linked Fort Bragg with Willits. Sailing ships and steam schooners collected milled lumber and other goods at landings all along the Mendocino coast, using wooden or wire chutes which lowered boards, shingles, railroad ties, and sometimes passengers from the bluffs down to ships resting at anchor below. During the early 1860s, some 300 schooners were in operation on the coast, each with a crew of five or six and a capacity of up to 100 thousand feet of lumber per trip.

Felling and transporting the huge logs of redwood trees, especially the old-growth trees first encountered by early loggers, presented a challenge even to those who arrived in California with logging experience from New England or elsewhere. Chopping down a redwood tree using an axe could take days of work; once down, the tree had to be stripped of its stringy bark, which was virtually impossible to saw, and then bucked into sections. The log segments were dragged out of canyons along skid roads made of logs laid side by side to mills located near the coast. Teams of oxen pulled the smooth logs, assisted after 1882 by Eureka resident John Dolbeer's invention of a steam engine, soon known as a "donkey," that winched the logs onto the skid roads.

Lumber mills were located in gulches the length of the Mendocino coast. Many of the towns that grew up around the mills, including Usal, Rockport, Hardy Creek, and Navarro, were later largely depopulated, after the lumber boom was over. Other mill sites, including Westport, Cleone, Fort Bragg, Noyo, Caspar, Mendocino City, Little River, Albion, Greenwood, and Gualala, developed into today's coastal towns and villages.

Loading lumber with donkey engine, Caspar Lumber Company, 1904

Dunes

MANY NORTHERN California beaches are part of a dune system. In these areas, wind is an important element in the beach dynamic. Dunes develop when onshore breezes carry loose, dry sand inland from the beach. Small surface irregularities from rocks, logs, or vegetation interrupt the airflow and lead to deposition of the sand particles in front of the obstacle. Eventually sand mounds or ridges form; some develop into ephemeral foredunes that are susceptible to removal by tides or storm waves. Others become permanent backshore dunes that support a beautiful array of vegetation.

As more complex root webs are established, the dunes trap increasing amounts of sand, forming hillocks and progressively higher dune ridges. These ridges orient at right angles to the prevailing winds, providing valuable habitat for a variety of plants and small animals. Deep-rooted plants such as beach strawberry, silvery beach sagewort, and yellow sand verbena grow on the foredunes. Coast buckwheat and silver bush lupine grow on the richer soils of the back dunes, in areas protected from direct wind and salt spray. Older, more mature dunes support woody shrubs and even trees. At Lanphere Dunes in Humboldt County, the dune system shows a broad range of vegetation, including mature Sitka spruce trees inland of the zone that was inundated three hundred years ago by a massive tsunami. Younger shrubs and herbaceous plants grow seaward of the tsunami run-up zone.

European beach grass, planted on dunes such as at Bodega Head in Sonoma County, is effective at trapping sand and stabilizing the dune, but it has negative impacts on native species. It develops into large uniform stands that crowd out plants that have evolved to endure the shifting sands of dunes.

Dunes function as a system, and a blowout or disturbance to one part of a dune can quickly propagate through the entire system. Loss of vegetation at the oceanfront of the dune system can expose the more landward dunes to higher winds, leading to further instability.

Ten Mile Dunes

The soft silvery **beach sagewort** *(Artemisia pycnocephala)* is one of the most abundant native dune plants. It has a yellow spike of tiny flowers with deeply divided foliage. The many tiny hairs on the leaves are an adaptation that helps protect it from the drying effect of the wind. In addition, the hairs capture moisture and reflect sunlight, further helping sagewort survive in this difficult environment. Beach sagewort is a psammophile, or "sand loving plant," and it blooms between June and August.

Beach sagewort

The wonderfully scented **yellow sand verbena** *(Abronia latifolia)*, a foredune or coastal strand species, has a very deep root system. This is one of many adaptations that allow it to survive in an arid environment. It is believed that Native Americans ate the stout roots of this species. The Cahuilla people called yellow sand verbena *nyuku*, or "earth cousin," because of the way the plant seems to hug the earth. Next time you see these flowers in bloom, bend down to get a whiff of their sweet perfume.

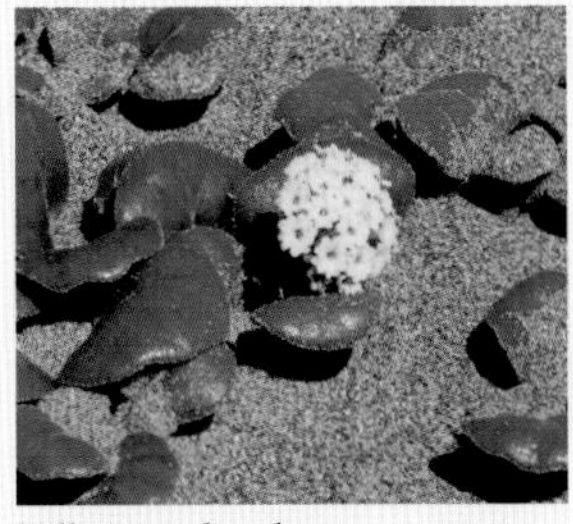

Yellow sand verbena

Found along sandy seashores of the Pacific coast, the **beach morning-glory** *(Calystegia soldanella)* has fleshy, kidney-shaped leaves, with a waxy outer layer to prevent them from drying out. Interestingly, this native California species has an international distribution, and also grows on beaches of Europe and South America. The showy, pink flower can be seen blooming in dunes along the coast between April and May. The flower is about two inches wide and has five white stripes in the shape of a star.

Beach morning-glory

The flowers of the **seaside daisy** *(Erigeron glaucus)* are pink or pale violet with yellow centers. The thick and fleshy leaves of the seaside daisy are spoon-shaped and are clustered together to form a dense mat. The genus name *Erigeron* comes from the Greek *eri*, meaning "early," and *geron*, meaning "old man." Thus, this plant is known as the "old man in the spring," referring to the fluffy white seed heads that develop after the flowers are done blooming.

Seaside daisy

Coast buckwheat

Coast buckwheat *(Eriogonum latifolium)* grows on coastal dunes and bluffs from Monterey into southern Oregon. The pom-pom-shaped flowers start out white and gradually turn pink. The gray-green leaves are fuzzy underneath and are a favorite meal for rabbits. Buckwheat is a popular nectar source for insects, including butterflies, wasps, bees, and flies. If you find buckwheat blooming in the dunes, take a few minutes to observe the pollinators that visit. You might be surprised by a showy little butterfly or the bombing about of a bumblebee.

Silver bush lupine

The **silver bush lupine** (*Lupinus chamissonis*) has soft, white hairs covering gray-green leaves. It flowers from March through July, bearing many stalks of showy blue-purple or violet spiky flowers. The flowers are quite fragrant, so be sure to take a moment to smell one if you discover one blooming. Watch the flowers closely, and you may witness the remarkable pollination mechanism that this plant has evolved. When a bumblebee lands in search of nectar, the weight of the bee causes a flower petal, called the keel, to lower and expose the flower's reproductive parts. The bee's belly is dusted with pollen as the bee collects nectar

Sea rocket

The four-petaled flowers of the **sea rocket** *(Cakile maritima)* are lavender. The plant has fleshy, succulent, and deeply lobed blue-green leaves. This European native was first documented in Marin County in 1935 and has since successfully colonized dunes all along the coast. You may find it blooming between June and September. Take a close look at the swollen two-jointed seed pod. The pod is designed to split in half, allowing a portion of the seeds to remain local and the other portion to disperse to distant shores. The pod has a corky wall, enabling it to float on the water and enhancing its ability to travel. The common name, sea rocket, refers to the ability of the pod to split in half and send one half into the great unknown.

Next time you take a stroll through the dunes on a warm summer day, notice all of the insects visiting flowers to gather food. Among the insects you will likely notice some bees. You may be familiar with honeybees and bumblebees, but have you ever heard of solitary bees? In contrast to social bees, solitary bees have no colony or hive and no stored resources. Instead, short-lived females work alone to build small nests in the soil. Some solitary bees utilize specific plant materials to line their shallow nests. For example, a **"woolcarder" bee** *(Anthidium palliventre)* gathers plant hairs from beach buckwheat to line its nest. Check out the leaves of beach buckwheat in April when *Anthidium* emerges, and you may notice bare spots where the "wool" has been removed, a sign that this bee is nesting nearby.

"Woolcarder" bee

Darkling beetles (*Eleodes* sp.) are medium-sized flightless beetles. Look for their diminutive tracks meandering through the dunes. Darkling beetles cannot fly because their wings are fused together to create a hard outer shell; this protects them from drying out and deters predators. Because they cannot escape via flight, they have developed a very interesting defense mechanism. They point their rear ends skyward when disturbed, and if handled roughly, they emit a foul-smelling, dark-colored fluid. Generally, however, the beetle just stands there, derrière in the air, until the danger passes. This behavior is enough to discourage all but the most determined predators.

Darkling beetle

The secretive and wary **brush rabbit** *(Sylvilagus bachmani)* has a small cottony tail. These small bunnies build tunnels through the brush of dune scrub. They are crepuscular (only active at dusk and dawn), so you may not see one during your trek through the dunes, but you are likely to see their sign. Look for a pile of scat or tracks that are in groups of four prints; the hind footprint is larger than the front. To hide themselves from predators, brush rabbits sometimes sit perfectly still. However, when they are threatened, they zig-zag speedily through the dunes.

Brush rabbit

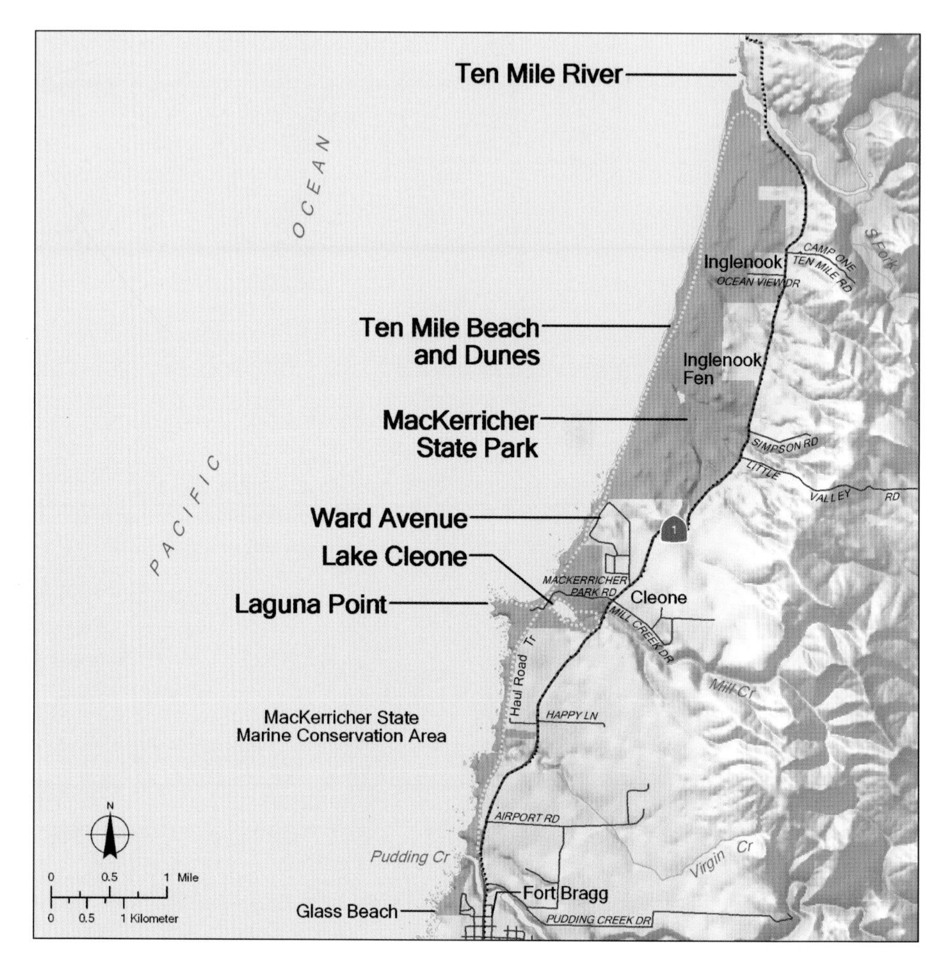

Marbled godwits at MacKerricher State Park

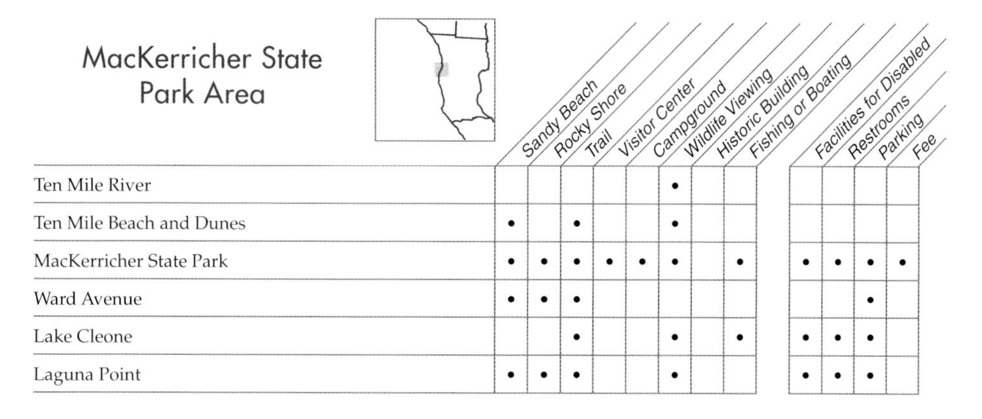

MacKerricher State Park Area

	Sandy Beach	Rocky Shore	Trail	Visitor Center	Campground	Wildlife Viewing	Historic Building	Fishing or Boating	Facilities for Disabled	Restrooms	Parking	Fee
Ten Mile River						•						
Ten Mile Beach and Dunes	•		•			•						
MacKerricher State Park	•	•	•	•	•	•		•	•	•	•	•
Ward Avenue	•	•	•								•	
Lake Cleone			•			•		•	•	•	•	
Laguna Point	•	•	•			•			•	•	•	

TEN MILE RIVER: *10 mi. N. of Noyo River.* So named in the 1850s because it is ten miles north of Noyo River, Ten Mile River and estuary constitute one of the county's largest wetlands. Seventy-five acres of salt marsh lie in back of the dunes at the river mouth. Common mergansers, or fish ducks, nest in the marsh at the river's mouth, and coho salmon, steelhead trout, and Pacific lamprey spawn in the north and south forks of Ten Mile River.

TEN MILE BEACH AND DUNES: *S. of Ten Mile River.* Ten Mile Beach, one of the longest stretches of dunes in California, extends from the river mouth south for 4.5 miles. A number of rare plants grow in the shifting sands of the dunes, and a wetland area known as the Inglenook Fen within the dunes supports a unique community of insects, spiders, and plants. On the beach are nesting sites for rare snowy plovers; watch for restrictions on public access that may be in place during the spring/summer breeding season. Ten Mile Beach, which is part of MacKerricher State Park, can be reached from any of the coastal access points within the park, including the west ends of Ward Ave. and Mill Creek Drive.

MACKERRICHER STATE PARK: *From Ten Mile River S. to Glass Beach in Fort Bragg.* Eight miles of beaches, dunes, and rocky shoreline characterize this state park, which includes

Ten Mile Dunes

a wide variety of day-use and camping facilities. Surf fishing and freshwater trout fishing are popular, along with hiking, bicycling, horseback riding, and wildlife viewing. The main park entrance is at Cleone, and other beach access points within the park include the ends of Ward Ave. and Mill Creek Dr., a parking area just north of Pudding Creek, the trestle over Pudding Creek, and Glass Beach.

The campground includes 139 developed and 10 walk-in sites set in a forest; some campsites are wheelchair accessible. Maximum length for RVs and trailers is 35 feet. There is a visitor center at the main park entrance, where binoculars can be rented for wildlife viewing. A humpback whale skeleton near the visitor center provides human visitors with a sense of scale. Nature walks and campfire programs are offered; a nature trail is wheelchair accessible. Horses can be rented locally for riding along the beach; see p. 115. Nearshore waters from Pudding Creek to north of Laguna Point are within the MacKerricher State Marine Conservation Area, where restrictions apply on removal of marine species; check Department of Fish and Game regulations. Call: 707-964-9112.

WARD AVENUE: *Just N. of MacKerricher State Park main entrance.* The northernmost access to the beach at MacKerricher State Park is at the northwest end of Ward Avenue. There are no facilities other than parking, but day-use visitors can explore miles of uncrowded beaches, backed by vast sand dunes.

A broad sandy beach lies at the base of the bluff west of the parking area on Ward Avenue. This is a popular and easily accessible surfing spot. The old Timber Haul road runs north and south along the shore here, providing pedestrian or bicycle access to miles of shoreline, although erosion has caused a gap in the Haul Road about a mile north of Ward Avenue. Beyond that point, pedestrian and equestrian access is possible, but note seasonal restrictions that may be in effect due to nesting habitat of the western snowy plover. Small pocket beaches line the shoreline between low rocky headlands.

Lake Cleone

Laguna Point

LAKE CLEONE: *End of Mill Creek Dr.* A trail 1.3 miles long rings Lake Cleone, accessible via the main park entrance or Mill Creek Drive. This former tidal lagoon is now a freshwater lake, which is stocked with rainbow trout. Birds seen year-round at the lake include great blue herons, ring-necked and mallard ducks, and osprey, which circle overhead before plummeting abruptly for fish. Around the lake, black-tailed deer are commonly seen, along with raccoons, gray foxes, rabbits, and gray and ground squirrels. The picnic area near the lake has tables, firepits with grills, restrooms, parking, and running water.

LAGUNA POINT: *End of Mill Creek Dr.* This headland extends west of the rest of MacKerricher State Park. Its seaward position made Laguna Point a natural transshipment point for the Cleone community in the late 19th century, when a wharf and loading chute brought lumber milled in the area to ships headed south for California's growing cities. Prior to that time, the Coast Yuki and nearby Pomo Indians took sea lions and other marine mammals here, fished for salmon, and gathered abundant abalone, barnacles, mussels, and other foods. Kelp forests grow in the ocean waters to the north of the point, and washed-up kelp is found on area beaches.

The tip of Laguna Point is a good place from which to see the harbor seals that haul out on the rocks, as well as gray whales that migrate along the coast between December and April. On a clear day the headland offers an unimpeded view to the north of the mountainous Humboldt County coastline. A wheelchair-accessible boardwalk leads around the point through fields of grass and seasonal wildflowers, past a series of interpretive signs.

Ponds and Lakes

PONDS AND LAKES are inland bodies of fresh water that are generally fed by streams, springs, precipitation, or groundwater. Many of the ponds that occur in California's coastal range were built by cattle ranchers, as a water source for their herds when ephemeral coastal drainages dried up in the summer. Whether formed naturally or by small dams, lakes and ponds provide essential habitat for wildlife. The margins of these water bodies provide important habitat for aquatic invertebrates. Mammals and birds use pools for watering, foraging, and resting, and in the spring, amphibians seek out small lake edges and shallow, ephemeral ponds to lay thousands of jelly-like eggs.

Northern red-legged frog

The **northern red-legged frog** *(Rana aurora aurora)* is so-called because of the reddish or pinkish tint usually present on the underside of the frog's hind legs and lower abdomen. Red-legged frogs prefer permanent shallow waters with vegetation along the edge; wooded wetlands, ponds, and wet meadows are ideal habitats. Mating usually begins in February. After breeding, females deposit 500 to 1,000 eggs on submerged aquatic vegetation, in a mass approximately the size of a cantaloupe. It generally takes about three weeks for the eggs to hatch.

Western pond turtle

Western pond turtles *(Clemmys marmorata marmorata)* can be found along rivers, ponds, and lake shores during the summer; look for turtles sunning themselves on rocks or logs emerging from the water. Pond turtles occupy the same stretch of river or lake for their entire lives, and they may live for 30 to 40 years. They grow quite slowly and may take up to eight years to reach sexual maturity. The western pond turtle was once quite common, but due to habitat loss and the introduction of large-mouth bass and bullfrogs, both of which prey on the young turtles, the numbers are quickly dwindling. The turtles are wary of people and getting a good picture can be difficult, but opportunities to see these native reptiles present themselves to those with patience.

The stately and dignified **great blue heron** *(Ardea herodias)* is the largest American heron. If you sneak up on one feeding at a pond's edge, pause for a moment. Notice how the heron stands motionless with its neck doubled into a flattened S-shape. There it will stay until its prey comes within striking distance, and then with a flash, the curved neck straightens out and the bill shoots downward, rarely missing the mark. If you spy a heron creeping slowly through the shallows of a pond, notice how carefully each foot is placed, with minimal disturbance to the water's surface.

Great blue heron

Many people mistake coots for ducks, because they are usually seen swimming together. However, the coot is actually a type of rail; its feet are not webbed like a duck. If you have a chance to see an **American coot** *(Fulica americana)* get out of the water, take a look at the feet, and you will see that on either side of each toe is a pair of scaly lobes that the coot uses to propel itself. Coots can be conspicuous, noisy, and aggressive during the breeding season. They are usually found in shallow ponds where reeds and rushes grow from the water.

American coot

The male **red-winged blackbird** *(Agelaius phoeniceus)* wears a bright red epaulette on each shoulder. In studies designed to investigate the significance of the red wing's badge, it appeared that the epaulette indeed carried status. The wing patch is puffed up during aggressive displays that the male makes to defend his territory during the breeding season. This species is unique in that it is one of the most highly polygynous of all bird species. This means that one male blackbird generally mates with more than one female. Males that have successfully claimed territories mate with three to six females. Up to fifteen females have been observed on the territory of a single male, although the male may not necessarily father all of the young on his territory.

Red-winged blackbird

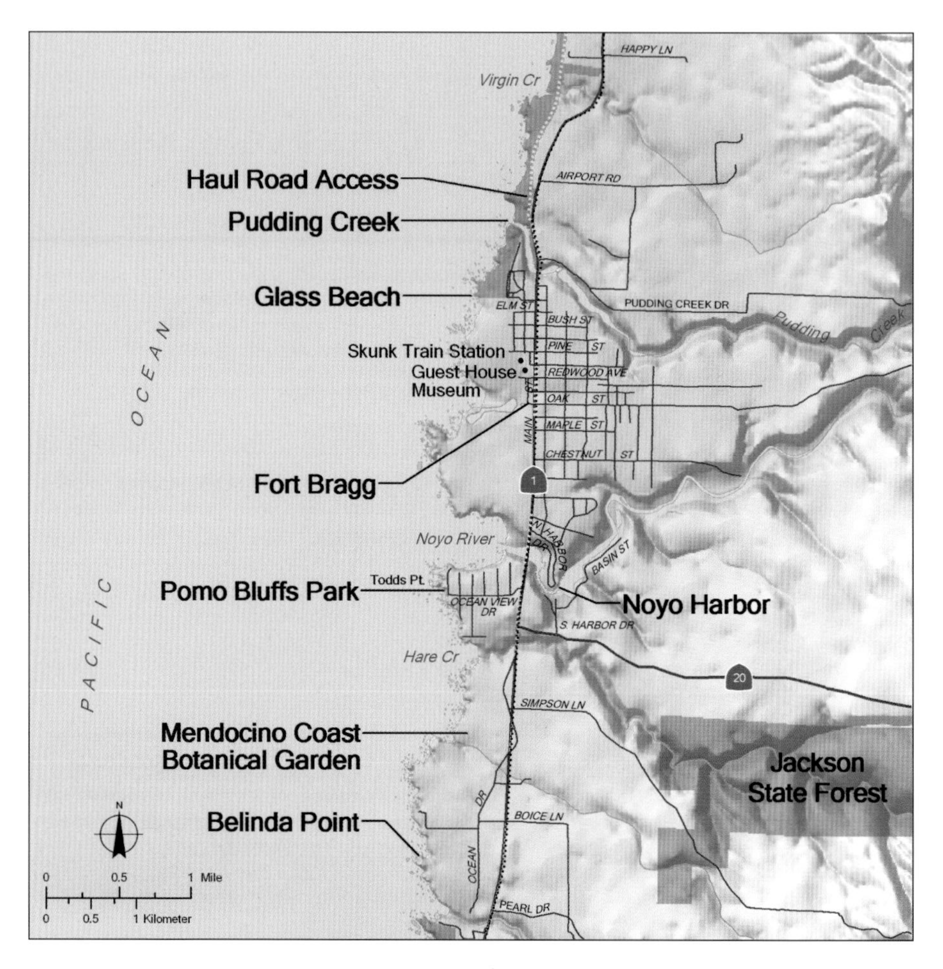
HAPPY LN
Virgin Cr
Haul Road Access
AIRPORT RD
Pudding Creek
Glass Beach
ELM ST
PUDDING CREEK DR
Pudding Creek
BUSH ST
PINE ST
Skunk Train Station
Guest House
Museum
REDWOOD AVE
OAK ST
MAPLE ST
CHESTNUT ST
MAIN ST
PACIFIC OCEAN
Fort Bragg
1
Noyo River
N HARBOR DR
BASIN ST
Todds Pt.
Pomo Bluffs Park
OCEAN VIEW DR
Noyo Harbor
S. HARBOR DR
Hare Cr
20
SIMPSON LN
Mendocino Coast Botanical Garden
Jackson State Forest
N
Belinda Point
BOICE LN
OCEAN DR
PEARL DR
0 0.5 1 Mile
0 0.5 1 Kilometer

Noyo Harbor

Fort Bragg

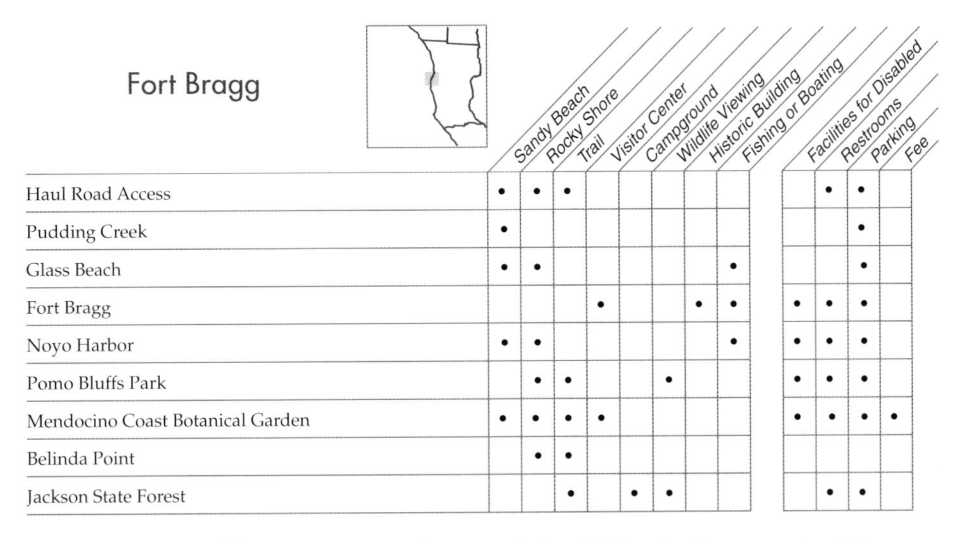

	Sandy Beach	Rocky Shore	Trail	Visitor Center	Campground	Wildlife Viewing	Historic Building	Fishing or Boating	Facilities for Disabled	Restrooms	Parking	Fee
Haul Road Access	•	•	•							•	•	
Pudding Creek	•										•	
Glass Beach	•	•						•			•	
Fort Bragg				•			•	•	•	•	•	
Noyo Harbor	•	•						•	•	•	•	
Pomo Bluffs Park		•	•			•			•	•	•	
Mendocino Coast Botanical Garden	•	•	•	•					•	•	•	•
Belinda Point		•	•									
Jackson State Forest			•		•	•				•	•	

HAUL ROAD ACCESS: *.1 mi. N. of Pudding Creek Bridge, Fort Bragg.* A large parking area between two motels provides access to the old lumber Haul Road that runs north and south along the shore, as well as to the grass-covered bluff and sandy pocket beaches. The Haul Road, once used to bring logs to the sawmill in Fort Bragg, now serves as an oceanfront trail linking MacKerricher State Park with the town of Fort Bragg. Located one-half mile north of the Haul Road parking area is a wide beach at the mouth of Virgin Creek, where breakers roll in on the sandy bottom at what is perhaps Fort Bragg's most popular surfing location.

PUDDING CREEK: *Hwy. One at Pudding Creek Bridge, Fort Bragg.* A sheltered, sandy beach is located under the old wooden trestle crossing Pudding Creek, north of the Hwy. One bridge. The State Department of Parks and Recreation intends to repair the trestle for pedestrian and bicycle use. A parking area next to the highway provides easy access to the beach and estuary, where shorebirds and migratory waterfowl may be seen.

GLASS BEACH: *W. end of Elm St., Fort Bragg.* Many decades ago, Glass Beach was a town dump, but now it is a visitor attraction for the surf-polished bits of blue, green, and red glass that can be found among the pebbles. The beach has been cleaned up, thanks to grants from the State Coastal Conservancy, the Integrated Waste Management Board, and the California Department of Transportation, and made a part of MacKerricher State Park. A seemingly endless supply of smooth glass chunks makes beachcombing fun. Hikers can head north along the low bluff, exploring small promontories that separate pocket beaches. Fishing from shore is possible here.

FORT BRAGG: *Hwy. One, 10 mi. N. of Mendocino.* Dominated for a century by its sawmills, by 2003 the town of Fort Bragg had seen the closure of all mill operations, including the huge oceanfront complex once run by the Union Lumber Company. The city government is looking at options for re-use of the area, which includes a mile of shoreline between Glass Beach and the Noyo River mouth; the city and the Coastal Conservancy are seeking a route for the Coastal Trail and public open space at the old mill site.

The old Union Lumber Company store on Main St. still stands, now converted for use by shops and restaurants, and other historical buildings are concentrated in the downtown area. The Guest House Museum at 343 N. Main St. has materials related to the history of the community and the logging industry; call: (707) 964-4251. Annual community events in Fort Bragg include Paul Bunyan Days, a Labor Day weekend event that celebrates the area's logging heritage, as well as the Salmon Barbecue, billed as the

Mendocino Coast Botanical Garden

world's largest and held around the Fourth of July. Call: Fort Bragg/Mendocino Coast Chamber of Commerce at 707-961-6300. The Skunk Train offers excursion trips through the redwoods between Fort Bragg and Willits; call: 866-457-5865.

NOYO HARBOR: *Off Hwy. One, 1.8 mi. S. of Fort Bragg.* Noyo Harbor is the main fishing port in Mendocino County. The fishing industry grew following construction in 1885 of the Union Lumber Company's railroad, which enabled Noyo fishermen to ship their fish, packed in ice, via Willits to San Francisco. A sandy beach under the Hwy. One bridge is accessible from North Harbor Drive, on the north side of the river; restrooms are at the parking lot. On the south bank of the river, the Noyo Harbor District provides two public boat ramps, one at the end of Basin Street and the other at the end of South Harbor Drive.

POMO BLUFFS PARK: *W. of Hwy. One, .2 mi. S. of Noyo River Bridge, Fort Bragg.* Turn west on Ocean View Dr., a quarter-mile south of the Noyo River Bridge. At the end of Ocean View Dr., turn north to a new park being developed by the City of Fort Bragg, with assistance from the State Coastal Conservancy. The bluffs offer a panoramic view of the mouth of the Noyo River and harbor

Charter boats at Noyo Harbor for salmon, crab, rock cod, or albacore fishing, as well as for whale watching in season:
All Aboard Adventures, 707-964-1881.
Noyo Fishing Center, 707-964-3000.
Telstar Charters, 707-964-8770.
Anchor Charter Boats, 707-964-4550.
Lady Irma II Charters, 707-964-3854.

Canoes, kayaks, and dive equipment for rent or sale:
Noyo Pacific Outfitters, 707-961-0559.

entrance. Trails, interpretive displays, restrooms, and parking are planned. For information, call: 707-961-2827.

MENDOCINO COAST BOTANICAL GARDEN: *1820 N. Hwy. One, Fort Bragg.* This is one of the few public gardens in the continental United States fronting directly on the ocean. Flowering shrubs and perennials are located in wind-sheltered glades among pine trees. In April and May, tree-sized rhododendrons are especially showy. Paths, some wheelchair accessible, lead through the woods to the ocean bluffs; electric carts are available for guests on a first-come, first-served basis. Picnicking on the grounds is allowed, and well-behaved dogs on leash are welcome. Docent-led tours are available year-round. Fee for garden entry; no fee for beach access. For information, call: 707-964-4352.

BELINDA POINT: *17410 Ocean Dr., S. of Fort Bragg.* Turn west from Hwy. One onto Ocean Dr. four-fifths of a mile south of the Hwy. 20 intersection. Continue another four-fifths of a mile to the signed trailhead on the right. A half-mile-long trail through the forest opens to a stunning view of rocky shore and along blufftop fingers overlooking a three-cove bay. The Mendocino Land Trust, which manages the accessway, plans future access improvements to a small cove beach. Call: 707-962-0470.

JACKSON STATE FOREST: *E. of Hwy. One off Hwy. 20.* This 50,200-acre demonstration forest of forest management practices features a self-guided nature trail (mostly along old logging roads). Managed by the California Department of Forestry and Fire Protection, the forest offers day use, including hiking, bicycling, horseback riding, swimming, and picnicking, and some 60 primitive campsites located along the Noyo River and tributaries. Camp One turnoff at milepost 5.9; other sites are scattered farther east. Campsites have picnic tables, fire rings, and pit toilets; some are for equestrian use. No water available. Campers must check in with camp host upon arrival. Pets allowed on leash. Hunting allowed in season; fishing prohibited. No fee for day use or camping. Contact Jackson State Forest, 802 N. Main St., Fort Bragg 95437, or call: 707-964-5674.

Belinda Point

Pygmy Forest

SEVERAL small forest areas known as pygmy forests are located on the coast of Mendocino and Sonoma Counties. These forests contain unusually dwarfed trees and shrubs that grow very slowly. The pygmy forest is underlain with severely leached and highly acidic soil known as podzol. In addition, the soil has an iron-rich hardpan layer that keeps roots from penetrating to more fertile soil and groundwater below. Due to both the soil chemistry and the hardpan, the trees that occur here are dwarfed. Much like bonsai, a century-old tree may be only three feet tall. There are several species of plants that are especially adapted to these conditions: pygmy cypress, Bolander pine, and bishop pine. However, all the plants within the pygmy forest are capable of growing to more normal heights in fertile soils. This unique vegetation type can be seen by hiking the ecological staircase trail at Jug Handle State Reserve, the self-guided trail at Van Damme State Park, or inland trails at Salt Point State Park.

Pygmy cypress

Pygmy cypress (*Cupressus goveniana pygmaea*) occurs in a narrow, discontinuous strip along the Mendocino County coast known as the "Mendocino white plains" or "pine barrens." Mendocino cypress cone production is abundant on dwarfed and mature trees, but is rare or absent on young vigorous trees. Staminate, or fertile, cones are not usually produced until the trees are six to seven years old. The dwarf appearance of these trees is due only to the harsh conditions of the soil. In good conditions, with fertile soil, this tree species can grow up to 50 feet tall.

Bolander or pygmy pine

Bolander or pygmy pine (*Pinus contorta* ssp. *bolanderi*) is endemic to the white plains of coastal Mendocino County. Bolander pine usually forms a dwarf forest with thickets of pygmy cypress and several endemic shrub species, including manzanita. Thickets of stunted trees often appear to be even-aged. This is possibly due to the fact that these trees require fire to reproduce. They have specially adapted cones that will not release seed until they reach high temperatures. After a fire does occur and the seeds are released, most of the propagules, or seedlings, will be approximately the same age.

Labrador tea (*Ledum glandulosum*) is a rhododendron-like evergreen shrub that grows four to five feet tall. Tiny, white, azalea-shaped flowers with a sweet, lemony fragrance appear in late spring or early summer. Native to boggy places, Labrador tea likes acidic conditions. Tea brewed from the leaves of the Labrador tea plant is said to be a tonic, having a pleasant odor and spicy taste. In addition, used as a wash, the tea is said to kill lice, and the leaves will keep away moths if scattered among stored clothing.

Labrador tea

The **glossy leaf manzanita** (*Arctostaphylos nummularia*) is a low-growing shrub that grows most abundantly in the pygmy forest but ranges as far south as San Francisco on poor soils. It has small, dark green leaves and red peeling bark. The leaves of the manzanita are positioned vertically, to prevent water loss by providing as little surface area as possible to the sun's rays. Pink urn-shaped flowers produce small apple-like fruit in the fall. The word *manzanita* means "little apple" in Spanish.

Glossy leaf manzanita

Fruticose lichen (*Usnea rubicunda)* is found on the branches of pygmy cypress. Don't let the name fool you; fruticose lichens don't look at all like fruits. The word fruticose is a technical term meaning "shrubby," and many fruticose lichens look like miniature shrubs an inch or so high. Each lichen species is technically two species, one of which is an alga and the other a fungus growing in mutualism, and behaving as one organism. The fruticose lichen is characteristic of the pygmy forest.

Fruticose lichen

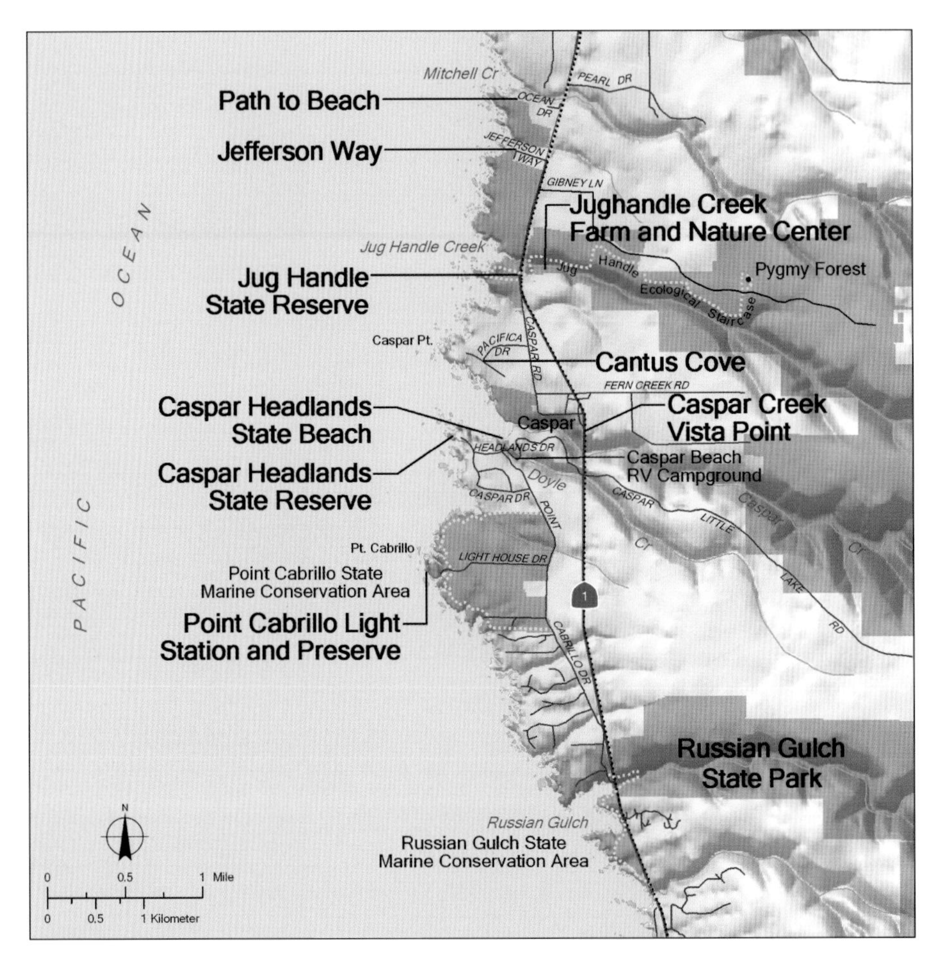

Russian Gulch State Park

Point Cabrillo Area

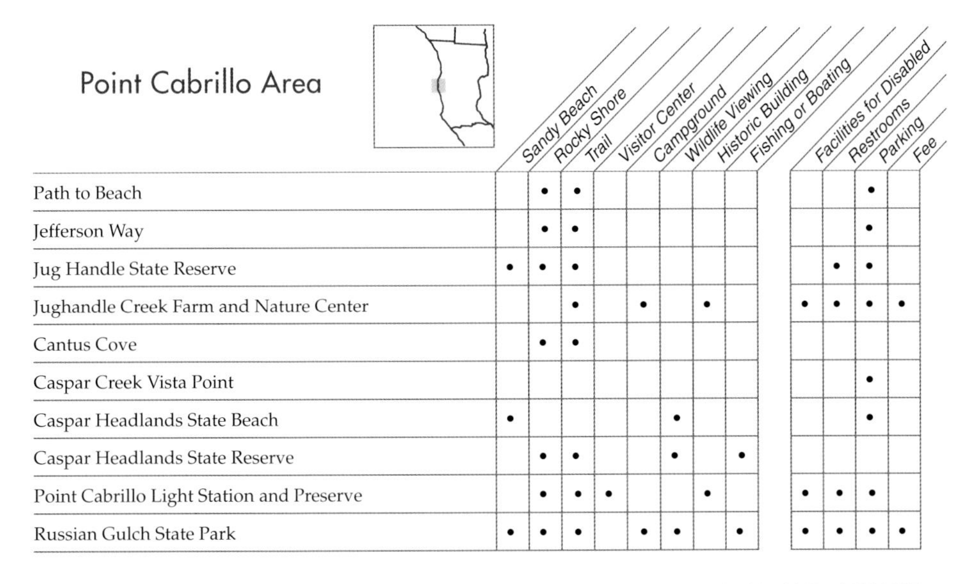

	Sandy Beach	Rocky Shore	Trail	Visitor Center	Campground	Wildlife Viewing	Historic Building	Fishing or Boating	Facilities for Disabled	Restrooms	Parking	Fee
Path to Beach		•	•								•	
Jefferson Way		•	•								•	
Jug Handle State Reserve	•	•	•							•	•	
Jughandle Creek Farm and Nature Center			•		•		•		•	•	•	•
Cantus Cove		•	•									
Caspar Creek Vista Point											•	
Caspar Headlands State Beach	•					•					•	
Caspar Headlands State Reserve		•	•			•		•				
Point Cabrillo Light Station and Preserve		•	•	•			•		•	•	•	
Russian Gulch State Park	•	•	•		•	•		•	•	•	•	•

PATH TO BEACH: *Ocean Dr. near Hwy. One, 2 mi. N. of Caspar Creek.* Park in an unpaved area near the Pine Beach Inn sign, and walk through groves of shore pines along a short level path to sea cliffs, part of Jug Handle State Reserve. In late spring and summer, the bluffs are dotted with Douglas iris and fiery red Indian paintbrush. No facilities.

JEFFERSON WAY: *W. of Hwy. One, 1.8 mi. N. of Caspar Creek.* Two pocket beaches within Jug Handle State Reserve are accessible from an unpaved parking area at the west end of Jefferson Way. Trails lead across the headlands, overlooking cove beaches. No facilities.

JUG HANDLE STATE RESERVE: *E. and W. of Hwy. One, 1 mi. N. of Caspar Creek.* This state park unit includes coastal terrace north and south of the Jug Handle Creek mouth, and forest extending inland over two miles. A sandy beach is located in a cove at the creek mouth. Five miles of trails lead through the Jug Handle Ecological Staircase, which illustrates five sequences in geological time. From the shoreline, these terraces support shore pine forest, bishop pine forest, coast redwood forest, grand fir/Sitka spruce forest, and pygmy forest; the oldest terraces are the highest and farthest inland. Brochures are available at the picnic area next to the parking lot for a self-guided tour. Call: 707-937-5804.

JUGHANDLE CREEK FARM AND NATURE CENTER: *E. of Hwy. One, .9 mi. N. of Caspar Creek.* This non-profit nature education center located next to Jug Handle State Reserve offers learning opportunities and overnight stays in a red Victorian farmhouse, three cabins, and a campground. Rooms are private, with shared bathrooms and kitchens. Room charges are discounted in return for a one-hour stewardship work contribution. Paths lead to the reserve's trail system and pygmy forest and to the beach at Jug Handle Cove. The farmhouse, campground, and some trails are wheelchair accessible. For information, call: 707-964-4630.

CANTUS COVE: *.1 mi. from Pacifica Dr. and Caspar Rd., Caspar.* In the village of Caspar, park on shoulder near stop-signed intersection of Caspar Rd. and Pacifica Dr. Walk west 200 yards to 45350 Pacifica Dr. The trailhead is unsigned; head north with the split rail fence on your right and the cypress grove on the left, then west through the woods to a scenic overlook managed by the Mendocino Land Trust. Call: 707-962-0470.

CASPAR CREEK VISTA POINT: *W. side of Hwy. One at Caspar Creek.* A roadside pullout overlooks the Pacific Ocean and heavily wooded Caspar Creek. A sawmill was built in 1861 at the mouth of the creek for what later became the Caspar Lumber Company.

The mill, once the largest lumber manufacturing operation on the Mendocino coast, operated until 1955.

CASPAR HEADLANDS STATE BEACH: *At Doyle Creek, off Pt. Cabrillo Rd., S. of Caspar.* A .2-mile-long sandy beach is located between the mouths of Caspar Creek and Doyle Creek, where Pt. Cabrillo Dr. (old Hwy. One) runs adjacent to the ocean. Skin diving and fishing are possible from the beach, which is frequented by foraging shorebirds. Harbor seals haul out on offshore rocks, south of the beach. The state park now includes over 100 acres of the Caspar headlands to the north of the beach; walk south from the end of Caspar Rd. to reach the newly acquired land, with its outstanding views of rocky shore. The Department of Parks and Recreation is planning additional expansion of the park to include more of the sandy beach, as well as the Sitka spruce forest to the east of Point Cabrillo Drive. Caspar Beach RV Campground, located at 45201 Point Cabrillo Dr., offers hookups for motorhomes up to 50 feet long and tent sites. Facilities include a convenience store, laundry, showers, and restrooms. Call: 707-964-3306.

CASPAR HEADLANDS STATE RESERVE: *End of Headlands Dr. off S. Caspar Dr., S. of Caspar.* Blufftop parcels totaling five acres are interspersed with private property. There is pedestrian access from Headlands Dr. to grassy finger-like headlands, with views of the coast in all directions and access to the rocky intertidal areas for fishing and abalone diving. A use pass is required, available at no charge from the state park district headquarters at Russian Gulch, located three miles south on the east side of Highway One. Call: 707-937-5804.

Point Cabrillo Light Station and Preserve

POINT CABRILLO LIGHT STATION AND PRESERVE: *Pt. Cabrillo Dr., 2 mi. N. of Mendocino.* The preserve occupies a dramatic headland jutting into the ocean. Park at the gate near the historic farmhouse rebuilt as a visitor center and walk one-half mile down the access road past three original light keepers' houses to the restored lighthouse with its operating Fresnel lens. Limited parking for those with special needs is available in front of the light station residences, one of which is being renovated as a museum. The lighthouse is open Friday through Monday from 11 AM to 4 PM, March through October; docent tours are offered on summer Sundays at 11 AM. The preserve grounds are open to pedestrians daily from sunrise to sunset. For information on the light station or parking, call: 707-937-0816. The ocean waters around the headland are within the Point Cabrillo State Marine Conservation Area, where no plants or animals may be removed; for information, call 707-964-9078.

RUSSIAN GULCH STATE PARK: *E. and W. of Hwy. One, 2 mi. N. of Mendocino.* A graceful high-arched bridge crosses Russian Gulch, carrying travelers along Hwy. One. The brief glimpse of coastline from the bridge only hints at the variety of sights within the state park, which is three miles long and up to a mile wide. On the headlands to the west of the highway is a blowhole that echoes with the sound of the waves when the tide is high. The hole was formed when the surf carved an inland tunnel, the roof of which later collapsed, leaving a hole 100 feet across and 60 feet deep. Trails wind around the perimeter of the headlands, through fields of wildflowers in spring. A picnic area is located on the north side of the gulch. Beneath and west of the Hwy. One bridge is a sea-level parking area and sandy beach, a good place to launch a kayak; an outdoor shower is nearby.

Facilities in the park east of Hwy. One include several campgrounds, a recreation hall, and ten miles of trails leading up the redwood-forested valley. Some trails are available to bicyclists and horses. Camping facilities include 30 family campsites, one of which is wheelchair accessible, a 40-person group campsite, a hike or bike area for up to 12 persons, and four primitive equestrian campsites for up to four persons. Reservations for equestrian sites are available only through State Parks district headquarters; for other campsites, call: 1-800-444-7275. Summertime reservations strongly recommended; campground closed from late October to early April. Fees for day use and camping. For information, call: 707-937-5804. The Russian Gulch State Marine Conservation Area includes ocean waters off Russian Gulch, where taking of aquatic plants and certain animals is banned; check Department of Fish and Game regulations.

The clipper ship *Frolic*, bound from Canton to San Francisco, struck an offshore rock and was lost off Pt. Cabrillo on July 25, 1850. The Boston-based owners of the *Frolic* had used her to smuggle opium into China from India, but news of California's gold rush prompted them to load the ship with bolts of silk fabric, china dishware, furniture, 84 cases of beer, and a prefabricated house, intended for sale in San Francisco. The goods were a mixture of ordinary household items and luxury goods, such as silver tinderboxes, that might catch the eye of a newly rich miner with gold dust in his pocket. As the *Frolic* sailed by dark, a day short of its destination, the crew kept a dim watch on distant mountains, while neglecting the dangerous rocks along the shore. The ship sank in shallow water only 75 feet from land, and the crew made it to shore safely. In the 1960s, sport divers found the wreck along with remnants of its contents, but shortly after the shipwreck much of the cargo had been removed by Mitom Pomo Indians who summered in the 1850s at the mouth of Big River, near the present town of Mendocino. An alert archaeologist excavating a Pomo village site in 1983 linked bits of porcelain ginger jars and bottle glass he found there to the cargo of the *Frolic*. Some of the items from the ship can be viewed at the Kelley House Museum in Mendocino, and others are at museums in Ukiah and Willits.

Rocks where the *Frolic* shipwrecked

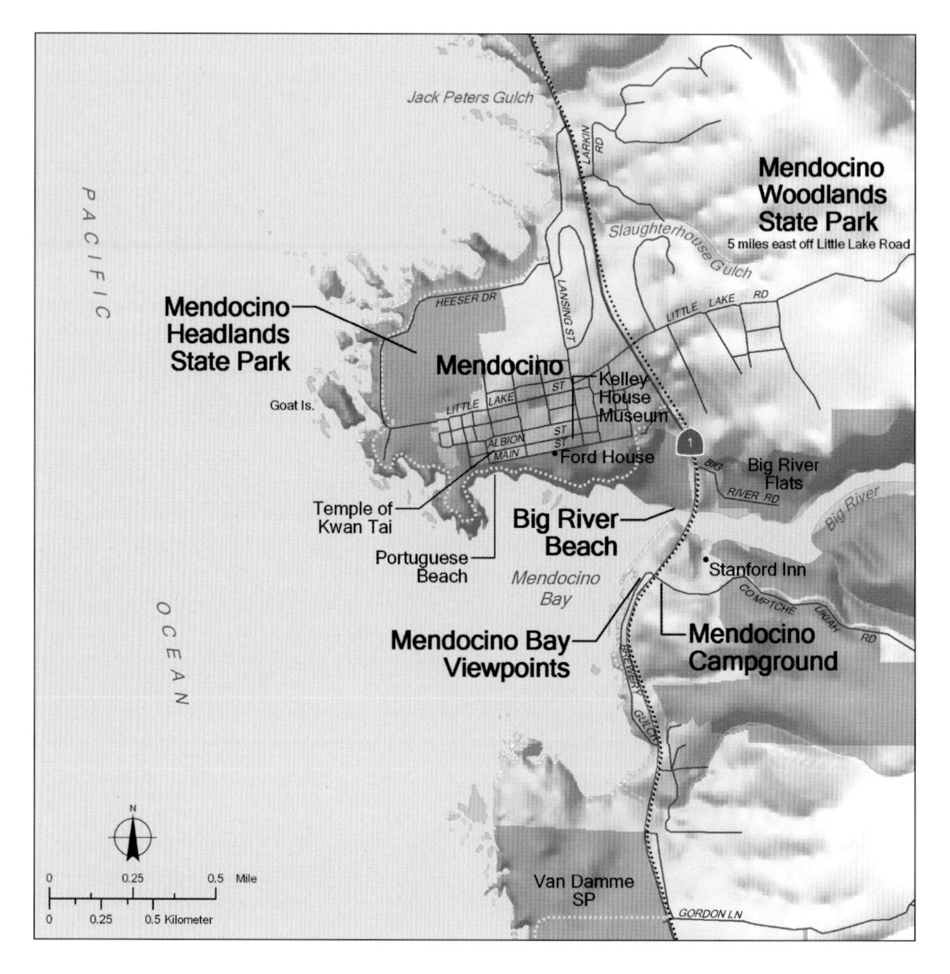

Town of Mendocino

Town of Mendocino

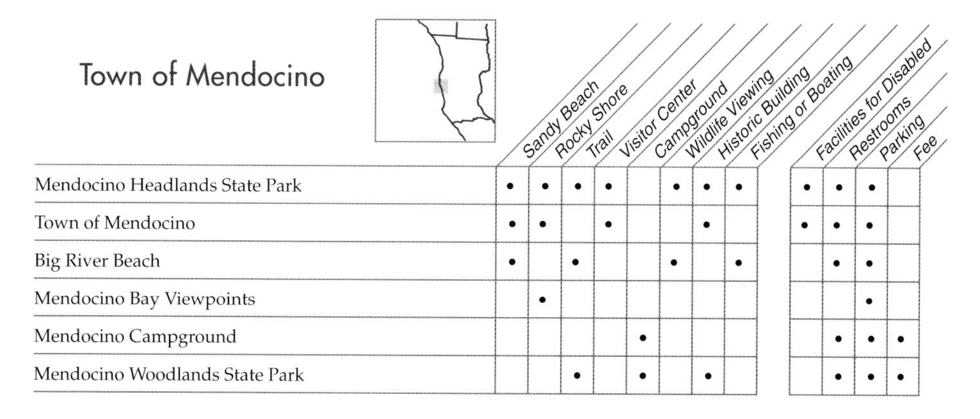

	Sandy Beach	Rocky Shore	Trail	Visitor Center	Campground	Wildlife Viewing	Historic Building	Fishing or Boating	Facilities for Disabled	Restrooms	Parking	Fee
Mendocino Headlands State Park	•	•	•	•		•	•	•	•	•	•	
Town of Mendocino	•	•		•			•		•	•	•	
Big River Beach	•		•			•		•		•	•	
Mendocino Bay Viewpoints		•									•	
Mendocino Campground					•					•	•	•
Mendocino Woodlands State Park			•		•		•			•	•	•

MENDOCINO HEADLANDS STATE PARK: *Seaward of Mendocino, from Lansing and Heeser Drs. to Big River.* The park is located on the ragged-edge marine terrace, surrounding the town of Mendocino on three sides. A trail winds along the circumference of the bluffs, which are up to 70 feet high, offering views of the rocky offshore islands, tidepools, and sandy beaches below; allow at least an hour for a complete circuit of the headland on foot. Never turn your back on the ocean; sleeper waves can reach higher on the rocks than you might expect. The edge of the headland can also be toured on bicycles or in vehicles along Heeser Dr., Main St., and intervening streets; there is no chance of getting lost, as the ocean is nearly always in view.

At the west end of Main St., a path leads onto a promontory where a blowhole can be seen, and a stair leads down to small, south-facing sandy Portuguese Beach. Opposite the west end of Little Lake St. is Goat Island, a flat-topped chunk of marine terrace only a few yards offshore; black oystercatchers, Brandt's cormorants, common murres, and pelagic cormorants are known to nest here. Along Heeser Dr., grasses are interspersed with spring wildflowers, including lupine and Indian paintbrush.

A visitor center at the historic Ford House has maps, books, and displays; open daily. Docent-led walks are available; picnic tables and restrooms are nearby. Another restroom is located on the north side of the Mendocino Headland, on Heeser Dr. near Lansing St. Call: 707-937-5804.

TOWN OF MENDOCINO: *Off Hwy. One, 10 mi. S. of Fort Bragg.* Mendocino's fortunes as a logging town waned with the decline of the industry in the 1930s; the community later became a favored residence for artists. The Mendocino Art Center was established in 1959 on the site of the former Dr. Preston mansion, used as a set in the 1954 film *East of Eden* but subsequently burned, and the flourishing artists' colony attracted an ever-increasing stream of visitors to its numerous

Town of Mendocino

galleries and shops. The Masonic Hall, built in 1865, is topped by a sculpture carved from a single block of redwood, with figures based on Freemasonry symbols. The whole town is on the National Register of Historic Places.

The Temple of Kwan Tai, built as early as 1867 with $12 worth of virgin redwood, was a center for the local Chinese community. Some accounts say the first settlers from China arrived directly in Mendocino, landing their junk on the beach at Caspar in 1854. The 1870 census counted 120 natives of China, out of a total population of about 2,900 living in coastal Mendocino County. The temple on Albion St. is dedicated to Guandi, a god associated with war but also literature, wealth, business, and social harmony. Visitors may view the interior of the Temple of Kwan Tai by appointment; write: P.O. Box 633, Mendocino, CA 95460.

BIG RIVER BEACH: *S. of Mendocino.* A broad white sandy beach, part of Mendocino Headlands State Park, extends along the north bank of Big River on both sides of Hwy. One. To reach the main parking area, turn east just north of the Big River Bridge. Boats can be launched at the east end of the parking area, and canoes, kayaks, and bicycles can be rented on the other side of the river at the Stanford Inn. Big River Beach can also be reached on foot via a stairway south of the historic Presbyterian church. The mouth of Big River is sometimes a good surfing spot.

Some 7,300 acres of the Big River watershed, much of it formerly logged, have been added to Mendocino Headlands State Park, including over eight miles of river frontage. No facilities have been developed yet, but boating, fishing, and wildlife viewing are possible on the river, which is a gentle Class I stream. Kayak or canoe trips are easiest if timed to go upstream on an incoming tide, then down on the outgoing tide. The river is habitat for otters, coho salmon, and steelhead, and the surrounding forest is home to bears, bobcats, and foxes. Kingfishers and osprey may be seen feeding on fish in the river. An old logging road, now a trail, departs the east end of the parking lot at Big River Beach. Hikers, equestrians, and mountain bikers permitted; no motor vehicles. Call: 707-937-5804.

MENDOCINO BAY VIEWPOINTS: *W. of Hwy. One, S. of Mendocino.* A fine vista of the much-photographed town of Mendocino, perched on its rocky headland, is avail-

Redwood lumber mill at Big River ca. 1850s

Mendocino was born as a lumber town, created to feed the huge demand for building materials in San Francisco and other new cities in mid-19th century California. Before local mills were developed, wood products for construction purposes came thousands of miles from the American East Coast, around South America. Henry Meiggs, with partners Jerome B. Ford and Stephen Smith, established the area's first redwood lumber mill in 1852 on Big River. Another partner, Edwards C. Williams, recalled years later their exploratory trip up Big River, assisted by early settler William Kasten:

In a rough canoe, which Mr. Kasten had fashioned from a redwood log, Mr. Ford and myself started out on a bright April morning with a fair light breeze from the ocean, and a flood tide to prospect the river for timber. For the first half-hour we were rather disappointed, but after that all that we hoped for was more than realized. The winter rains had not wholly ceased and the river bank full, its slight ripples meeting the verdure of the shore, the tall redwoods with their great symmetrical trunks traveling toward the skies, with the bright colors of the rhododendrons profusely scattered over the hills forming the background, the clear blue sky above reflected in the placid river and over all the hush and solitude of the primeval forest – all combining to impress upon our minds the beauty and truth of the opening of Bryant's Thanatopsis, "The groves were God's first temples," and as I recall the beauty of the picture, I cannot but regret the part it appeared necessary for me to enact in what now looks like a desecration.

The environmental effects of logging were severe, including the clogging of salmon streams with mud and the destruction of habitat of birds and animals that depend on old-growth forest. The logging industry seemed determined simply to remove as much timber as possible, as fast as possible. The speculative ventures of lumberman Henry Meiggs brought him profits, but also overwhelming debts, which apparently prompted him to abruptly depart San Francisco one October night in 1854 for Chile, by way of Tahiti. The mill operations in Mendocino continued, in the hands of others. The first mill was joined in 1853 by a second larger sawmill, which at its peak turned out two million board feet of lumber per month. The Page mill had 200-foot-tall smokestacks, which collapsed in the 1906 earthquake. The flats east of the Hwy. One bridge, where open beach and parking are located today, were the site of the Page mill and its rows of workers' cottages. Big River, as the town of Mendocino was known in its first years, drew immigrants from other forested locales; in 1860, nearly a quarter of the area's 638 inhabitants were born in Maine, Canada, or Scandinavia. The last mill in town closed in the 1930s after processing more than seven billion board feet of redwood.

Logger's home in the forest, 1904

able from a blufftop area managed by the Mendocino Land Trust. Turn west off Hwy. One on Brewery Gulch Road, just south of the Big River Bridge; shoulder parking. To contact the Mendocino Land Trust, call: 707-962-0470. A second vista point is located one-half mile farther south, where Brewery Gulch Rd. rejoins Hwy. One.

MENDOCINO CAMPGROUND: *Comptche-Ukiah Rd. and Hwy. One.* The privately run Mendocino Campground with 60 sites is located just south of the Big River. Open April to October; reservations required holidays and weekends; dogs allowed. For information, call: 707-937-3130.

MENDOCINO WOODLANDS STATE PARK: *E. of Hwy. One, off Little Lake Rd. S. of Fort Bragg.* Reservation-only camping for groups of between 30 and 200 people is available some five miles east of Mendocino, in a redwood forest and stream corridor setting. Cabins, tent cabins, showers, and kitchen/dining halls are available.

Reservations can be made up to two years in advance; contact Mendocino Woodlands Camping Association, P.O. Box 267, Mendocino 95460; 707-937-5755. Swimming holes and trails for hiking and mountain biking are available also for drop-in day-use visitors; no fee.

Mendocino Headlands State Park

The coast of Mendocino County has been featured in many Hollywood productions, for both television and theaters. Oddly, for a community as visually distinctive as Mendocino, the town rarely plays itself when filmmakers come to town. *Johnny Belinda* was set on Canada's Cape Breton Island, but filmed on the Mendocino coast in 1948. The production left its name behind: a bluff south of Fort Bragg is still called Belinda Point. The film is sentimental hokum, although Jane Wyman won an Academy Award for her wordless performance as a young deaf woman living on a farm by the sea. Today's film viewer might watch for scenes of Mendocino's landmark 1867 Presbyterian church, Fort Bragg's Noyo Harbor, and an unintentionally humorous scene when the fog blows in off the ocean, causing the characters to go into a panic at the prospect of the impending storm.

Still captivating, with very strong performances, is *East of Eden*, based on the novel by John Steinbeck. The 1954 film directed by Elia Kazan and starring James Dean, Julie Harris, and Raymond Massey is set partly in Monterey, for which Mendocino stands in. Background to the opening scenes of the film are views of Mendocino's Main Street, characteristic water towers, and the bay south of town, none of them resembling at all that other picturesque coastal town, Monterey. The brief presence in town of the movie company in the spring of 1954 left its mark, resulting in a book of reminiscences, *James Dean in Mendocino*, on sale at the Ford House Visitor Center.

James Dean as Cal Trask in *East of Eden*
EAST OF EDEN © Warner Bros. Pictures, Inc. All Rights Reserved
TM/© 2005 James Dean, Inc., by CMG Worldwide, www.JamesDean.com

Colors of the Coast

WHILE the northern California coast can be gray and foggy at times, visitors willing to look carefully will see an entire palette of colors, sometimes subtle, often striking. Like the colors reflected in a piece of abalone shell, a rainbow of color can be found along the coast. Each hue has a unique story to tell.

Indian paintbrush

RED

Have you ever walked out along the coastal bluffs on a spring day and seen the beautiful red color of the **Indian paintbrush** (*Castilleja* sp.)? The red that you see does not occur on the petals of the flower, but instead on modified leaves called "bracts," which protect the flower and attract pollinators. The true flowers are actually yellow. If you happen upon one, part the red bracts to get a look at the snapdragon-shaped flower hiding within. Indian paintbrush is "hemiparasitic," meaning that the plants get some of their nutrients by tapping into the roots of nearby plants.

Chert

Chert is a rock that comes in many colors. Particularly beautiful is the red chert that can be seen at many locations on the Marin Headlands. Chert forms on the sea floor from the accumulation of siliceous skeletons of microscopic organisms called radiolarians and diatoms. These accumulations occur in deep ocean basins, far from other sources of sediments that might dilute the rain of skeletons. Small amounts of iron are intermixed with the skeletal material, and as these constituents accumulate slowly on the ocean floor, the iron interacts with oxygen in the water to yield a beautiful red color.

Sticky monkeyflower

ORANGE

Sticky monkeyflower *(Mimulus aurantiacus)* is a drought-tolerant shrub that occurs in coastal scrub habitat. It produces abundant trumpet-shaped orange flowers throughout the summer. The beautiful flowers attract hummingbirds and insect pollinators. One of the unusual features of this plant is that the stigma (the sticky white lips that receive pollen) actually close in response to touch (e.g., pollination). The closed stigma prevents the plant's own pollen from being deposited as the pollinator is leaving. You can witness the stigma closing by gently touching it with your finger or a twig.

The beautiful orange color of the **monarch butterfly** *(Danaus plexippus)* serves to warn predators that their intended meal might be their last. The larvae of the monarch butterfly dine on plants in the milkweed family, and these plants are known to contain poisons that are toxic to vertebrates. The toxins are stored in the body of the caterpillar and passed on to the butterfly stage. Each fall thousands of these black-and-orange beauties make the long journey south to wintering grounds with mild weather. On the California coast, monarchs may be observed at Stinson Beach, Muir Beach, and other locations south of Marin County.

Monarch butterfly

YELLOW

Common yellowthroats *(Geothlypis trichas)* are stunning warblers that frequent marshy habitat. Between May and September they sing their distinctive song, *wichety, wichety, wichety*, from the cover of cattails and dense, brushy vegetation. During the breeding season, the male yellowthroat performs an aerobatic display during which he shoots out of the brush, ascends rapidly, and utters a jumble of high-pitched notes. Next time you hear one call, wait, and you may get a glimpse of its brilliant yellow throat and black mask.

Common yellowthroat

The **American goldfinch** *(Carduelis tristis)* has been described as "an active little bird, always in the best of spirits . . ." Goldfinches live in brushy thickets and weedy fields and feed mainly on the seeds of plants in the aster family. Male goldfinches get their bright yellow feathers in spring. A goldfinch in flight can be said to be "riding the waves of a stormy sea, giving, as it rises to each crest, its little phrase of four happy notes." These small gregarious birds are rarely seen outside the company of others. Keep an eye out for their deeply undulating flight. They appear in the sky like a burst of little yellow rockets.

American goldfinch

Maidenhair fern

GREEN

Maidenhair fern *(Adiantum capillus-veneris)* is found in forested canyons and rock walls near seeps and waterfalls. Its delicate, bright green fronds stand out in the dark shade of the forest. Native Americans used the black leathery stems of the maidenhair fern to make baskets. Next time you see this seemingly fragile denizen of the forest understory, take a closer look at the strong, fibrous stems that support the beautiful green leaflets.

Rocks including jade

Jade has been prized for millennia for its translucent green color. Actually two minerals, jadeite and nephrite, jade is formed at very high pressures but relatively low temperatures. These conditions are found deep within subduction zones, regions where an oceanic plate dives beneath a continental plate. Until about 30 million years ago, the coast of California was all part of a subduction zone (and still is, north of Cape Mendocino). Rocks containing jade were brought to the surface through the uplift processes that have shaped so much of the California coast. Jade can sometimes be seen interspersed with other pebbles on beaches in Marin, Sonoma, and Mendocino Counties, but it is rare, so look carefully. You can tell jade from the more abundant green chert pebbles because it is much harder. Jade pebbles are very difficult to break with less than a full swing of a hammer.

Brandt's cormorant

BLUE

The **Brandt's cormorant** *(Phalacrocorax pencillatus)* develops a striking cobalt-blue throat pouch and slender white plumes on the face during the breeding season. These loon-like birds can be seen on coastal or offshore rocks and in coastal waters. Brandt's cormorants often gather in flocks of several hundred where their prey is plentiful. The name "cormorant" is derived from the Latin words *corvus marinus,* meaning marine crow. This is an apt name, given the bird's large size and sooty black feathers.

The **western pygmy blue** *(Brephidium exile)* is the most diminutive butterfly in California. With a wing span of only one-half to three-quarters of an inch, it can easily be overlooked. Look for pygmy blue butterflies between July and September at Bolinas Lagoon and other coastal salt marshes in Marin and Sonoma Counties. You will find them fluttering weakly, close to the ground. When the wings are closed, the butterfly is copper–brown, with a white fringe and a row of black spots at the outer margin; when the wings are open, you may catch a glimpse of beautiful ultramarine blue, centered near the thorax and surrounded by a rich chocolate brown.

Western pygmy blue

VIOLET

The striking purple **Douglas iris** *(Iris douglasiana)* usually grows within sight of the ocean. It is fairly common on coastal bluffs and grassy hillsides. The fibrous and leathery grass-like leaves of the Douglas iris provided important rope-making fiber for Native Americans. Rope making was a very difficult and time-consuming process; one report states that it took about six weeks to make 12 feet of rope. Due to its beauty and ease of cultivation, the Douglas iris is used widely in the nursery trade. Next time you think about doing some landscaping, consider planting native species.

Douglas iris

The **purple sea urchin** *(Strongylocentrotus purpuratus)* has relatively short spines and is light purple in color. It is commonly found on exposed and semi-protected rocky areas on the Pacific coast. This odd little invertebrate has the ability to create shelter from the heavy surf by carving caves out of rocks with its teeth and spines. However, you will usually see sea urchins out in the open, slowly moving in search of food. Their diet consists mainly of seaweed and small organisms. If you find an empty sea urchin shell, or even just a piece of one, look closely at the little holes and bumps. Moveable spines and tube feet extend from these holes and bumps, allowing the sea urchin to move across the rocky shore.

Purple sea urchin

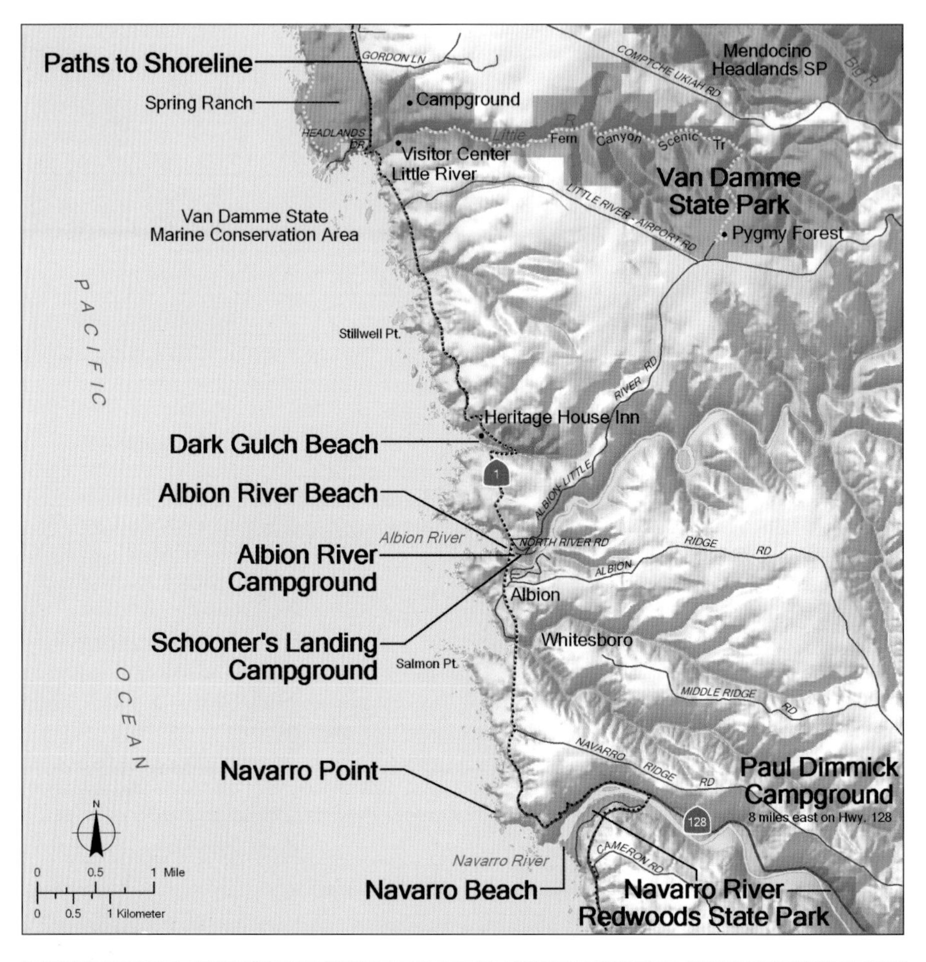

Van Damme State Park

Van Damme to Navarro River

	Sandy Beach	Rocky Shore	Trail	Visitor Center	Campground	Wildlife Viewing	Historic Building	Fishing or Boating	Facilities for Disabled	Restrooms	Parking	Fee
Paths to Shoreline / Van Damme State Park		•	•			•					•	
Van Damme State Park	•	•	•	•	•	•		•	•	•	•	•
Dark Gulch Beach		•	•								•	
Albion River Beach	•					•		•				
Albion River Campground					•			•	•	•	•	•
Schooner's Landing Campground					•			•	•	•	•	•
Navarro Point		•	•								•	
Navarro Beach	•				•	•		•	•	•	•	•
Navarro River Redwoods State Park	•				•	•		•	•	•	•	•
Paul Dimmick Campground					•			•	•	•	•	•

PATHS TO SHORELINE/VAN DAMME STATE PARK: *Hwy. One at Gordon Lane, 1.5 mi. S. of Mendocino.* The old Spring Ranch, located along the ocean north of Little River, is now an undeveloped addition to Van Damme State Park, with trails offering fine views of rocky coves and pocket beaches. Two trails lead from Hwy. One to the sea, the first starting at a dirt parking area opposite Gordon Lane. The second trail starts at an opening in the fence about 50 feet north of the inn called Auberge of Mendocino; there is parking on the south side of Headlands Dr. just south of the inn, off Hwy. One. The two paths descend toward the ocean, across the sloping terrace, and they are connected by a north-south trail, about one-half mile long. From the bluff, visitors may see sea lions or harbor seals hauled out on the rocks, a variety of shorebirds, and perhaps a northern harrier on the bluffs. Wildflowers dot the meadows in spring and early summer.

VAN DAMME STATE PARK: *E. and W. of Hwy. One at Little River, 3 mi. S. of Mendocino.* Rocky coast, a sheltered beach, ten miles of trails, and forests ranging from redwood to pygmy make this one of Mendocino's most popular parks. Divers, kayakers, hikers, picnickers, and campers enjoy the park's facilities, which include nearly three square miles extending along the Little River and north along the coast. The community of Little River subsisted economically on the lumber industry between the 1850s and the end of the 19th century, but today recreation and tourism are mainstays.

The small sandy beach has paved parking next to Highway One. There are fire rings, restrooms, and an outdoor shower adjacent

Pygmy forest trail, Van Damme State Park

to the Little River mouth. Offshore is a particularly popular site for divers because of the wealth of marine resources and the ease of transporting equipment from the sea-level parking area. The Van Damme State Marine Conservation Area includes nearshore waters, where the taking of aquatic plants and certain animals is banned; check Department of Fish and Game regulations.

East of Hwy. One, within the state park, is a picnic area and a campground with 74 sites, ten environmental sites, and one group site for up to 75 people; campers and trailers limited to 35 feet in length. Hike or bike campsites are located within the family campground. Restrooms and showers available. Campground open all year; reservations strongly recommended during summer; call 1-800-444-7275. Twenty-four enroute campsites are available in the beach parking lot if the regular campground is full. Fees are charged for camping and park day use; no fee for beach day use.

The visitor center has a diorama depicting the underwater environment visited by divers, along with other interpretive materials; open weekends year-round. There is a campfire center, and a bog trail around a wetland. The Fern Canyon trail extends along the Little River several miles upstream, where it connects to other paths; bicycles are allowed on some of the trails. A self-guided trail through a pygmy forest is located in the eastern part of the park. The pygmy forest trail, a wheelchair-accessible boardwalk, can also be reached from Little River Rd., off Hwy. One. Go east past the county airport to a parking area on the north side of the road. For information, call: 707-937-5804.

DARK GULCH BEACH: *Hwy. One, Dark Gulch, milepost 45.24.* Just south of Heritage House is an accessway leading to a secluded cove beach. There is limited roadside parking on the seaward side of Hwy. One along the wooden fence. From the road, the path leads through a corner of the Heritage House

Spring Ranch Cove, Van Damme State Park

Albion River

grounds, down several flights of stairs, and across a 30-foot-long boardwalk over Dark Gulch Creek. The accessway is managed by the Heritage House Inn.

The beach is paved with rounded cobbles, sheltered by the surrounding headlands. Sea anemones are plentiful among the rocks in the intertidal zone, and black oystercatchers may be seen feeding along the shore. A thick growth of willows lines the stream, and wildflowers including seaside daisies and monkeyflowers grow on the cliffs.

ALBION RIVER BEACH: *Albion flats, W. of Albion River Campground.* A sandy river beach faces the ocean near the mouth of the Albion River, which remains open year-round. Park near the café, then walk under the bridge.

A signal on a rock in the channel sounds a warning to boaters entering the little harbor, which is frequented by harbor seals and, farther upstream, river otters. Eelgrass grows along both sides of the channel, and the estuary is inhabited by surfperch, starry flounder, and Dungeness crab. Great blue herons build their nests along the river, and osprey can be spotted fishing at the river mouth. The river is an important sport fishing area for coho salmon and steelhead trout.

ALBION RIVER CAMPGROUND: *Hwy. One, N. side of Albion River.* Two campgrounds are located on the flats along the river, below the historic wooden trestle bridge that crosses high over the Albion River. North of the bridge, turn east on North River Rd., which winds down the hill to an active small harbor. Albion River Campground offers both RV (with or without hookups) and tent camping. Also available are a boat launch, bait and tackle, a café, restrooms with showers, and a pumpout station for RVs. Fees for camping or day use; no charge for pedestrians headed for Albion River Beach. For information and reservations, contact Albion River Campground, P.O. Box 217, Albion 95410; or call: 707-937-0606.

Hollow log at Navarro River Redwoods State Park

SCHOONER'S LANDING CAMPGROUND: *Hwy. One, north side of Albion River.* Also located on the flats beneath the Albion River Bridge is Schooner's Landing, which offers RV camping with hookups, boat launch and dock, and picnicking. There are restrooms with showers. Fee for camping or day use. For reservations and information, contact Schooner's Landing, P.O. Box 218, Albion 95410; or call: 707-937-5707.

NAVARRO POINT: *Off Hwy. One, 1.7 mi. N. of the junction of Hwy. 128.* The Mendocino Land Trust plans in mid-2005 to open a new blufftop viewing area on a spectacular 55-acre parcel of land on the Navarro Headlands. At milepost 41.83, an eight-car parking area is planned, and trails will extend north and south, forming a loop approximately one mile long. The meadow on the bluff contains wetlands, native bunchgrasses near the bluff edge, and resplendent wildflowers in spring. For information, call: 707-962-0470.

NAVARRO BEACH: *W. end of Navarro Bluff Road.* South of the Hwy. One bridge over the Navarro River, turn west onto Navarro Bluff Rd. to a very wide sandy beach strewn with driftwood that stretches a half-mile beneath the bluffs at the mouth of the river. There is a picnic area at the south end of the beach. Ten campsites have picnic tables and firepits; no water available. RVs and trailers can be accommodated. Fee for camping; no reservations taken. Call: 707-937-5804.

Sea lions and occasionally dolphins may be sighted from the beach, and pelicans and osprey feed near the river mouth. In the winter months, migratory waterfowl use the estuary. Offshore rocks are roosting sites for seabirds, and migrating gray whales pass nearby. The riparian vegetation along the river supports many bird species. When the flow of water in the river is low, the sandbar extends all the way across the river mouth, allowing access to the beach on the north side and a sea cave in the bluff. Plans are under way to restore as a visitor center the historic Navarro-by-the-Sea buildings on Navarro Bluff Road.

NAVARRO RIVER REDWOODS STATE PARK: *Hwy. 128, 2 mi. E. of the junction of Hwy. One.* A redwood forest stretches for ten miles along Hwy. 128 on the lower reach of the Navarro River. Navarro River Redwoods State Park takes in both redwood forest and the beach at the mouth of the river, and swimming, boating, camping, and beachcombing are popular. In summer, the park often offers a range of summer temperatures, from cool and potentially very windy at the beach to warm among the redwood groves along the river, as well as a variety of activities. Scattered pull-outs on the south side of Hwy. 128 provide parking and access to trails through redwood groves to the banks of the Navarro River. Dense second-growth redwood forest towers overhead, and ferns, wood violets, and horsetails grow along the paths. No facilities.

PAUL DIMMICK CAMPGROUND: *8 mi. E. of Hwy. One on Hwy. 128.* This unit of Navarro River Redwoods State Park offers both camping and day use opportunities in a redwood grove along the Navarro River. Twenty-five campsites are located close to Highway 128 on a paved road. All sites include picnic tables and fire rings. Non-potable running water is available. Campers up to 30 feet; trailers up to 22 feet. The campground is a popular spot for kayaking, canoeing, and winter steelhead fishing; catch-and-release strongly encouraged. Lush vegetation along the river includes willows and big-leaf maple, and the redwood forest contains an understory of sword ferns, redwood sorrel, and thimbleberry. Campground host available in summer only. For information, call: 707-937-5804.

Big-leaf maple

Offshore rocks (Bird Island, Marin County)

© 2004, Tom Killion

The Ocean

A GREAT PARADOX of the sea is that it shows us a nearly featureless surface, while concealing high mountains, deep valleys, and a great richness of life. The ocean still contains many mysteries, and its rules cannot be understood through superficial analogies with terrestrial systems. One might imagine that warm equatorial seas, like tropical forests, would be teeming with life, but northern California's brisk ocean supports greater biological diversity and abundance. Visitors may shiver when dabbling their toes at the beach, but marine animals and plants thrive in those nutrient-laden, cold waters.

The marine world off northern California is strongly influenced by ocean currents and a phenomenon known as upwelling, which is characteristic of only a handful of the world's shorelines. The dominant current along the California coast is the California Current. It is part of a large ocean gyre, or circular current, that travels westward across the equatorial Pacific, north past Japan (where it is called the Kuroshio Current), east across the northern Pacific and then south along the Washington, Oregon, and California coasts. The California Current is some 100 miles wide, and it carries approximately 10 million cubic meters of water per second along the coast. One reason that California coastal waters are so much cooler than coastal waters at a comparable latitude on the Atlantic shore is that the California Current is carrying cold arctic water south, while on the Atlantic, the Gulf Stream is carrying warm tropical waters north. In the warmest months of the year, the ocean at Mendocino reaches an average high temperature of 54 degrees Fahrenheit, whereas the ocean at Atlantic City, New Jersey, almost due east, averages 70 degrees.

Ocean circulation off California is partly due to the Coriolis effect, which is the apparent deflection of a moving body as it moves toward or away from the equator. In the northern hemisphere, this effect results in a slight deflection to the right, and in the southern hemisphere, a slight deflection to the left, responsible for (among other things) the cyclonic motion of hurricanes. The Coriolis effect also causes a slight offshore movement of water along the California coast. In the spring and summer months, winds augment this movement through a process called Ekman transport that results in water flow

Surf off Rodeo Beach, Marin County

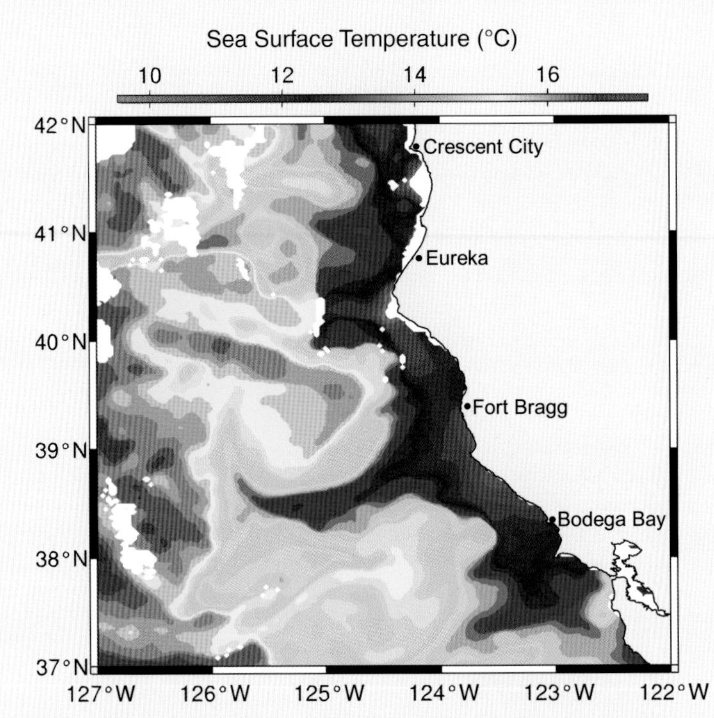

NOAA/NASA AVHRR, processing by John Ryan, Monterey Bay Aquarium Research Institute

that is 90 degrees to the right of the dominant wind direction. The combined result is that surface waters are pushed offshore, and deeper, nutrient-rich bottom-waters are brought up to the surface through upwelling. This process is also responsible for the natural air-cooling fog system that can catch summer beach visitors unawares.

Wind can actually be considered the main force for coastal upwelling since the ocean currents themselves derive their energy from surface winds. There are complex interactions between the ocean and the atmosphere, but scientists speculate that if the wind were to stop, surface currents would die down and the ocean surface would be completely flat within about three years.

While upwelling occurs along much of the central and northern California coast, the intensity of upwelling depends strongly on the topography of the land and offshore bathymetry, or underwater topography. As currents pass a headland, they can draw water away from the area down-current of the headland. For example, Point Arena is a strong upwelling region. Along northern California, the continental shelf (less than about 200 meters deep) is relatively narrow, extending seaward only four or five miles. Beyond that, the sea floor drops dramatically. Although less precipitous than the Monterey Canyon near Monterey Bay, the Noyo Canyon off Fort Bragg reaches depths of 1,000 meters only some 12 miles from shore.

Upwelling is a major factor in the nearshore food chain. Mixed with the colder water is a soup of organic matter from decaying plants and animals that have sunk to the ocean floor. Upwelling brings this mixture to the surface, where sunshine warms it, and a population explosion begins of minuscule single-celled algae called phytoplankton.

Tiny animals known as zooplankton feed on the phytoplankton, as well as each other, creating the bottom rung of the food chain. Shrimp-like krill and the larvae of fish and crabs are eaten by larger fish, which in turn are eaten by birds and marine mammals. The abundance of food sources, both large and small, results in a relatively dense assemblage of marine life in the nearshore waters off Northern California. Beyond the continental shelf, by contrast, the ocean waters are clearer and bluer, and biological resources are sparser.

The process of upwelling varies by season and year. The process is heightened from about March to September, which happens to coincide with the longest days of the year, when the period of maximum sunshine is available for photosynthesis. In the fall, upwelling tends to break down, and the growth of the sea's phytoplankton slows, while beachgoers find more comfortable air and water temperatures. In winter, northwesterly winds are sometimes replaced with those from the southwest, which tend to push surface waters toward shore, causing a condition of "downwelling." At the same time, the Davidson Current often develops, traveling from south to north. This current brings with it warmer, less biologically rich ocean water, and it pushes the California Current farther offshore.

Every few years, the pattern is disrupted by an eastward flood of equatorial warm water known as El Niño. The California Current is shifted farther offshore, its strength diminishes, less cold water is carried down from the polar regions, and the upwelling of cold, nutrient-rich water is suppressed. As a result, during an El Niño episode, sea surface temperatures are high and nutrients are low, with a sometimes devastating impact on the fish stocks that normally rely on the upwelling system. Effects of an El Niño onshore may include violent storms and exceptionally heavy rains.

There are several ways to get a look at northern California's marine resources. From promontories such as Point Reyes, Bodega Head, Point Arena, and Patrick's Point, coastal visitors can see migrating gray whales between November and mid-May. The harbor porpoise may be seen sometimes at Drake's Bay in Marin County or other shallow areas. Boat excursions to view the ocean's resources close up are available at Bodega Bay, Noyo Harbor, Humboldt Bay, Crescent City, and other locations. Seabirds, including the albatross, that rarely come close to shore may be sighted on ocean-going trips. One destination of ocean trips is the Cordell Bank, where shallow depths and an abundance of food draw sea birds and animals; see p. 232. Locations where marine mammals and birds can be seen are noted throughout this guide.

Gray whale

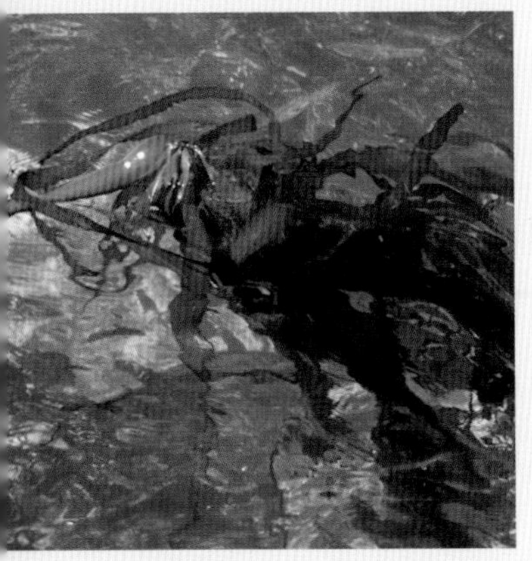
Bull kelp

Bull kelp (*Nereocystis luetkeana*) is an annual species of algae and one of the fastest growing seaweeds in the world, achieving its adult length of up to 100 feet in its single growing season. It can grow as much as six inches in one day. The surface canopy of kelp beds provides an important resting place for otters, herons, and gulls. Below the water's surface, kelp forests shelter a great variety of fishes and invertebrates, such as abalone and other snails, crabs, sea urchins, and sea stars. Bull kelp attaches to rocks with a "holdfast," a gnarly twisted hand-like growth that can sometimes be seen washed up on the beach. A single robust stipe extends from the anchoring holdfast to the bulbous air-filled float from which the long trailing blades extend. The blades grow close to the water's surface, where they can collect the sun's radiant energy. Much of the brown color in bull kelp comes from tannins, which help discourage the growth of parasites.

Red abalone

Red abalone (*Haliotis rufescens*), which is protected by the kelp forest, is the largest of California's marine snails. Most of the abalone's body under its oval-shaped shell consists of a "foot" for attaching itself to a rock surface and for locomotion, although a mature abalone is generally quite sedentary. Abalone feed on algae, and the red or coral color of the outer surface of a red abalone shell results from consumption of red algae. The holes along the edge of the shell serve several functions: they release eggs or sperm, discharge metabolic wastes, and allow water to flow out after passing through the animal's gill chamber.

China rockfish

The **china rockfish** (*Sebastes nebulosus*) is an attractive solitary fish with mildly venomous spines located on the dorsal and anal fins. The china rockfish most commonly occurs in 30 to 300 feet of water. The juveniles live in the open ocean, but the adults are sedentary, associated with rocky reefs or cobble, where they may travel less than three feet from their home range. The adults are generally found resting on the bottom or hiding in crevices, where their large pectoral fins enable them to support themselves on the floor of a cave and to maneuver in crevices. The china rockfish's diet consists of brittle stars, crabs, and shrimps. The fish have been found to live up to 26 years.

No discussion of California's northern offshore waters would be complete without mention of the **white shark** (*Carcharodon carcharias*). The world's largest predatory fish, the white shark feeds on sea lions, seals, whales, sea otters, and sea turtles. The waters off California offer a rich bounty of food for white sharks, and every summer and fall they actively feed in nearshore areas where their prey are found. The white shark can grow to be about 21 feet long, and it has approximately 3,000 razor-sharp teeth. The shark's acute sense of smell and its ability to detect electric charges helps it to detect its prey. Despite the white shark's reputation as the oceans' most ferocious predator, white shark numbers are dwindling, and the species may be close to extinction in some areas. The sharks have been hunted for their teeth and jaws, and many are also accidentally netted as by-catch in fishing nets. In some areas, white sharks have been taken for shark fin soup. In 1993, California became the first state to protect the white shark. The population along the California coast may number only 100 individuals. Mature females are believed to give birth to only a small litter every two to three years, meaning that in areas where they are still hunted, shark reproduction may not keep up with the hunters.

White shark

Pacific white-sided dolphins (*Lagenorynchus obliquidens*) are very energetic, frequently leaping, belly-flopping, and somersaulting above the ocean's surface. They are strong, fast swimmers and enthusiastic bow-riders, often staying with moving vessels for extended periods. They are very social animals and swim in groups that generally number between 50 and 100 individuals. White-sided dolphins reach a length of seven to eight feet and weigh 300 pounds. They have sharp teeth that are ideal for gripping their slippery food, especially squid and fish.

Pacific white-sided dolphin

Cabezon

In Spanish, the word *cabezon* means big-headed or stubborn, and, proportionally, the massive head of this fish is its largest feature. **Cabezon** (*Scorpaenichthys marmoratus*) have a marbled or mottled appearance, which can be reddish, greenish, or bronze. They occupy rocky reef areas and kelp beds, normally near shore; their depth range extends from the intertidal zone to 335 feet. In shallower water, cabezon move in and out with the tide to feed, and as they get older and larger, they tend to migrate into deeper water. Cabezon are long-lived and have been determined to reach a maximum age of 17 years for males and 16 years for females. They are not sexually mature until they are three or four years old. Females spawn their eggs on algae-free rocky surfaces, primarily in crevices and under rocks. The protective male guards the nest during the two-to three-week period that it takes the eggs to mature. Cabezon are considered "sit-and-wait" predators. Their mottled coloration allows them to blend into their environment. A cabezon may sit motionless waiting for its next meal to swim by, and then quickly lunge after unwary prey, engulfing it in its large mouth.

Surf scoter

The **surf scoter** (*Melanitta perspicillata*) is a type of sea duck that dives for mollusks and crustaceans in shallow ocean waters subject to breaking waves. The surf scoter has a large bill that is swollen on the top, giving the bill a knobby appearance. The decidedly odd-shaped bill has earned this species a multitude of descriptive informal names, including goggle-nose, horse-head coot, plaster-bill, snuff-taker, blossom-billed coot, bottle-nosed diver, and mussel bill. The bill of the surf scoter is hard and stout, ideal for seizing shellfish underwater. A large portion of the scoter's diet consists of clams and mussels, but they also eat other invertebrates and algae.

The **sooty shearwater** (*Puffinus griseus*) belongs to a family of birds that have a unique skimming flight pattern, consisting of a flap and a rapid glide close to the surface of the sea. During the glide phase, these birds can soar above or skim just over the waves at sea, banking and careening with remarkable agility using their long, stiff wings. Sooty shearwaters nest in the sub-Antarctic, particularly on islands off Tierra del Fuego and New Zealand. During August and September the sooty shearwater is one of the most abundant marine birds along the California coast. Thousands of sooty shearwaters have been observed in seemingly endless winding rivers of birds flowing just above the ocean's surface. Sometimes they follow huge schools of fish in toward shore and feed in monstrous flocks. Regarding the sooty shearwater, Charles Darwin wrote in *The Voyage of the Beagle*, "I do not think I ever saw so many birds of any other sort together, as I once saw of these.... Hundreds of thousands flew in an irregular line for several hours in one direction. When part of the flock settled on the water the surface was blackened, and a noise proceeded from them as of human beings talking in the distance."

Sooty shearwater

The only albatross that regularly visits the West Coast is the **black-footed albatross** (*Phoebastria nigripes*). It is most commonly observed during the summer, when it can be found following shrimp or fishing boats. These large and beautiful seabirds have a wingspan ranging from six to seven feet, wing tip to wing tip. Their large wings allow them to travel great distances with little effort. The albatross is monogamous, and only one chick is produced per breeding season. While rearing their chicks on islands in the Hawaiian archipelago, some individual birds have been tracked by satellite to California's continental shelf where they foraged for prey to feed their young waiting thousands of miles away. They make this lengthy trip because ocean currents around Midway Island offer little to eat, and upwelling currents along the continental shelf in California are rich in prey. The prey is stored in the belly in a very viscous rich oil, and upon arrival back at the nest, it is regurgitated for the chick. Due to its low reproductive rate and losses of the birds as by-catch on commercial fishermen's long lines, the species is considered "Endangered" by the International Union for the Conservation of Nature (IUCN).

Black-footed albatross

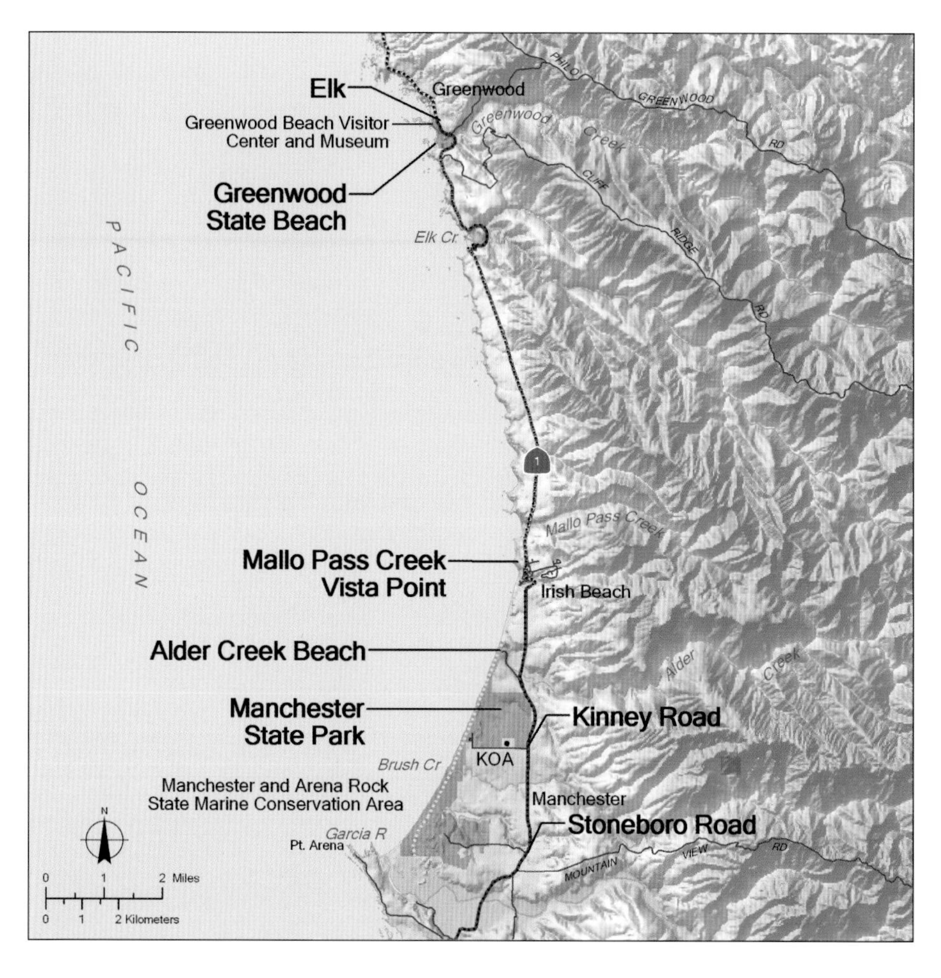

Osprey at Alder Creek

Elk to Manchester State Park

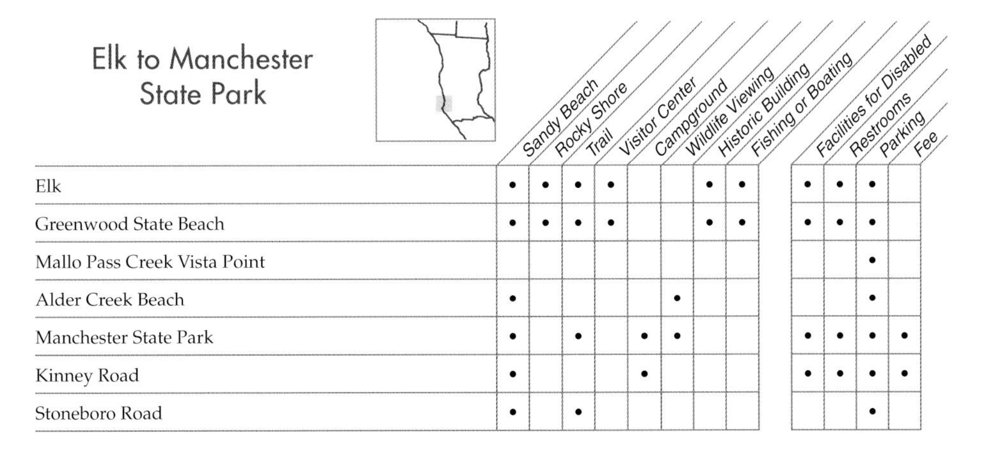

	Sandy Beach	Rocky Shore	Trail	Visitor Center	Campground	Wildlife Viewing	Historic Building	Fishing or Boating	Facilities for Disabled	Restrooms	Parking	Fee
Elk	•	•	•	•			•	•	•	•	•	
Greenwood State Beach	•	•	•	•			•	•	•	•	•	
Mallo Pass Creek Vista Point											•	
Alder Creek Beach	•					•					•	
Manchester State Park	•		•		•	•			•	•	•	•
Kinney Road	•				•				•	•	•	•
Stoneboro Road	•		•								•	

ELK: *Hwy. One, 6.2 mi. S. of junction with Hwy. 128.* The quiet present-day village of Elk was a bustling lumber center in the late 19th century, with a sawmill at nearby Greenwood, a railroad that brought logs from the forest east of town, and wharves to load wood products onto waiting schooners. The Greenwood Beach Visitor Center and Museum, open weekends from 11 AM to 1 PM, mid-March through October, is located in the former mill office of the Goodyear Lumber Company. The museum offers an eclectic collection of artifacts pertaining to the history of the local area; for information, call: 707-877-3458.

GREENWOOD STATE BEACH: *Hwy. One at Greenwood Creek, Elk.* A bluff some 150 feet high backs a wide sandy beach at the mouth of Greenwood Creek, just south of the community of Elk. Next to the unpaved parking area on Hwy. One are picnic tables, fire rings, and restrooms. A trail leads north along the bluff one-quarter mile to the Greenwood State Beach visitor center. An old road, now a wide but fairly steep path, leads one-quarter mile from the parking area down to the sandy beach, where there are picnic tables and fire rings. Visitors roll small boats on carts down the road to the sheltered beach, a good launching spot for kayaks. Day use only; dogs must be leashed. For information, call: 707-937-5804.

Rock promontories border the long curving beach, divided by marshy Greenwood Creek. Coho salmon and steelhead spawn in the creek after winter rains open the sandbar that forms in summer. The creek is a wintering site for migratory geese and ducks, and offshore rocks are nesting sites for cormorants and pigeon guillemots. Sea caves and tunnels riddle the offshore rocks, some of which were once linked with elabo-

The village of Elk

rate trestles that brought cargo seaward to load onto waiting ships.

MALLO PASS CREEK VISTA POINT: *W. of Hwy. One, 4.2 mi. N. of Manchester.* Just south of Mallo Pass Creek, this large paved parking area provides elevated views of the rocky ocean shore to the west and the dense, riparian forest to the east. No facilities.

ALDER CREEK BEACH: *End of Alder Creek Rd., W. of Hwy. One, 7 mi. N. of Point Arena.* No sign marks Alder Creek Rd., which intersects Hwy. One just south of the bridge over Alder Creek. There is limited parking and no facilities at the end of the half-mile-long road, where a short, steep path leads down to the northern end of Manchester State Park. Smelt spawn in the mouth of Alder Creek, and migratory waterfowl winter in the freshwater habitat that extends inland along the stream. Horses allowed; dogs must be leashed; day use only. For information, call: 707-882-2463.

MANCHESTER STATE PARK: *W. of Hwy. One, from Alder Creek to just N. of Point Arena.* This state park includes more than one square mile of beaches, sand dunes, and grasslands, including more than three miles of driftwood-strewn, often-windy shoreline. The long sandy beach, located a mile west of the coastal highway, offers a sense of solitude. There are seasonal wildflowers, including sea pinks, poppies, lupines, baby blue eyes, and irises. The park provides habitat for tundra swans, and two streams, Brush Creek and Alder Creek, feature salmon and steelhead fishing. From Hwy. One, Manchester State Park can be reached on Alder Creek, Kinney, and Stoneboro Roads. Call: 707-882-2463. The Manchester and Arena Rock State Marine Conservation Area includes nearby ocean waters, where the taking of aquatic plants and certain animals is banned; check Department of Fish and Game regulations.

KINNEY ROAD: *End of Kinney Rd., W. of Hwy. One, 6 mi. N. of Point Arena.* The main entrance to Manchester State Park has day-

Cuffey Cove, north of Elk

Alder Creek

use and camping facilities. A picnic area is located at the end of the road, along with restrooms and a quarter-mile flat trail to the beach. The campground, set among trees, offers tent and RV camping, including 46 primitive sites, a 40-person group site, and one hike or bike site. There are fire pits, chemical toilets, running water, and RV dump station; no showers are available. RVs up to 30 feet allowed; trailers up to 22 feet. One mile east of the main park entrance are an additional nine primitive walk-in sites. Horses not allowed; dogs must be leashed, and additional restrictions on beach use may be in effect during the breeding season of snowy plovers. For information, call: 707-882-2463.

A privately operated KOA campground is located east of the state park on Kinney Road. Facilities include 124 tent and RV sites, some with partial or full hookups, cabins and cottages, showers, pool and spa, game room, laundry, and store. Summer reservations recommended; call: 1-800-562-4188. For information, call: 707-882-2375.

STONEBORO ROAD: *End of Stoneboro Rd., W. of Hwy. One, 4.5 mi. N. of Point Arena.* The southernmost access to Manchester State Park is at the end of Stoneboro Rd., 1.5 miles west of Hwy. One. There is a dirt parking area, from which a half-mile sandy path leads up and over a series of dunes to the beach; the going is moderately strenuous

Stunning views can be had of Point Arena and the lighthouse to the south, and a seemingly endless strand to the north. To find your way back through the dunes, covered with European beach grass, take note of the pole at the ocean end of the trail. No facilities. Horses allowed; dogs must be leashed; day use only. Call: 707-882-2463.

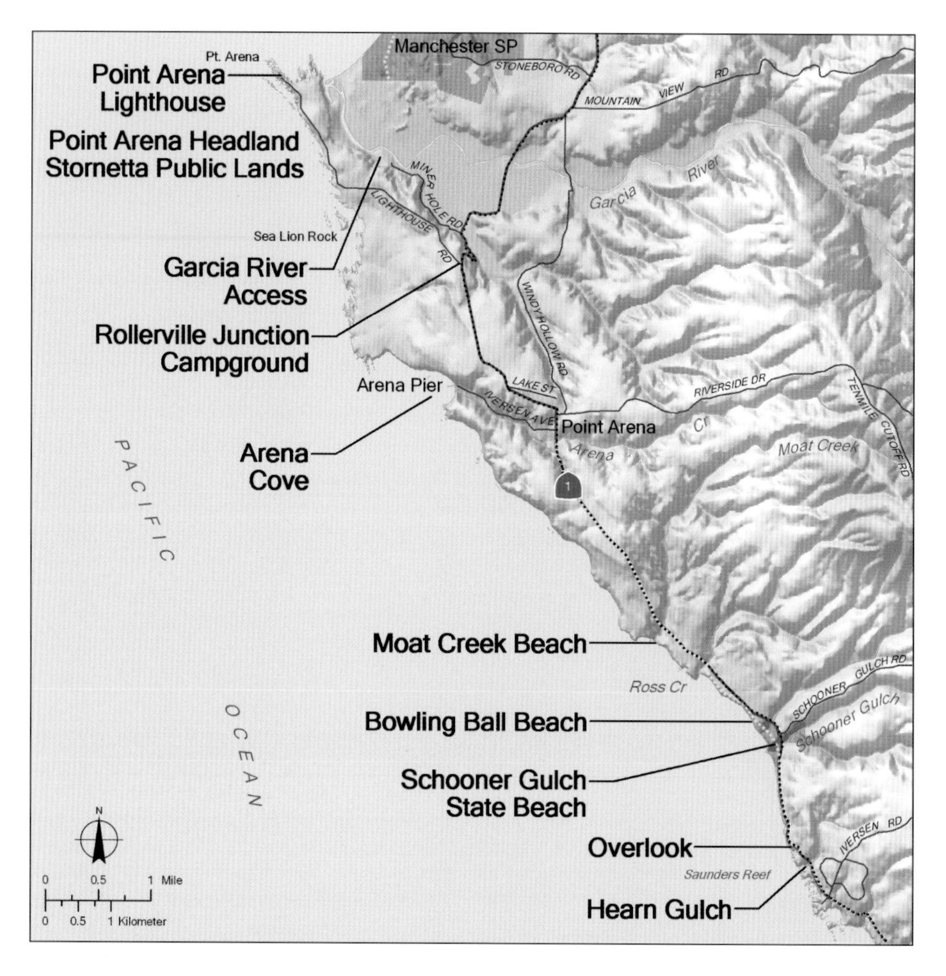

Arena Cove harbor

Point Arena to Hearn Gulch

	Sandy Beach	Rocky Shore	Trail	Visitor Center	Campground	Wildlife Viewing	Historic Building	Fishing or Boating	Facilities for Disabled	Restrooms	Parking	Fee
Point Arena Lighthouse				•			•		•	•	•	•
Point Arena Headland / Stornetta Public Lands		•	•			•				•	•	
Garcia River Access			•					•			•	
Rollerville Junction Campground					•				•	•	•	•
Arena Cove		•						•	•	•	•	
Moat Creek Beach		•	•						•	•	•	
Bowling Ball Beach	•	•	•									
Schooner Gulch State Beach	•	•	•					•		•	•	
Overlook						•					•	
Hearn Gulch	•	•				•						

POINT ARENA LIGHTHOUSE: *End of Lighthouse Rd., W. of Hwy. One.* The 115-foot-tall lighthouse was built in 1870 and severely damaged in the 1906 earthquake on the nearby San Andreas Fault. The rebuilt tower has one of the most powerful lights on the coast, and there are panoramic views from the top. Guided tours available; visitor center and gift shop, with saltwater tank housing tidepool creatures. Due to its westward location, the lighthouse is a popular place to observe gray whales, which migrate near the shore between December and April. The lighthouse is open 10 AM–4:30 PM daily, 11 AM–4:30 PM March through October. The facility is owned and operated by the Point Arena Lighthouse Keepers Association, a non-profit organization. Fee for entrance or parking. Call: 877-725-4448.

POINT ARENA HEADLAND / STORNETTA PUBLIC LANDS: *2 mi. N. of Point Arena, W. of Hwy. One.* In 2004, a dramatic addition to publicly accessible coastal land was made through the transfer of 1,132 acres of the historic Stornetta Brothers Ranch to the Bureau of Land Management with funding from the State Coastal Conservancy, U.S. Fish and Wildlife Service, State Wildlife Conservation Board, and the Mendocino Coast Audubon Society. Two miles of shoreline are included, on both sides of the Point Arena Lighthouse, along with riparian corridors, wetlands, ponds, cypress groves, meadows, sand dunes, and Sea Lion Rock offshore. Migratory waterfowl, shorebirds, raptors, and several special-status species may be found here.

Access to the Stornetta Public Lands is available from Miner Hole Rd. or all-weather Lighthouse Road, which is narrow and does not allow shoulder parking. Visitors may park at the Point Arena Lighthouse (fee

Point Arena Headland / Stornetta Public Lands

The geology of the North Coast is much in evidence around the rugged promontories of Point Arena. The steeply dipping beds of the Monterey Formation are eroded into twisted forms, riddled with sea caves formed by erosion along the weaker beds. The Monterey Formation consists of shale, sandstone, and chert, with abundant volcanic ash in places. These rocks vary tremendously in their resistance to erosion, so the surf has carved them into a series of points and embayments. Caves form along fractures or in especially susceptible rocks.

Just north of the Point Arena Lighthouse, the Garcia River enters the Pacific Ocean. As can be seen on a map of the area, the Garcia River runs along a razor-straight valley for many miles to the southeast. This valley marks the San Andreas fault, along which the relentless grinding of the North American and Pacific plates has fractured the rock, leaving it vulnerable to erosion. The upper Garcia River, like the South Fork of the Garcia River to the south, has carved this valley in these weaker rocks.

When standing at the lighthouse, you are on the Pacific plate, atop a sliver of rocks that have been transported hundreds of miles to the north over the past 30 million years. To the east of you, across the valley of the Garcia River, lie rocks of the Franciscan Complex, on the North American plate.

View northwest to Point Arena Lighthouse

applies); land managers plan a new parking facility off Lighthouse Rd., accommodating autos, RVs, and vehicles with horse trailers. Restrooms are also planned. No fee will be charged for parking. Public uses of the Stornetta Public Lands include hiking and equestrian use, picnicking, and wildlife observation. Motor vehicles, bicycles, and camping are prohibited. Call: 707-468-4000.

GARCIA RIVER ACCESS: *2.4 mi. N. of Point Arena, W. of Hwy. One.* The river is known for its steelhead and coho salmon, although the numbers of fish have declined due to past logging, road building, and gravel mining. Unpaved Miner Hole Rd., impassable in winter, leads from Hwy. One along the south bank of the river to a turn-around just short of the road's end. Park at the turn-around, and look on the north side for a half-hidden path through a willow thicket to the riverbank, a popular fishing spot. Private ranch lands are located across the river; do not trespass. On the south side of the pull-out there is access through a fence to former Stornetta Ranch lands, now administered by the Bureau of Land Management. No facilities are available.

ROLLERVILLE JUNCTION CAMPGROUND: *Hwy. One at Lighthouse Rd.* This privately owned facility offers 46 campsites, most with full RV hookups; separate tent sites and cabins available. Facilities include a grocery store, café, laundry, hot showers, swimming pool, and spa. Call: 707-882-2440.

ARENA COVE: *End of Iversen Ave., 1 mi. W. of Hwy. One.* At the base of high cliffs is a small, rocky beach. The Arena Pier is a municipal wharf, where fishing without a license is permitted, although catch limits apply. Facilities at the pier include picnic tables, public restrooms, outdoor showers, fish cleaning tables, and a boat launch, one of the few between Fort Bragg and Bodega Bay. A bait shop, restaurant, gift shop, and lodging are also available. The cove is one of the most popular surfing spots north of San Francisco, and divers often launch here. For information and current conditions at the cove, call: 707-882-2583.

The town of Point Arena is the smallest municipality in Mendocino County. The community bustled during the lumber boom of the late 1800s, and today, its pocket downtown includes art galleries, gift shops, restaurants, and an Art Deco–style theater hosting films and live performances.

MOAT CREEK BEACH: *W. of Hwy. One, 2 mi. S. of Point Arena.* A small, rocky beach, backed by high cliffs, is located 200 yards from the parking area. The cove is sheltered

Moat Creek Beach

For a striking view of bedded sedimentary rocks, visit Bowling Ball Beach at a very low tide; check local tide tables for when minus tides occur. When the tide drops to about 1 foot below mean lower low water (that is, well below an average low tide), a vast flat bench cut into sandstones and shales of the Schooner Gulch Formation is exposed. The strata are tilted steeply on end, dipping to the west, and the high sea cliff above the bench is formed of the dipping beds. The bench has been eroded into a series of parallel grooves and ridges several feet wide and hundreds of feet long, resembling a series of bowling lanes. Nature has even provided the bowling balls: nearly spherical boulders found scattered about at the base of the cliff. These oddly spherical boulders are concretions—unusually well-cemented nodules found in the sedimentary rocks making up the cliffs. Harder than the surrounding rocks, the concretions remain in the surf after the surrounding matrix has been worn away.

Bowling Ball Beach

Harbor seals at Point Arena

from prevailing northwest winds, and especially in springtime is a popular surfing destination; the accessway is managed by a joint venture of the Moat Creek Management Agency. A restroom is near the parking area. The trail to the beach follows Moat Creek to the shore, where black oystercatchers can be seen on nearby rocks. To the south, stairs lead up onto the grass-covered bluff, strewn with wildflowers in spring, where a trail leads 1.4 miles to Bowling Ball Beach.

BOWLING BALL BEACH: *W. of Hwy. One, 3.7 mi. S. of Point Arena.* Bowling Ball Beach is located midway between Moat Creek Beach and Schooner Gulch Beach, with no direct public access of its own. To reach the unusual reef at Bowling Ball Beach, you must walk from parking areas at Moat Creek or Schooner Gulch along the bluff, or along the beach at low tide. From Moat Creek Beach, take the blufftop trail south 1.4 miles to Bowling Ball Beach. Or park facing south on the shoulder of Hwy. One, opposite Schooner Gulch Rd., and take the trail at the north end of the pull-out. At the restroom, take the right fork and follow the trail along the bluff, through Schooner Gulch State Beach, turning toward the ocean when the trail forks again. A steep staircase descends to a sandy beach; head north one-half mile to Bowling Ball Beach. At minus tides, bowling ball-shaped rocks are visible along the surf line. The coastal bluffs are high, and some lie at a shallow angle, with wildflowers, grasses, and bluff lettuce growing on the slopes.

SCHOONER GULCH STATE BEACH: *W. of Hwy. One at Schooner Gulch Rd., 3.8 mi. S. of Point Arena.* Park facing south on the shoulder of Hwy. One, opposite Schooner Gulch Rd. Take the quarter-mile trail at the north end of the pull-out downhill through a patch of redwood forest to the beach; restrooms are located near the bottom of the hill. A small creek spills onto the sandy beach, which stretches south for 100 yards to the base of eroding bluffs. Picnicking, fishing, and surfing are popular here. No camping or fires; dogs must be leashed. From Schooner Gulch Beach, a six-mile round trip hike to Moat Creek Beach is possible, either along the blufftop or, at very low tides, along the beach.

OVERLOOK: *W. of Hwy. One, milepost 10.52.* A paved parking area with no facilities is located on the bluff at milepost 10.52. Fine 180-degree views of the shoreline, with huge kelp beds visible offshore. Cormorants nest in spring on the sheer faces of nearby sea stacks.

HEARN GULCH: *W. of Hwy. One, N. of Iversen Rd., 5.5 mi. N. of Anchor Bay.* An undeveloped headland, with a small beach at the base, was acquired by the Redwood Coast Land Conservancy in 2001. The Conservancy plans future access improvements; in the interim, limited parking is available on an unpaved pull-out on the west side of Hwy. One at milepost 10.0, there is also parking at the overlook at milepost 10.52. From the blufftop, there are dramatic views of rocks and waves. A very steep 200-yard-long trail leads down to a white-sand beach washed by turquoise waves, a launch point for experienced kayakers and divers to explore nearby sea caves. Offshore is Saunders Reef, a popular surfing destination.

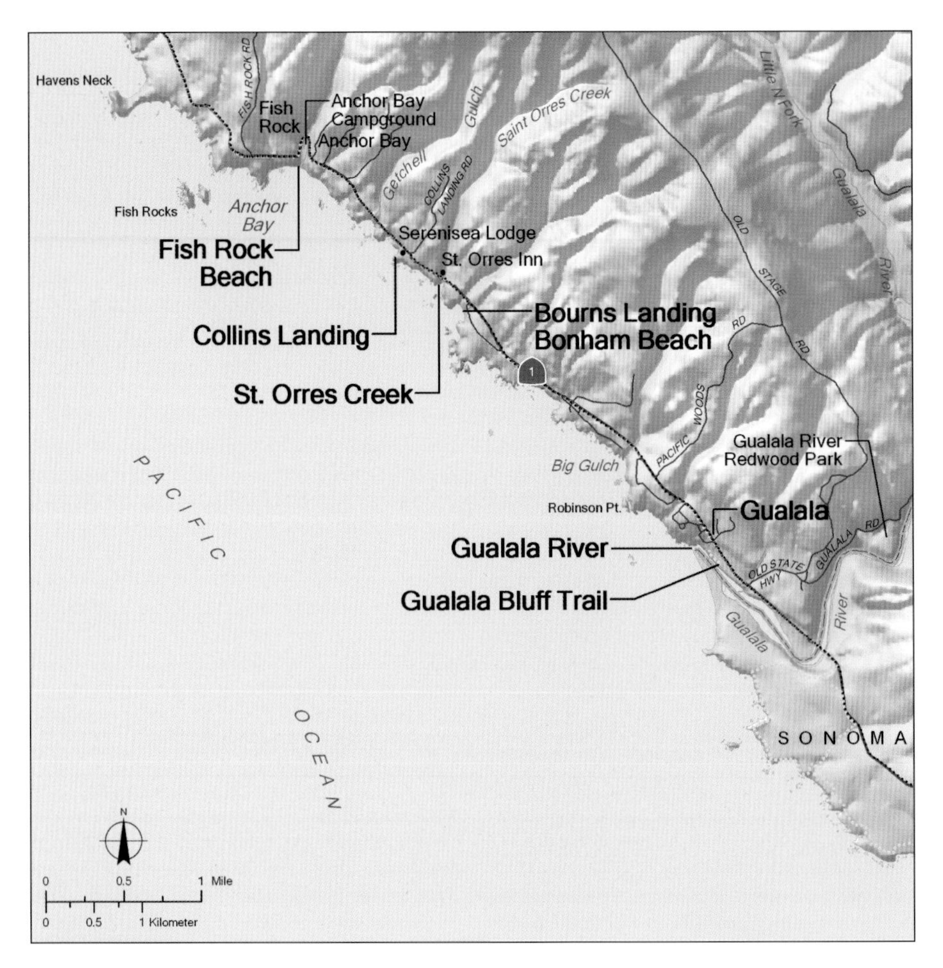

Beach at St. Orres Creek mouth

Anchor Bay to Gualala

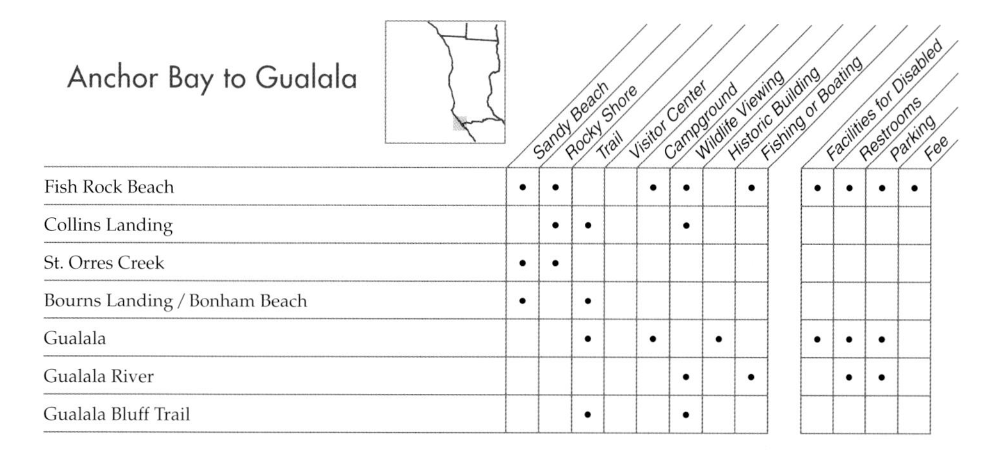

	Sandy Beach	Rocky Shore	Trail	Visitor Center	Campground	Wildlife Viewing	Historic Building	Fishing or Boating	Facilities for Disabled	Restrooms	Parking	Fee
Fish Rock Beach	•	•			•	•		•	•	•	•	•
Collins Landing		•	•			•						
St. Orres Creek	•	•										
Bourns Landing / Bonham Beach	•		•									
Gualala			•		•		•		•	•	•	
Gualala River						•		•		•	•	
Gualala Bluff Trail			•			•						

FISH ROCK BEACH: *At Anchor Bay Campground, .2 mi. N. of town of Anchor Bay.* Public day use and camping are available at the privately owned Anchor Bay Campground, which is situated in a narrow forested gulch west of Hwy. One. At the mouth of the valley, flanked by cliffs, lies a sandy beach with scattered rocks and tidepools. The campground has 36 overnight sites with picnic tables and firepits; most have water hookups, and some have electrical hookups. There are restrooms, hot showers, a septic dump station, a fish-cleaning house, and a dive-gear wash area.

Fishing, abalone diving, and birding are popular activities. Fish Rocks, located offshore, are a nesting site for black oystercatchers, cormorants, pigeon guillemots, and other seabirds. A short, steep trail from the Anchor Bay Campground climbs the bluff to the village of Anchor Bay, where there is a general store, café, and laundromat. Fee for day use or camping; call: 707-884-4222.

COLLINS LANDING: *At Serenisea Lodge, .7 mi. S. of town of Anchor Bay.* A curving, rocky cove with tidepools is accessible by a steep trail and stairway beginning at the north end of the Serenisea Lodge complex. The trail overlooks Fish Rocks, from which sea lions may be heard barking. Cliffs backing the beach are composed of tilted planes of rock, with wildflowers, live-forevers, and sword ferns growing on the slopes. Use of the accessway requires the permission of the

Fish Rock Beach

Bourns Landing / Bonham Beach

Serenisea Lodge management; the lodge office, open 9 AM–8 PM daily, is adjacent to the accessway. Call: 707-884-3836.

ST. ORRES CREEK: *Hwy. One, 2.1 mi. N. of Gualala.* A small beach is located at the mouth of St. Orres Creek, down a steep slope from Hwy. One. Shoulder parking on Hwy. One southbound, just north of the onion-domed St. Orres Inn; at milepost 3.33, an unimproved path down an eroding bluff leads to the cove. No facilities. The Redwood Coast Land Conservancy plans to improve the public access trail to the cove and beach, starting on the road shoulder at milepost 3.30. Call: 707-785-3327.

BOURNS LANDING / BONHAM BEACH: *Hwy. One, 2.3 mi. N. of Gualala.* Bonham Beach, formerly known as Cooks Beach, is a wind-sheltered, wide curve of sand enclosed by cliffs, some with residential development. There is lots of driftwood, some piled into shelters. The beach is accessible from the northern end of the old Coast Highway, which forms a loop on the seaward side of Hwy. One south of St. Orres Inn at milepost 3.14. A short path splits in two: one fork leads along the bluff, offering views of Bonham Beach and offshore rocks, and the second fork winds downhill among pine trees to the beach. No facilities; call: 707-785-3327. The Redwood Coast Land Conservancy plans to construct a blufftop viewing platform and improvements to the beach trail.

GUALALA: *Hwy. One, N. side of Gualala River mouth.* Once a logging center, Gualala is now a destination for outdoor recreation and cultural activity. The Gualala River to the south of town draws visitors for fishing, kayaking, camping, and hiking. The Gualala Arts Center offers a variety of cultural and community events, including the "Art in the Redwoods Festival," which takes place every year on the third weekend in August. Visitor information is available at the Dolphin Gallery, located in the shopping center next to the Gualala Hotel.

Gualala River Redwood Park is a privately owned campground with 120 RV or tent sites, all with water, electricity, picnic tables, and fire rings. From Hwy. One, one-half mile north of the Gualala River, turn east on Old State Rd. and go one mile to the campground. Some sites overlook the river. Convenience store and children's playground available; call: 707-884-3533.

GUALALA RIVER: *W. of Hwy. One at Gualala.* The Gualala River marks the boundary between Sonoma and Mendocino Counties. The lower river is a protected estuary much of the year, when a sandbar forms across the river mouth. From the Mendocino County side, access to the river is available on a gravel road leading west, just north of the Hwy. One bridge. The quarter-mile-long road provides direct vehicle access to the gravel bank of the river and is a good place to put in a boat or kayak, which can be rented locally.

GUALALA BLUFF TRAIL: *Hwy. One, at Gualala.* A level path, some two hundred yards long, runs along the bluff on the river side of Gualala's commercial strip, offering panoramic views of the Gualala River mouth and ocean, benches, and picnic tables. Pedestrian access from Hwy. One is via the driveway through the Surf Motel; the trail can also be reached through the Sea Cliff Center complex. Call: 707-785-3327.

Winter surf at Collins Landing

Islands and Offshore Rocks

AS YOU DRIVE along the coastal highway in northern California, you will notice that offshore rocks and small islands take a variety of forms. Storm waves buffet the shore and whittle away the coastal cliffs, leaving isolated remnants of the former coastline. Depending on their size and shape, the windswept rocks and islands support varying types of vegetation. Most of the smaller, pointy rocks have little or no vegetation, while the larger, flatter islands may develop a thin layer of soil where grasses and herbaceous plants, or even trees, persist.

With their cold-water surroundings and lack of shelter, northern California's offshore rocks and islands are not the isles of daydreams. But even though they seem somewhat barren, rocks and islands provide a very important resource for marine mammals and seabirds. Sea lions, harbor seals, and northern elephant seals haul out regularly on offshore rocks, and Steller sea lions breed on a few of the more remote rocks. Nineteen species of birds, some of them endangered, use offshore rocks for breeding in California. The rocks offer a mix of crevices, soil, and open areas, required by different species for nesting, and they are without terrestrial predators. Intrusion by humans is very limited on rocks and islands, and this allows sensitive bird species to nest without too much disturbance. If black oystercatchers, pelagic cormorants, or pigeon guillemots are frightened off their nests, even briefly, eggs or small chicks may be gobbled by western gulls.

The rocks provide not only a place to nest but also food resources; shorebirds, such as black turnstones and surfbirds, feed on the margins of offshore rocks. Intertidal species such as barnacles, mussels, and periwinkles are present, and on the larger islands, land snails, butterflies, and other invertebrates can be found. Many species of birds and animals that used mainland beaches and cliffs for feeding or nesting prior to California's rapid population growth are now dependent on offshore sites for these purposes.

Islands and offshore rocks have been used by humans for thousands of years. Native Americans have traditionally used them for hunting, gathering of seaweed and shellfish, and as temporary landing sites. Offshore rocks were also important in navigation, and they played a part in the mythology of some tribes. In the 19th century, rocks were often made part of landings, which were used to ship out lumber products, as can be seen by remnant footings embedded in the rocks at Elk in Mendocino County.

In the year 2000, the California Coastal National Monument was created to protect and foster public appreciation of California's coastal islands, rocks, and surrounding resources. At least 20,000 rocks, small islands, and exposed reefs are included in the monument, with a combined land area of about 1,000 acres spread the length of California. Only reefs or pinnacles above the level of mean high tide are included in the monument; the tiniest rocks with less than about 40 square feet of surface are excluded.

This is a national monument like no other. Few humans ever visit it, and yet its features are among the most photographed and well-loved of all California's landscapes. Plans for management of the monument are being prepared, and programs are expected to emphasize protection of scenic views, of bird, animal, and plant habitats, and of traditional uses by Native Americans. Recreational activities on the rocks are not encouraged, but nearby uses that do not disturb wildlife, such as kayaking and photography, are possible. The visual allure of the islands presents an opportunity for education, and interpretive facilities, such as wayside exhibits, may be constructed at key mainland parks and overlooks. Offshore rocks and islands are a mirror of the mainland coast, on a small scale, offering in some cases a look at biological resources that existed before human disturbance. Evidence suggests that some islands may even contain unique plants that are found nowhere else.

Not every islet and rock is included in the California Coastal National Monument. The U.S. Coast Guard took title many years ago to rocks on St. George Reef northwest of Crescent City and to Sugarloaf Island off Humboldt County for lighthouses that were never built. A few islets, such as a group north of Cape Vizcaíno in Mendocino County, remain in private hands.

Offshore rocks are scattered along most of the northern California shore, with particular concentrations north of Pelican Bay and west of Crescent City in Del Norte County; from Patrick's Point to Trinidad and along the Lost Coast in Humboldt County; along most of the Mendocino and Sonoma coasts except opposite dune fields such as the Ten Mile Dunes; and, in Marin County, off the tip of Point Reyes and south of Stinson Beach. For more information on the California Coastal National Monument, write: 299 Foam St., Monterey 93940; or call: 831-372-6115.

Early summer is a great time to view the **harbor seal** (*Phoca vitulina*), California's most common marine mammal, which can be seen hauled up on rocks offshore or silently watching you from the water just beyond the wave line. Harbor seals have spotted coats in a variety of shades from silver-gray to black or dark brown. Their mottled fur allows them to practically disappear against the rocks and sand. Loosely translated, their scientific name means "sea dog," and after you have seen them, you will understand why. They are generally very shy; please watch them from a distance.

Harbor seals

Steller sea lions (*Eumetopias jubatus*) have a bulky build and a very thick neck, which resembles a lion's mane, hence the name "sea lion." They can be easily distinguished from harbor seals by their great size. In addition, sea lions have external ears, while harbor seals do not. Steller sea lions are the largest of the "eared seals"; adult males weigh an average of 2,000 pounds and reach up to nine feet in length. These massive creatures are protected pursuant to the Federal Endangered Species Act as a threatened species. One suspected cause of their decline in numbers is the overfishing of their prey species, which include walleye pollock, mackerel, salmon, Pacific cod, herring, flounder, squid, and rockfish. Listen for their loud barking on offshore islands.

Steller sea lions

California sea lions

California sea lions (*Zalophus californianus*) range from British Columbia to the Baja California peninsula. These are the very vocal sea lions that you may see swimming around fishing piers and harbors. This playful, noisy pinniped is the trained seal of the circus. An adult male can weigh up to 800 pounds and has a sagittal crest, or bump, on the top of the head; the female is considerably smaller, averaging 250 pounds. You may see a sea lion in the water with a flipper held up in the air, as if waving. The reason for this is to regulate body temperature; a wet flipper in the air causes evaporation, cooling down the sea lion. Sea lions do not breed until they are four or five years old.

Pigeon guillemot

Pigeon guillemots (*Cepphus columba*), or "sea pigeons," are medium-sized, black-and-white sea birds with an upright posture and short wing span. In some ways they are similar to penguins, although they are not even closely related. They nest on sea cliffs and "fly" underwater to catch fish. Due to their short wingspan, guillemots have to flap their wings very fast in order to fly. They nest in little dens or cavities that they excavate in coastal cliffs and islands. Listen for their high thin whistles and squeaks. In all seasons you can identify a pigeon guillemot by its brilliant red legs and feet.

Tufted puffins

Rocky islands along the California coast host a variety of nesting seabirds each June, but none is as flamboyant as the **tufted puffin** (*Fratercula cirrhata*). Its most distinctive feature is its large and colorful bill. During the breeding season, look also for the tan-colored feathers that extend from above and behind its eyes down to the nape of its neck. These tufts are what give this puffin its name. These features, along with vermilion spectacles and orange legs and feet, give this unique bird the appearance of a caricature. One of the largest colonies of tufted puffins in California is found at the Farallon Islands. A pelagic bird trip to see these striking birds and other offshore species is a very worthwhile adventure. Due to population declines, the tufted puffin is designated by the State of California as a "Species of Special Concern."

During the breeding season, **rhinoceros auklets** (*Cerorhinca monocerata*) develop an unusual horny growth at the base of the bill, used to excavate the nest burrow. Once nesting has begun, the burrow is aggressively defended. These colonial dove-sized seabirds are nocturnal, and there is not much activity at the burrows during the day. At night, however, they compete for nesting spots, dig their burrows, lay eggs, and feed their young, causing quite a racket at the nesting colony. Rhinoceros auklets were once plentiful in California, but presently only three offshore islands (Castle Rock, the Farallon Islands, and Año Nuevo Island) provide nesting habitat for approximately 95 percent of the California breeding population of rhinoceros auklets, totaling only about 2,000 birds.

Rhinoceros auklet

Cassin's auklets (*Ptychoramphus aleuticus*) come ashore only during the nesting season. Like many nocturnal birds that need to find their mates and young at night, Cassin's auklets are vocal in their colonies. In the small hours of the morning in May and June, they set up a chorus reminiscent of spring peepers and other swarming frogs. Once the young leave the nest, the nightly choruses cease, and only the occasional tentative signal from late breeders is heard. By late July, the colonies are deserted and silent. During the non-breeding season, these little birds are found on the open ocean at the edge of the continental shelf, and they can be observed during pelagic bird trips or whale-watching outings.

Cassin's auklet

Brown pelican

During summer and fall, the **brown pelican** (*Pelecanus occidentalis*) flies northward from breeding grounds on the Channel Islands in southern California and other islands off the Baja California peninsula. Brown pelicans can be observed along the North Coast gathered along the beach with other birds, or diving for fish out in the open water between May and October. The bill of a pelican has an enormous pouch attached to it that is used like a net. The pelican dives from the air with wings partly folded and plunges into the water to catch its prey, using its huge bill and pouch to scoop up fish and water. After the bird rights itself, it strains out the water from the side of its bill, and tips its head back to swallow the fish. The brown pelican was listed as endangered in 1970 due to a precipitous decline in numbers. This decline was caused primarily by a reduction in eggshell thickness and concomitant low reproductive success, a result of the ingestion of pesticide residues, specifically DDT and PCBs, in the birds' prey species. In 1972, the use of DDT was banned in the United States, and since then there has been a corresponding increase in the eggshell thickness and reproductive success of the brown pelican.

Brown pelican

Page opposite: Pebble Beach, Sonoma County

Sonoma County

To Hwy. 101
Clear Lake
MENDOCINO
Cloverdale
Gualala R
Gualala Point Regional Park
Sea Ranch
Lake Sonoma
Geyserville
Gualala R
Stewarts Point
S Fk
Healdsburg
Kruse Rhododendron SR
Salt Pt.
Salt Point SP
Austin Creek SRA
SONOMA
Fort Ross SHP
Guerneville
Russian
RIVER RD
Jenner
Russian R
Sonoma Coast SB
BOHEMIAN HWY
Sebastopol
Santa Rosa
COLEMAN VALLEY RD
Sonoma Coast SB
Bodega Bay
BODEGA HWY
Mussel Pt.
Bodega Head
Bodega Bay
VALLEY FORD RD
BODEGA AVE
TOMALES RD
PETALUMA
Petaluma
Sonoma
PACIFIC
OCEAN
MARIN
San Pablo Bay
SAN FRANCISCO
N. Farallon Is.
Farallon Is.
200
600
1000
1400
1800
2200
2600
N
0
10
20 Miles
0
10
20 Kilometers
Depth Contours Shown in Meters

Sonoma County

THE SONOMA COAST has a history of human habitation that stretches from the native Kashaya Pomo through Spanish exploration, Russian enterprise, Mexican land grants, and American ranches and timber operations. Bishop pine forest, grazing lands, and coastal prairie cover the steep slopes north of the Russian River, while a chain of sandy and pebble beaches, separated by rocky headlands, stretches south from the Russian River to Bodega Head. Highway One hugs the shoreline along nearly the entire county, and numerous pull-outs offer views of cove beaches and distant mountains.

Between Jenner and Fort Ross, Highway One climbs to an elevation above sea level of over 600 feet. This is perhaps the most vertiginous point on California's coastal highway, where the ocean seems an endless blue plane and the land is too steep to allow more than a glimpse of waves and beaches. The Sonoma coast is unstable ground; Highway One has been closed by landslides many times.

Sonoma County's shoreline is subject to intense summer upwelling, the process by which colder subsurface waters rise to the surface. Tourists on shore may shiver from the cool breezes, but marine animals and plants thrive on the nutrient-laden cold deep-sea waters. Diving for abalone or to view anemones, sea urchins, and rockfish is a popular activity on the Sonoma coast, such as at Salt Point and Fort Ross State Parks. Here the seafloor is rocky, and underwater visibility is better than in sandy or muddy areas.

On the rocky shore, fishing for crabs or rockfish is popular, as is poke-poling in crevices for the eel-like fish known as the monkeyface prickleback, said to be good eating. In the mudflats of Bodega Harbor, digging for gaper, littleneck, and Washington clams is popular.

The Bodega Bay Fishermen's Festival is held annually in April. Events include an arts and crafts show, activities booths, music, boat rides, children's activities, and the Blessing of the Fleet, honoring fishermen lost at sea. Bodega Bay Visitor Center, call: 707-875-3866 or 877-789-1212 (toll free). Russian River Chamber of Commerce, call: 707-869-9000.

Surfboard, boogie board, skim board, and wet suit rentals:
Bodega Bay Surf Shack, 707-875-3944.

Diving equipment and boogie board rentals:
Bodega Bay Pro Dive, 707-875-3054.

Salmon, albacore, and Dungeness crab fishing charter boats and whale-watching trips:
Bodega Charters, 707-463-3618.

Horseback riding on the beach:
Chanslor Ranch, Bodega Bay, 707-875-3333.

Kayak rentals and tours:
Bodega Bay Kayak, 707-875-8899.

Kayak rentals on the lower Russian River:
Lotus Kayaks, Jenner, 707-865-9604.
Russian River Outfitters, Duncans Mills, 707-865-9080.

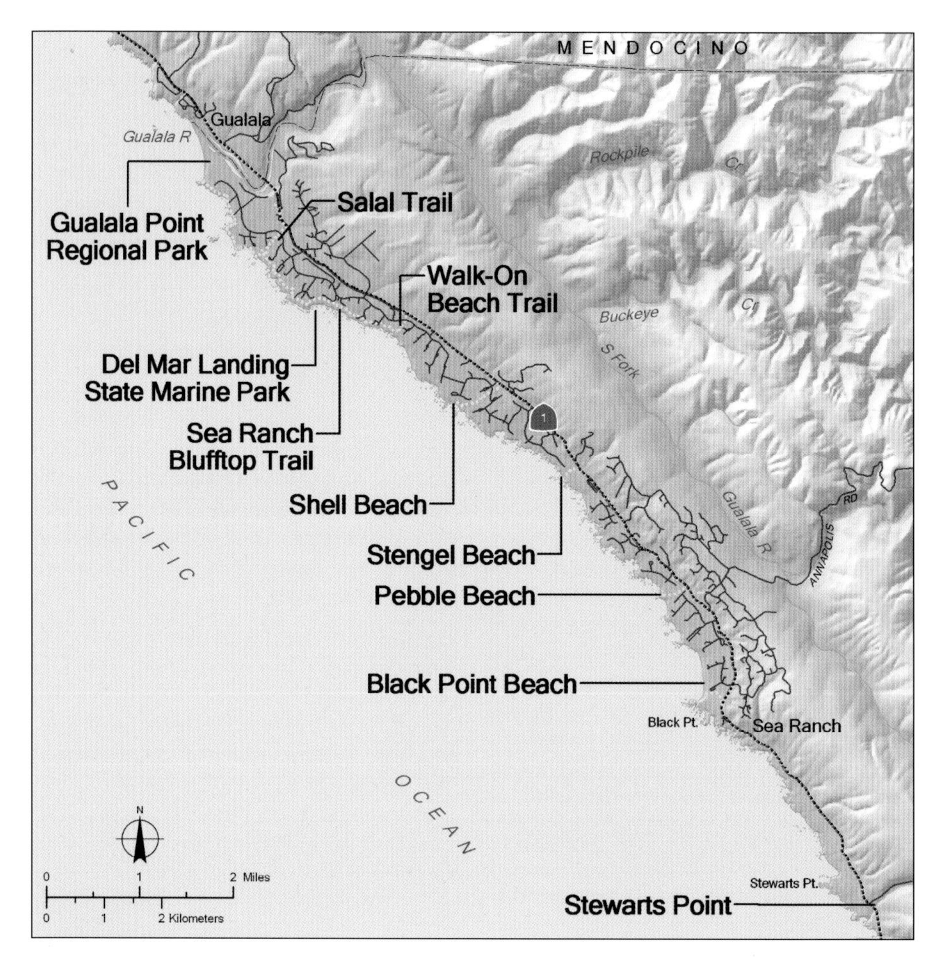

Gualala River mouth

Northern Sonoma County

	Sandy Beach	Rocky Shore	Trail	Visitor Center	Campground	Wildlife Viewing	Historic Building	Fishing or Boating	Facilities for Disabled	Restrooms	Parking	Fee
Gualala Point Regional Park	•		•	•	•	•			•	•	•	•
Salal Trail			•								•	
Del Mar Landing State Marine Park		•				•		•				
Sea Ranch Blufftop Trail		•	•			•						
Walk-On Beach Trail			•							•	•	•
Shell Beach	•	•	•							•	•	•
Stengel Beach	•	•	•							•	•	•
Pebble Beach	•	•	•							•	•	•
Black Point Beach	•		•							•	•	•
Stewarts Point							•					

GUALALA POINT REGIONAL PARK: *Hwy. One, 1 mi. S. of Gualala.* This popular park offers access to the beach at the mouth of the Gualala River and the blufftop trail that leads south through part of the Sea Ranch residential community. West of Hwy. One there are day-use areas with river and mountain views, while east of the highway the park offers camping facilities in the forest.

A volunteer-staffed visitor center with changing exhibits is open weekends from Memorial Day to Labor Day. A wheelchair-accessible trail leads west across meadows where swallows bank and turn in the spring. The north fork of the trail leads to the sandy, driftwood-strewn beach, and the south fork connects to the Sea Ranch blufftop trail.

Once a portion of the northernmost Mexican land grant in California, Rancho German, this area was used for cattle grazing and ranching activities until the 1960s. The rows of Monterey cypress trees along the beach trail are a remnant of the windbreaks planted by ranchers. Its location at the mouth of the river gives Gualala Point Regional Park a great diversity of wildlife habitats. Great blue herons, pygmy owls, hawks, and hummingbirds are resident here, and seabirds can be seen on the rocks offshore, while ospreys fly overhead on fishing expeditions. California gray whales can be seen offshore in winter and spring, and deer roam the meadows that are thick with grasses, bracken ferns, and lupine. Pacific rhododendron and sword ferns grow in the redwood forest along the river.

Camping facilities at Gualala Point Regional Park include picnic tables, firepits, and a trailer sanitation facility. Camping fees are charged for improved and hike or bike campsites. Call: 707-785-2377.

SALAL TRAIL: *Hwy. One, 1 mi. S. of Gualala.* The trail starts at Hwy. One near the entrance to Gualala Point Regional Park, where parking is available, and leads along a thickly vegetated creek to the bluff above the ocean, intersecting with the Blufftop Trail.

DEL MAR LANDING STATE MARINE PARK: *Ocean W. of Sea Ranch, 2 mi. S. of Gualala.* California's state marine parks are created to protect or restore areas for rare plants, animals, and habitats, as well as for historical or archaeological preservation or research. The Del Mar Landing State Marine Park includes an area of ocean adjacent to the Sea Ranch residential community. Fishing, scuba diving, wildlife viewing, and snorkeling are allowed; the taking of marine plants and invertebrates is prohibited. The reserve can be reached from Gualala Point Regional Park.

Stengel Beach

SEA RANCH BLUFFTOP TRAIL: *Hwy. One at Gualala Point Regional Park.* A trail winds along the bluff edge south from Gualala Point Regional Park to the Walk-On Beach Trail. Spectacular views of rocky coast and pocket beaches. Sea Ranch residences are nearby; respect private property.

WALK-ON BEACH TRAIL: *Hwy. One, 2.5 mi. S. of Gualala.* A trail leads from the parking area to the bluff near Walk-On Beach. Due to bluff erosion, Walk-On Beach may not be accessible to the public, although visitors can head north on the Sea Ranch blufftop trail to the Salal Trail and Gualala Point Regional Park.

The Sea Ranch, established in the 1960s, is a planned residential community with architecture that is characteristic and much-copied. The emphasis on native plant landscaping has also been influential. The developer's original plan to keep the development entirely private, with no public access to some ten miles of shoreline, had a different kind of influence. The proposal helped to propel the coastal protection movement, leading to creation of the California Coastal Commission and the State Coastal Conservancy. Ultimately, limited public shoreline access at selected points was incorporated into the Sea Ranch development plan.

Beach access facilities at the Sea Ranch include Gualala Point Park and the Salal Trail, both originally proposed by the subdivision's developer. Later negotiation with the Coastal Commission led to creation of five trails from parking lots on Highway One to the shoreline at Walk-On Beach, Shell Beach, Stengel Beach, Pebble Beach, and Black Point Beach, as well as the Blufftop Trail that runs along three miles of shoreline. The bluff at the southern end of the Blufftop Trail has been eroded, temporarily interrupting public access near Walk-On Beach.

Public access rights on the beaches at Sea Ranch extend to the base of the bluff or the first line of vegetation. Trails within the development open to public use are marked, and others are private; no dogs or bicycles allowed. All roads in the Sea Ranch are private; do not trespass, and please observe signs. Call: 707-565-2041.

SHELL BEACH: *Hwy. One, 3.7 mi. S. of Gualala.* The trail is half a mile long, leading along a dense thicket of shrubs including coffeeberry and western azalea, which blooms pale pink in the late spring. As the trail continues through a meadow, the dark forest covering the ridge is visible to the east. At the end of the trail is a fine sand beach, backed by a low bluff. Offshore rocks shelter the ocean waters here, where harbor seals gather.

STENGEL BEACH: *Hwy. One, 5 mi. S. of Gualala.* A short trail leads from Hwy. One along a small creek. Beach access is via a wooden staircase that rests on a huge tilted rock formation. Rock outcroppings and cliffs create parallel planes, diagonal to the beach. A small creek spilling over the bluff edge creates a miniature waterfall.

PEBBLE BEACH: *Hwy. One, 6.7 mi. S. of Gualala.* A small parking lot is located next to Hwy. One; a quarter-mile-long trail winds through groves of trees before passing into a meadow overlooking the sea. A small secluded beach is paved with fine pebbles; tidepools are scattered among rocks and small seastacks.

BLACK POINT BEACH: *Hwy. One, 8.1 mi. S. of Gualala.* A well-maintained trail, one-fifth of a mile long, leads to the beach through a meadow browsed by deer. A high wooden staircase provides access down the vertical cliff, which borders a long curving sandy beach. South winds and swells sometimes draw surfers.

STEWARTS POINT: *Hwy. One, 11 mi. S. of Gualala.* Stewarts Point is an excellent example of a 19th century coastal settlement, with little alteration of its historic buildings. The Stewarts Point School (circa 1860), constructed in the Greek Revival style, is typical of the one-room schoolhouses that were found throughout Sonoma County before the turn of the 20th century. At present, the general store (built in 1868) is the only historic building accessible to the public. Founded in 1857 by A. L. Fisk, Stewarts Point became one of the major doghole lumber ports of the Sonoma coast. (A doghole port, used by ships in 19th-century coastal commerce, is a cove just big enough for "a dog to turn around in.") Cut lumber was sent by an aerial chute or cable from the clifftops to schooners in the cove for shipment to San Francisco.

Black Point Beach

Coastal cliffs and beach (Usal Beach, Mendocino County) ©2004, Tom Killion

Coastal Prairie

GRASSLAND is one of those major plant community types that we all can recognize instantly without any scientific training. Prairies and tree-studded savannas are found throughout the world, covering perhaps 20 percent of the earth's land surface. The native grasslands of California fall into two broad categories: northern coastal prairie and valley grassland. Northern coastal prairie is scattered along the coast from Oregon to Monterey; within this area, grassy fields and bluffs were originally populated by bunchgrasses such as oatgrass and the graceful purple needlegrass. Valley grassland was most common in the Central Valley and includes purple needlegrass, as well as other perennial species such as the sod-producing creeping wild rye.

Prior to the arrival of Europeans in the 18th century, grasslands covered more than a quarter of California and were relatively undisturbed and unmanipulated, although it is thought that Native Americans used fire to manage these habitats. This situation changed rapidly with the introduction of exotic grasses, herds of domesticated animals, and cultivation. The grasses that defined much of the native California prairie were predominantly long-lived perennials, whereas the plant "colonists" from the world's other Mediterranean climates were mostly annual species that produce a thick thatch and copious amounts of seed, and are thought by many to be better adapted to cope with the intensive cattle grazing that became widespread in the 19th century.

The early arrival of exotic plants was cleverly documented by a botanist named George Hendry. Hendry obtained adobe bricks from buildings of known age built between the early 1700s and about 1840, dissolved the adobe in water, and then identified the remains of the plants that had been used as a binder. Many of the "non-native" grasses and forbs that are common in today's landscape have been part of California's flora for well over 200 years, and most of the dominant annual grasses have been naturalized for more than a century.

Bodega Head

Mixed grasses, Bodega Head

There is no consensus as to the ecological mechanisms underlying the historical changes in California's grasslands, but the evidence is all around us that these plant colonists effectively displaced the native grass communities and their associated annual species of wildflowers. Although the constituent species are still found in many habitats, coherent native grassland communities no longer exist, except in a few relict stands. However, scattered patches of natives mixed in various proportions with introduced species are still common. In Marin, Sonoma, Mendocino, Humboldt, and Del Norte Counties, remnant native prairie exists in mixed stands along the coastal strip, often sandwiched between the humid coastal forest and the ocean bluff. Point Reyes is an excellent place to see native coastal prairie, but small stands are also present at Sonoma Coast State Beach and Fort Ross and Salt Point State Parks.

The drastic decline in native biodiversity in coastal grasslands has been attributed to habitat loss from cultivation, increased grazing pressure, and the elimination of annual fires. Unfortunately, the actual composition and extent of the pristine coastal prairie is largely unknown, the only hints coming from the journals of early travelers and settlers, including the occasional botanist. As a result, today's land stewards are faced with a challenge as they struggle to choose the most responsible land management techniques. For example, when coastal terraces are colonized by shrubs and trees from the adjacent forested mountainside, managers need to know if this is a return to the natural state prior to European settlement or a modern aberration caused by the environmental impacts of those settlers and their descendents. In this case, ecological sleuths have come up with an answer in the form of microscopic mineral bodies called phytoliths. Phytoliths are formed in plant leaves and other tissues from the silica in ground water, and they remain in the soil after the plant dies and decomposes. Since different types of plants form different shapes of phytoliths, plant ecologists can get an idea of the pristine vegetation in an area by analyzing the phytoliths in the soil column.

This has been done at various locations along the coast, including near Sea Ranch and Salt Point in Sonoma County, and it shows that the coastal terrace was indeed dominated by coastal prairie and not by forest. Since the evidence suggests that the pristine vegetation of the coastal terrace was grassland, ecologists at Salt Point State Park are removing trees and shrubs to reduce competition and to encourage the continuance of coastal prairie vegetation.

Although the coastal prairie is defined by perennial bunchgrasses, many associated annual species provide a striking spring wildflower display. Now often a subtle beauty, these wildflowers must once have painted whole sections of the coast with reds, purples, and yellows. The keen eye may still find patches of such unrestrained floral exuberance. Hidden within the grassy thatch are populations of rodents that wax and wane with the weather. During years of plentiful rainfall and grass growth, there will be explosions of chubby voles that are eaten by birds of prey.

Look for the **white-tailed kite** *(Elanus leucurus)* (formerly known as "black-shouldered kite") in open grassy areas along the coast. This is a mostly white raptor with a gray back and a black patch on the shoulder. This streamlined hunter is especially easy to spot when it is hovering in one spot above the ground, searching for prey. If you are lucky, you might spy one tucking in its wings and diving down for the catch. In the 1930s, this beautiful bird of prey was hunted almost to extinction. Farmers believed that white-tailed kites threatened their chickens, and kites were shot on sight. However, the diet of the white-tailed kite is made up primarily of small mammals such as the California vole, and fortunately the kites were able to make a comeback in the latter half of the 20th century, after legal provisions were made for their protection.

© 2004 Tom Greer

White-tailed kite

Like many small mammals, the **California vole** *(Microtus californicus)* is a burrower, but it also forms surface runways. When you walk through coastal prairie habitat, look for the tiny trails meandering through the grass. These are the trails of the California vole and other small burrowing mammals. However, voles are nocturnal, and it is unlikely that you would see one during your coastal foray. In the spring and summer, California voles feed on grasses and other green vegetation. In the winter, their diet consists mostly of roots. In one study, it was shown that the California vole made up 85 percent of the white-tailed kite's diet.

California vole

American badger

The **American badger** *(Taxidea taxus)* is a nocturnal burrowing mammal with a black-and-white striped face. These fierce predators are adapted for digging up burrowing rodents, which are their main prey item. In particular, they are connoisseurs of the pocket gopher. On their front feet they have two-inch claws and webbing between the toes to help them scoop dirt away from where they dig. Look for their eight-inch-wide elliptical burrows in coastal prairie habitat. The badgers dig along the sides of the burrow, making it wider than it is tall, and sometimes claw marks are visible along the walls.

California oatgrass

California oatgrass *(Danthonia californica)* is a lovely, native grass species that grows from 12 to 30 inches tall. It is a tuft-forming perennial with an open, branching flower stalk. The flowerhead is sparse and has only two to five spikelets, one each at the ends of the branches. It has a narrow, flat, and somewhat in-rolled leaf. When the species is not in flower or fruit it can be found by searching for the clump of leaves that form the "tuft."

Purple needlegrass

Perhaps California's best known native bunchgrass is **purple needlegrass** *(Nassella pulchra)*. Purple needlegrass thrived in California grasslands long before Europeans arrived. It is thought that the grass once spread across 25 million acres, but today it is found on less than 100,000 acres. Producing deep-rooted, basal-leaved tufts, this species can live for over 100 years. The tuft of this bunchgrass stays green most of the year, because its roots, which can extend to a depth of 20 feet, tap into deep sources of water.

Sky lupine

Sky lupine *(Lupinus nanus)*, an annual plant in the pea family, is often found in grassy areas along the coast. Sky lupine can grow to a height of about two feet, depending on environmental conditions. It has green, palmate leaves and blue flowers that turn pink after pollination. The leaves are perfectly shaped to direct morning condensation or fog drip down to the base of the plant.

Blue-eyed grass *(Sisyrinchium bellum)* is neither a grass nor is it blue-eyed. This diminutive member of the iris family has a yellow center and striking blue petals. The flowers close up at night and open after the sun is up. It is best not to look for them on foggy mornings, as they will likely still be waiting for the sun's rays to trigger their unfurling. This species is widely distributed in California and can be found in open grasslands and coastal bluffs. Look for blue-eyed grass at Point Reyes National Seashore starting in March.

Blue-eyed grass

California poppy *(Eschscholzia californica)* is a charming wildflower that seems to radiate light from its golden flowers. California's state flower is named for Johann Friedrich Eschscholtz (1793–1831), a surgeon and naturalist with Russian expeditions to the Pacific coast in the early 1800s. Spanish explorers called the California poppy *copa de oro,* meaning "cup of gold." In the spring, the four glistening petals of the poppy unfurl and push off the pointed green calyx that appears much like a small dunce cap. Once the cap comes off, the beautiful bright blossoms open during the day and close up at night. Native Americans used the green foliage as a mild pain-killer, crushing the leaves and placing them inside the cheek to alleviate the pain of an aching tooth.

California poppy

California poppies at Bodega Head

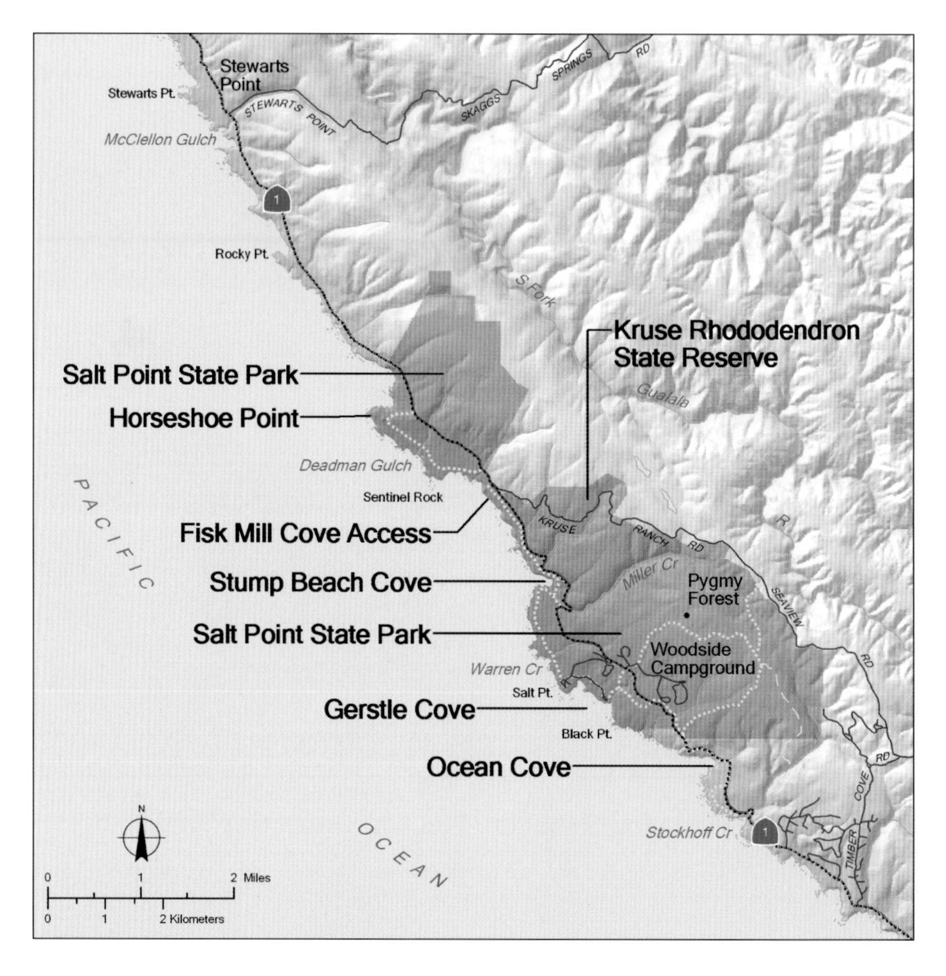

Diver off Sonoma coast

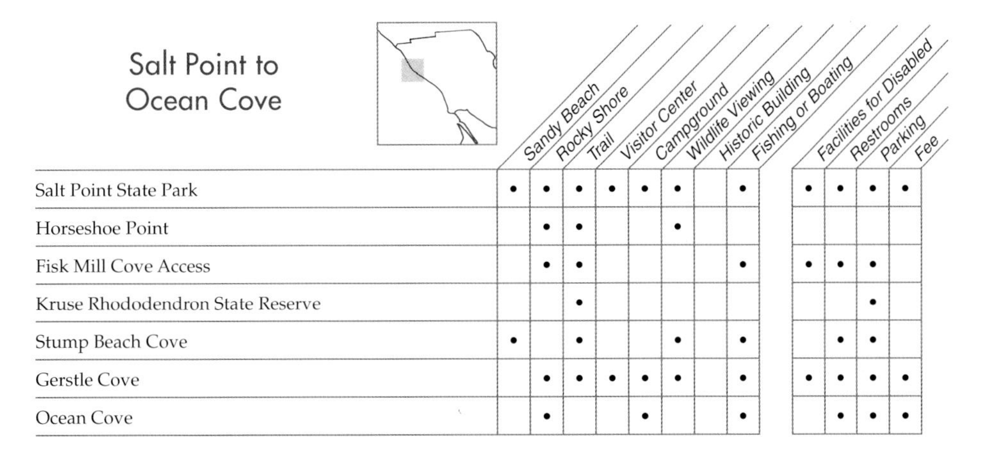

Salt Point to Ocean Cove

	Sandy Beach	Rocky Shore	Trail	Visitor Center	Campground	Wildlife Viewing	Historic Building	Fishing or Boating	Facilities for Disabled	Restrooms	Parking	Fee
Salt Point State Park	•	•	•	•	•	•		•	•	•	•	•
Horseshoe Point		•	•			•						
Fisk Mill Cove Access		•	•					•	•	•	•	
Kruse Rhododendron State Reserve			•								•	
Stump Beach Cove	•		•			•		•		•	•	
Gerstle Cove		•	•	•	•	•		•	•	•	•	•
Ocean Cove		•			•			•		•	•	•

SALT POINT STATE PARK: *Hwy. One, 18 mi. N. of Jenner.* The park includes 6,000 acres of mixed conifer forest, grasslands, rocky shore, sandy beach, and even an underwater park, where divers hunt for red abalone or spear fish. Salt Point State Park includes all the land between Hwy. One and the sea for a distance of six miles, as well as much of the area inland to the top of the ridge. Besides dense conifer forest, the slopes to the east of Hwy. One feature an open prairie, once grazed by elk, and a stand of pygmy forest. This botanical oddity is home to cypress, pine, and redwood trees that are stunted by poor soil drainage and nutrient deficiency. Some 20 miles of trails in the park are used by hikers, mountain bikers, and equestrians, and there are two campgrounds and several picnic areas. Call: 707-847-3221.

HORSESHOE POINT: *Hwy. One, .5 to 1.5 mi. N. of Fisk Mill Cove park entrance.* North of Fisk Mill Cove, several Hwy. One pull-outs provide access to trails that lead to the ocean bluff through forest dominated by bishop pine. These pines form closed cones, which remain smooth, intact, and usually tightly attached to the tree for years. Seeds are released sometimes only through the effects of fire. In the forest, listen for woodpeckers, drumming or calling hoarsely. On the bluffs, exposure to winds and salt spray create miniature wildflower gardens, composed of sea thrift and tidytips. Pockmarked tafoni formations are found along the shore here. Horseshoe Cove has a surf break known as Secrets, said to be good during wintertime south winds and swells.

Rhododendron at Salt Point State Park

FISK MILL COVE ACCESS: *2.7 mi. N. of Gerstle Cove park entrance.* A thick growth of bishop pines along Hwy. One screens the rocky shore that comes dramatically into view from the bluff and trails at the Fisk Mill Cove area of Salt Point State Park. Thick stands of fern and huckleberry surround the trees; listen for chickadees calling their name as they dart overhead. From the northernmost of three paved parking lots, walk one-tenth of a mile to the viewing platform atop Sentinel Rock to gain a view of rugged coastline from above the treetops. Wind-sheltered picnic tables, barbecue grills, drinking water, and restrooms.

KRUSE RHODODENDRON STATE RESERVE: *.3 mi. E. of Hwy. One on Kruse Ranch Rd.* This small reserve located adjacent to Salt Point State Park offers two miles of loop trails through a dense mixed-conifer forest and, in late spring, a vibrant display of pink-flowered native rhododendrons. The blooms are particularly splendid in this location due to a major fire that burned the area a century ago. The rhododendrons grew back rapidly afterward, benefiting from the sudden input of sunshine. Now second-growth redwoods, Douglas-firs, grand firs, and tanbark-oaks are gradually shading the forest floor, as part of the natural succession of plant species. Kruse Ranch Rd. is a one-lane dirt road with few places to turn around.

STUMP BEACH COVE: *Hwy. One, 1.2 mi. N. of Gerstle Cove park entrance.* A broad sandy beach ringed by sandstone cliffs is reached by a short trail and easy steps leading from the parking area. The beach is a fine place to view the breeding behavior of pelagic cormorants. With field glasses, look on the north side of the cove for nests in niches on

The rocks exposed along the shoreline at Salt Point State Park reveal dramatic changes through time. These rocks were deposited 40–60 million years ago in a deep marine basin located some 350 miles south of their current position. They were transported to where we see them today by movement along the San Andreas Fault over the last 15–20 million years. The rocks consist of sediments that were cemented together by heat and pressure. The nature of the rocks depends on the size of the sediment grains: pebbles and gravels become conglomerate, sand becomes sandstone, and mud becomes shale.

The sediments were carried into the basin by underwater mudslides and debris flows known as turbidity currents, a kind of submarine landslide traveling at speeds of as much as 60 miles per hour. The sediments accumulated into huge fan-shaped piles, known as submarine fans, built up by these currents. The conglomerates and sandstones represent channels that carried these turbidity currents out across the fan, and the finer sandstones and shales are from areas between channels where sediment overflowed the channel banks. The sediments were converted to rock and were uplifted above sea level by compression along the San Andreas fault, which accompanied their northward movement. These layered rocks, tilted during their upward journey, are well exposed at Stump Beach cove.

Weathering of the sandstones has resulted in a beautiful, honeycomb-like formation known as tafoni, which seems to be formed by a combination of processes. Leaching of minerals from the interior of rock and their deposition near the surface leads to a hardened exterior shell. Where this shell is breached, weathering and erosion, enhanced by salt spray, is concentrated and works quickly to excavate holes that coalesce to form the distinctive box-like pattern. Tafoni features are especially well developed at Gerstle Cove.

the cliffs. The birds skim low in search of fish over the protected waters of the cove, rising sharply to alight on what appear to be sheer vertical walls. Picnic tables and restroom near the parking lot.

GERSTLE COVE: *Hwy. One, 7 mi. N. of Fort Ross.* Salt Point State Park's main day-use and camping facilities are located at Gerstle Cove and at nearby Woodside Campground on the inland side of Hwy. One. The cove is reachable by a paved road that winds down the bluff. Offshore, clumps of sea palms look like limp mops in the surf, while harbor seals can be seen lounging on the rocks. On land, look for honeycomb tafoni formations. Facilities include a visitor center overlooking the sea, picnic areas, trails, and a small boat launch. The nearby kelp forests make Gerstle Cove particularly attractive to divers. Call 707-847-3222 for recorded ocean conditions. The inner cove is a marine reserve where no form of marine life may be taken or disturbed.

From the parking area at Gerstle Cove, a 1.2-mile trail leads north along the ocean bluff to Stump Beach Cove past honey-colored rock formations that take on fantastic forms, like dinosaur vertebrae or castles tilting into the sea.

Other paths connect to the much larger area of the state park that lies east of Hwy. One. Trail maps available at the park entrances at Gerstle Cove and Woodside Campground. Camping facilities at Salt Point State Park include at Gerstle Cove: 30 improved family sites and a 40-person group camp; and at Woodside Campground: 79 improved sites, 10 walk-in sites, and 10 unimproved hike or bike sites. For camping reservations, call: 1-800-444-7275.

OCEAN COVE: *Hwy. One, 5 mi. N. of Fort Ross.* A privately owned bluff and beach offers fishing and diving along the rocky shore, a boat ramp, and camping. Some 100 campsites are located on the ocean bluff or in the trees, with firepits, picnic tables, restrooms, water, and hot showers. Reservations taken for group campsites. Pay camping and day-use fees at the grocery store. For information, call: 707-847-3422.

Tafoni formations at Salt Point State Park

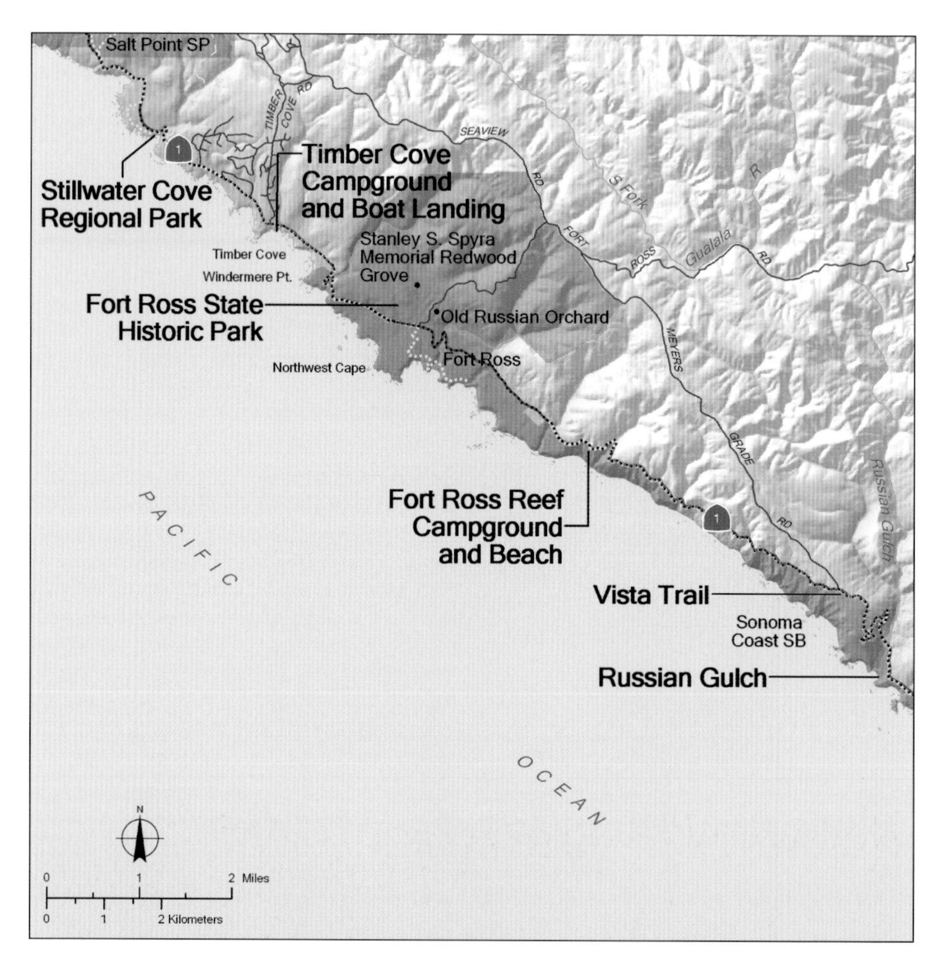

Stillwater Cove Regional Park

Stillwater Cove to Russian Gulch

	Sandy Beach	Rocky Shore	Trail	Visitor Center	Campground	Wildlife Viewing	Historic Building	Fishing or Boating	Facilities for Disabled	Restrooms	Parking	Fee
Stillwater Cove Regional Park		•	•		•	•	•	•	•	•	•	•
Timber Cove Campground and Boat Landing		•			•			•		•	•	•
Fort Ross State Historic Park	•	•	•	•			•	•	•	•	•	•
Fort Ross Reef Campground and Beach		•			•				•	•	•	•
Vista Trail			•						•	•	•	
Russian Gulch	•		•							•	•	•

STILLWATER COVE REGIONAL PARK: *Hwy. One, 4 mi. N. of Fort Ross.* This 210-acre park offers picnic and camping facilities among forest and meadows, and ocean access at a cove beach. Divers set out from the pebbly cove, where small boats can be launched. An unpaved trail leads to the beach from the day-use parking area, passing through riparian forest of grand fir, red alder, rhododendron, and western azalea. Horsetails and ferns line the path.

Spring wild flowers on the bluff include deep-blue Douglas iris. Ospreys nest in the tops of fir and redwood trees along the ridgelines, and these fish-eating hawks can be seen from the shore. A half-mile trail leads to the restored one-room Fort Ross schoolhouse, built around 1885. Campsites are scattered through the forest, some with filtered views of the sea. Some sites improved for disabled access; hike or bike site available. Restrooms, coin-operated showers for use by campers, electrical outlets, and a dump station. For information, call: 707-847-3245.

A pull-out on the west side of Hwy. One just north of Stockhoff Creek is located adjacent to an unimproved day-use portion of Stillwater Cove Regional Park. A path leads along the grassy bluff edge, providing vistas of the tiny coves below; Indian paintbrush grows from the steep slopes.

TIMBER COVE CAMPGROUND AND BOAT LANDING: *Hwy. One, 1 mi. N. of Fort Ross.* Day use and camping with hook-ups at this privately owned facility. Boat rentals, boat launch, and scuba rentals are available; fishing licenses, bait and tackle, and propane. Information: 21350 Hwy. One, Jenner 95450; 707-847-3278.

The Timber Cove Inn at Timber Cove is the site of a Beniamino Bufano statue, *Peace*. The 85-foot-high statue, a Sonoma County Historic Landmark, was made of concrete and mosaic by Bufano in 1960 and was his last finished work.

Fort Ross schoolhouse

FORT ROSS STATE HISTORIC PARK: *Hwy. One, 11 mi. N. of Jenner.* The park includes over 3,000 acres of coastal bluff, beaches, and forest, along with the settlement site of native peoples and Russian traders. The focus of the park is historical, centering on the Russian occupation in the early 19th century, but recreation and natural resource protection are also featured.

Fort Ross State Historic Park includes the reconstructed fort, within which the Rotchev House is original construction. Other structures have been rebuilt, including the chapel that collapsed in the 1906 earthquake that devastated San Francisco, and plans are under way to reconstruct the Russian-era fur warehouse for interpretive purposes. Presentations by costumed park interpreters are offered within the fort. There is a visitor center nearby with exhibits on local history and the Kashaya people, and a bookstore with a large collection on Russian history; for information, call: 707-847-3437.

Day-use beach access is available near the fort in the coves where the Ross Colony workers built ships and where ranchers once shipped out bulk cargo using a 180-foot-long chute from the blufftop. Visitors can picnic near the visitor center, in the fort, or next to the historic Call ranch house, where a colorful flower garden originally planted by George W. Call's Chilean-born wife, Mercedes, draws hummingbirds and song sparrows. Red abalone and rockfish may be taken in season, and scuba divers can explore the wreck of the 19th century S.S. *Pomona* in an underwater portion of the state park. Inland of Hwy. One, Fort Ross Rd. climbs 1,400 feet up the ridge, past the old Russian orchard of apple, plum, and pear trees and the Stanley S. Spyra Memorial Redwood Grove. The state park also extends north along Hwy. One as far as Windermere Point, seaward of the Fort Ross Store, although there are no facilities. For general information, call: 707-847-3286; for educational opportunities, call: 707-847-4777.

FORT ROSS REEF CAMPGROUND AND BEACH: *Hwy. One, 10 mi. N. of Jenner.* A campground is located 1.6 miles south of the main entrance to Fort Ross State Historic Park. Twenty primitive campsites are reached by a narrow road leading down into a sheltered canyon. Tables, stoves, and food lockers; restrooms and drinking water available. No reservations; no dogs allowed. Large RVs not advisable, due to limited turning space. From the Hwy. One turn-off, a separate unpaved road leads across the bluff to a day-use parking area. A steep trail leads down to a long, curving rocky beach.

VISTA TRAIL: *Hwy. One, 4.5 mi. N. of Jenner.* Particularly expansive views of the shoreline from a paved wheelchair-accessible loop trail some 600 feet above the sea. On spring afternoons, cliff swallows swoop above the steep slopes. Facilities include picnic tables, restrooms, and a parking lot.

RUSSIAN GULCH: *Hwy. One, 3 mi. N. of Jenner.* A large cove beach bounded by nearly vertical cliffs is reached by a short trail. Thick riparian vegetation lines Russian Gulch Creek, providing habitat for songbirds. Parking lot; restroom; fee. For information, call: 707-875-3483.

Call Ranch House

Long before the 19th century, the Kashaya Pomo people lived along the coast between the rivers now known as the Gualala and the Russian. The Kashayas spent winters on the high ridge above the sea and summers close to the coast, where the ocean's food supplies of abalone, mussels, fish, and sea plants were readily available. The people collected sea salt for their own use and to trade with others, and they produced fine baskets. The Russians were apparently the first Europeans to be encountered by the Kashayas.

The Ross Colony was established in 1812 by the Russian-American Company, a commercial venture chartered by Tsar Paul in 1799 with a monopoly over Russian enterprises in North America. The settlement on the Sonoma coast was undertaken to raise wheat and other food crops for Sitka and other settlements in Alaska, then part of the Russian Empire.

The colony was also intended to provide a base for hunting sea otters and for attempting trade with the Spanish who controlled California from San Francisco Bay southward. The silky, lustrous fur of the sea otter was highly prized for capes and other garments among the wealthy in China, where each pelt might bring the equivalent of $55, or several months' wages for a workingman in the U.S. at the time. The enormous profits to be made from the sea otter trade attracted not only the Russians but also other foreigners to California, including Americans later involved in the take-over of California from Mexican sovereignty. Although the numbers of otters taken for the 19th century fur trade were never large, perhaps a few thousand a year at most, the creatures became noticeably scarce almost at once.

Russian citizens were outnumbered by others at the Ross Colony, where most of the work of farming, hide tanning, blacksmithing, and boat building was done by the local Kashayas and by Alutiiq people from Alaska. By 1841, the Russians turned to the Hudson's Bay Company outposts farther north to supply their Alaskan settlements, and they sold Fort Ross to John Sutter of the New Helvetia settlement in the Sacramento Valley. The Ross Colony site became part of the 15,000-acre George Washington Call ranch, which produced timber and dairy products for California's growing population and continued as a Call family enterprise until 1973.

Chapel at Fort Ross

Rocky shore (Westport, Mendocino County)

©2004, Tom Killion

Rocky Shore and Nearshore Waters

THE PACIFIC COAST has miles of rocky shores, where nooks and crannies, surge channels, and tidepools provide homes for specialized plants and marine animals. Waves and tides together shape the rocky shore and influence which species can survive there. Intertidal organisms cope with pounding waves, extreme temperatures, and variable salinity. They are adapted to dealing with flooding and drying caused each day by the tides. They also have to avoid being eaten during low tides by specialized birds, mollusks, and crabs, and during high tides by fish and other marine life.

The **finger limpet** *(Lottia digitalis)* can be found high up on the rocks in the intertidal zone where ocean spray is abundant. Many finger limpets cluster along cracks where moisture collects. Finger limpets are basically conical but have an off-center peak that curves over. Their diet is made up of microscopic algae. The limpet breathes by drawing water from its left side, over its gills, and pushing it out the shell on the other side.

Finger limpet

The **California mussel** *(Mytilus californianus)* is one of the most common creatures on California's rocky shores and in tidepools. Mussels can form massive beds, sometimes several feet across, on surf-exposed rocks and wharf pilings. They are generally grayish black, and have very hard shells that are difficult to pry open. Mussels attach themselves to rocks very tightly. From time to time, the mussel sneaks out its foot (the part inside the shell) and touches the rock, secreting a special thread of cement. After doing this several times, these threads of cement hold the mussel to the rock. Once a mussel has found a home, it opens its paired shells to let in seawater and food particles. California mussels are edible only in the colder months, because in summer they may ingest large amounts of small floating algae or phytoplankton whose poisonous excretions accumulate in the mussels' tissues and can cause sickness and death in humans.

California mussel

Gumboot chiton

The **gumboot chiton** *(Criptochiton stelleri)* is a type of mollusk that has eight internal butterfly-shaped plates along the back. The plates provide the chiton with a backbone-like structure. Gumboot chitons can be found attached to rocks in the low intertidal zone. They are the largest chitons in the world, growing to nearly a foot long. The mantle, or back, of the gumboot chiton is reddish brown, thick, and leathery. The underside is a large foot that acts as a suction cup to keep the chiton attached to the rocks.

Purple shore crab

Next time you have a chance to look among the tidepools in the seaweeds and rock crevices, you will likely come across the **purple shore crab** *(Hemigrapsus nudus)*. Aptly named for its red-purple spotted shell, it is easily distinguished from the slimmer, striped shore crab by the red or purplish spots on its claws. It grows up to two inches wide and has no hair or spines on shell or legs (hence the species name *nudus*). This crab feeds primarily on the film of small algae on rocks, but is also a scavenger of animal matter. Shore crabs store water in specialized gill chambers that enable them to remain active out of the water for extended periods of time.

Blue-banded hermit crab

The **blue-banded hermit crab** *(Pagurus samuelis)* is a type of crab that does not have a very hard shell. Not a true crab, it uses other animals' old shells for protection. Most hermit crabs like to live in snail shells. As the hermit crab grows in size, it must find a larger shell. Hermit crabs carry their shell homes on their backs and tuck themselves inside it for protection. They use the large right claw for protection and holding food, and the small left claw for eating. Hermit crabs are mainly scavengers and can often be seen digging for food, preying on smaller organisms, or scrounging for scraps in tidepools along shores.

The **black turban snail** *(Tegula funebralis)* gets its name from its turban shape, which, along with its color and size, helps it to blend very effectively with rocks. The turban snail can crawl amazingly fast to evade predators. If it is on a sloping surface, the snail may use another method of escape and release its grip on the rock to roll to safety. One of its main predators is the ochre sea star. The black turban snail has the ability to detect a chemical scent in the water emitted by predatory sea stars, and the snail attempts to seek higher ground, above the predator's foraging range. Black turbans often congregate in great clusters in crevices or on the sides of small boulders. Juvenile black turbans occur in the higher tidal zones, where they live for five to seven years. As they grow larger, they slowly migrate to lower levels.

Black turban snails

The carnivorous **giant green sea anemone** *(Anthopleura xanthogrammica)* lives on rocks in surge channels and lower tidepools. This huge anemone species reaches a maximum size of eight to ten inches in diameter. The reason the anemones are green is that green algae live within their tissues. If you are brave enough and very careful, you can gently touch the green sea anemone to see what it feels like. You might even feel a tingling sensation. Do not worry, though; they are harmless to humans.

Giant green sea anemone

Ochre sea stars *(Pisaster ochraceus)* come in many different colors: brown, yellow, purple, orange, or reddish. They are generally abundant in the intertidal zone. They have no head, tail, or brain. Most ochre stars have five arms, which are dotted with rough-feeling, white, blunt spines that form lined patterns. These small bumps are actually pincers that keep other animals off the slow-moving sea stars. The bottoms of their arms are covered with many suction cup–like tube feet. Ochre sea stars eat mussels, snails, barnacles, chitons, limpets, and sea urchins. A star grips its prey with its tube feet and pulls open the shell. A sea star actually extrudes its stomach outside of its mouth and digests the prey externally. The process can take two or three days. Like many other sea stars, the ochre star can regrow a lost arm. This regeneration can take up to a year.

Ochre sea star

Giant Pacific octopus

The brain, sense organs, and central nervous system of the **giant Pacific octopus** *(Enteroctopus dofleini)* are considered among the most highly developed of those of any invertebrate. The giant octopus is a master of disguise and can quickly change the color of its skin to match its surroundings. Giant octopuses reside in crevices or under rocks between the low intertidal zone and about 200 meters deep. During the day they remain in their dens, coming out at night to feed. Octopuses often eat inside the safety of the den and then discard the shells and bones of their prey in a pile outside. They use their parrot-like beaks to feed on shrimp, crabs, abalone, scallops, and clams. The giant octopus is the largest of all the octopus species, growing to as much as 23 feet long and weighing up to 100 pounds. These gentle giants are endemic to the northwest and northeast Pacific Ocean.

Sea palm

The **sea palm** *(Postelsia palmaeformis)* is a brown alga resembling a miniature palm tree. It is locally abundant in the upper to mid-tidal zones from Vancouver Island to Morro Bay. It is restricted to rocks exposed to heavy surf. Currently, it is illegal to harvest sea palm in California. However, in the past when the species was more abundant, some people considered the tender blades a delicacy. Sea palm is an annual kelp, meaning that it only survives for one year. It thrives in dense stands, where its dispersal and germination are local and highly influenced by seasonal disturbance. Sea palms compete with mussels and other algae for space. The palms produce many sporelings, which attach to competitors and act as drags, causing competitors to rip off the rock surface during storms. The remaining sporelings then grow to maturity in these cleared spaces.

Shorebirds as a group are foremost among the earth's greatest travelers, and the **black turnstone** *(Arenaria melanocephala)* is no exception. Black turnstones breed along the coast of Alaska and winter as far south as Mexico. The stocky little birds can be found along the North Coast's rocky shores during May as they migrate north and after early August as they wend their way south. They have a unique wedge-shaped bill that allows them to toss aside debris and rocks as they look for invertebrates underneath. It is in search of these small meals that the turner of stones operates.

Black turnstone

The **surfbird** *(Aphriza virgata)* is one of the characteristic winter shorebirds of the Pacific Coast. It is a dark-gray plover-like bird that has a broad black band on a white tail. The markings of the surfbird may not be all that distinct, but this dull coloration operates like camouflage, allowing the bird to forage along the rocky shore unnoticed. This species is poorly known from its remote mountain breeding grounds of Alaska and the Yukon Territory; its nest and eggs were first discovered only in 1926. The threat of oil spills along this species' wintering grounds, together with increased human development along the Pacific coast, makes the surfbird a species of conservation concern.

Surfbird

Black oystercatchers *(Haematopus bachmani)* are shorebirds that feed in the rocky intertidal zone. They are permanent residents of the Pacific coast. In the spring, you can find them by following their loud and raucous mating call. Despite its name, the black oystercatcher rarely eats oysters. It dines on creatures that cling to rocks below the high tide line. Next time you spy one, take a close look at the long and stout red bill. This chisel-shaped bill allows the birds to pry or hammer open bivalve shells or to remove limpets and other shellfish from the rocks.

Black oystercatchers

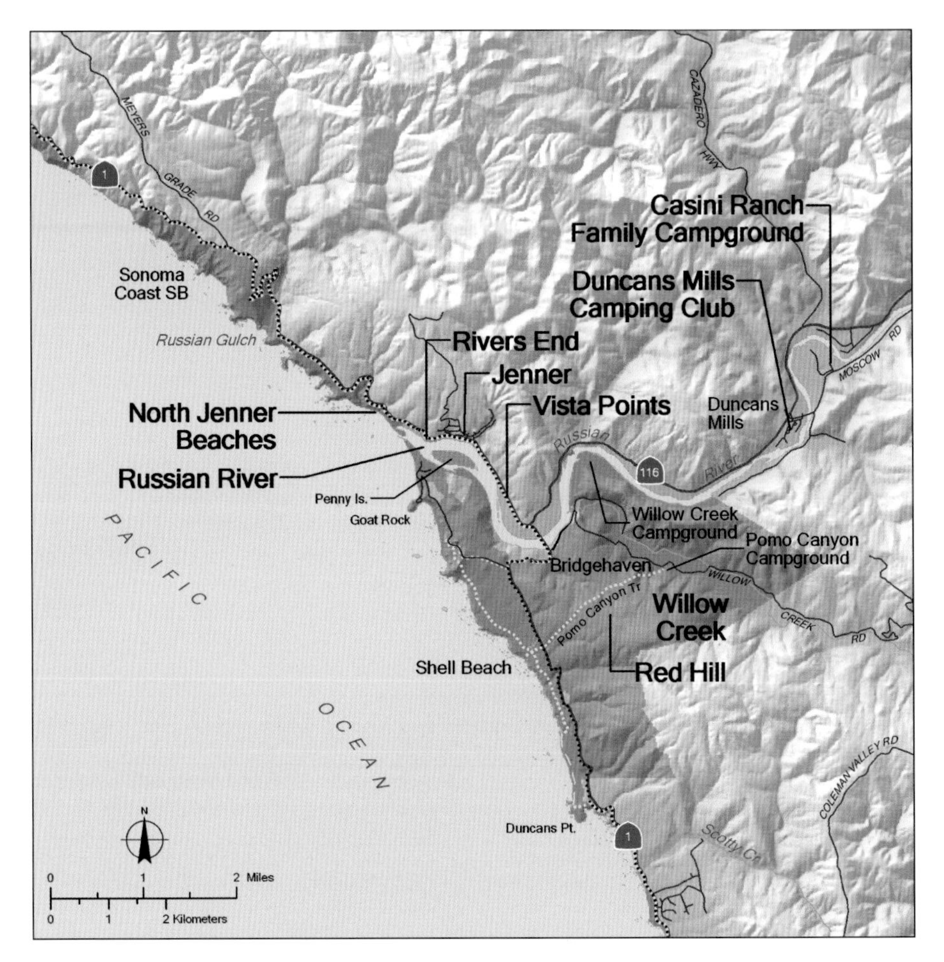

Goat Rock, south of Russian River mouth

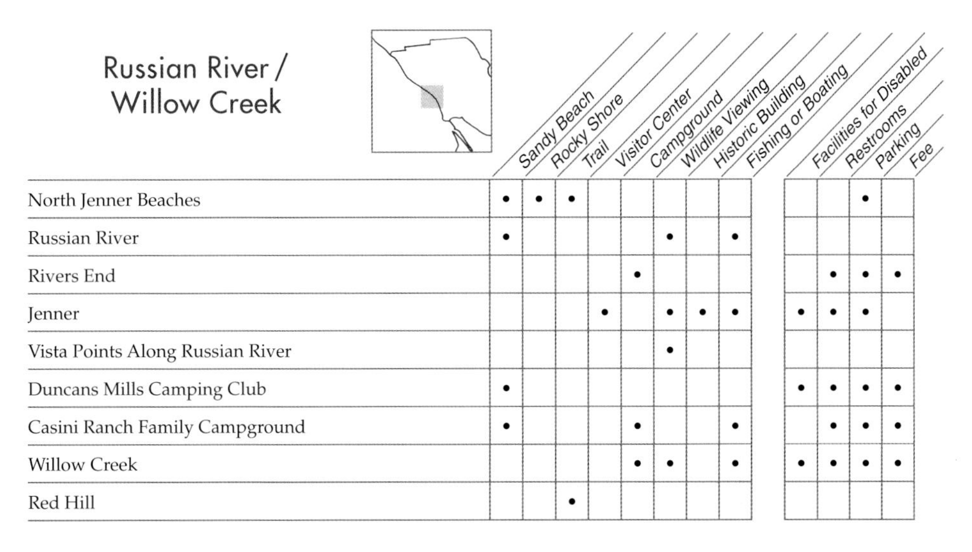

Russian River / Willow Creek

	Sandy Beach	Rocky Shore	Trail	Visitor Center	Campground	Wildlife Viewing	Historic Building	Fishing or Boating	Facilities for Disabled	Restrooms	Parking	Fee
North Jenner Beaches	•	•	•								•	
Russian River	•					•		•				
Rivers End					•					•	•	•
Jenner				•		•	•	•	•	•	•	
Vista Points Along Russian River						•						
Duncans Mills Camping Club	•								•	•	•	•
Casini Ranch Family Campground	•				•			•		•	•	•
Willow Creek					•	•		•	•	•	•	•
Red Hill			•									

NORTH JENNER BEACHES: *Hwy. One, from Russian Gulch to Jenner.* State park lands include the stretch of coast south of Russian Gulch to the mouth of the Russian River at Jenner. No facilities, but pull-outs along Hwy. One lead to several trails that cross the bluff. Steep trails down the eroding bluffs provide access to several beaches.

RUSSIAN RIVER: *Hwy. One at Hwy. 116.* The 110-mile-long Russian River, one of the largest North Coast river systems, drains a watershed of approximately 1,500 square miles. A major resort area, the Russian River valley offers a sunny and warm contrast in summer to the generally cool ocean coast. Inns, campgrounds, canoe rentals, and other small-scale visitor facilities are spread along some 15 miles of Hwy. 116.

The slopes above the river are largely forested with Douglas-fir, redwood, and live oak; manzanita and chaparral are also widespread. The riparian corridor along the water's edge includes white alders, box elders, and rushes. Gravel bars exposed during the summer months support dense stands of willow and cottonwood.

The earliest known inhabitants of the area were the Kashaya Pomo Indians, who called the river "Shabaikai" (long snake). Russian colonists explored the river area in 1809, naming the river "Slavianka" (Slav woman), before establishing a settlement at Fort Ross; the Spanish later called the river "Rio Ruso" (Russian River), referring to the Russians in the area. Settlement of the area increased rapidly during the early 1850s as the lumber industry was established. Farming, logging, and mining activities developed as towns sprang up in the river basin. In 1877, North Pacific Coast Railroad service was extended to the Russian River at Duncans Mills, and trains began to bring vacationers from San Francisco to the river.

RIVERS END: *Hwy. One, Jenner.* This privately owned resort provides day use, camping, and fishing. Boat launch and ramp, cabins, restaurant, and bar. Call: 707-865-2484.

JENNER: *Hwy. One at mouth of Russian River.* The Jenner area marks the transition between the steep rugged headlands that extend north from the Russian River and the wide marine terraces of the southern Sonoma coast. Located on a steep slope above the river, Jenner is a small, compact coastal town. It became a summer resort area around the turn of the 20th century and contains several bungalows and cottages dating from the early 1900s. The volunteer-run Jenner Visitor Center on the bank of the river has displays about the Sonoma coast and a small bookshop. Other facilities include parking, a small boat launching ramp, picnic tables, and restrooms. The Visitor Center is wheelchair accessible. Call: 707-865-9433.

VISTA POINTS ALONG RUSSIAN RIVER: *Hwy. One, Jenner to Bridgehaven.* Several pullouts along Hwy. One offer views of the Russian River estuary and Penny Island. Ospreys nest in the tops of trees along the river banks and can be seen fishing at the mouth of the river. The shallow waters in the river estuary attract large numbers of birds, sometimes running to the thousands. Penny Island, opposite Jenner, serves as a feeding and resting area for several species of birds, including brown pelicans. Jenner Pond, visible from the intersection of Hwy. One and Hwy. 116, is a small freshwater marsh used extensively by shorebirds, coots, rails, and ducks, notably the cinnamon teal.

DUNCANS MILLS CAMPING CLUB: *Hwy. 116, Duncans Mills.* The Duncans Mills Camping Club, a private, membership-only campground located on the north bank of the Russian River, allows public day use of the river beach. Fee charged; no more than four people per vehicle permitted. Park outside the campground and walk in; restrooms and picnic tables available. Call: 707-865-2573.

CASINI RANCH FAMILY CAMPGROUND: *Moscow Rd., .75 mi. E. of Hwy. 116, Duncans Mills.* This privately owned facility offers camping and river beach access on the south side of the Russian River. In summer when the river is low, there is a broad pebbly beach along a curve in the river with a backdrop of forested hills. There are picnic tables and restrooms at the beach. The campground includes 160 tent sites and 214 trailer sites with hookups; a general store offers food, fishing licenses, and bait and tackle. Kayak and canoe rentals during the summer, boat launch, playground, picnic areas, and recreation hall. Call: 707-865-2255.

WILLOW CREEK: *Willow Creek Rd., off Hwy. One.* Turn east off Hwy. One just south of the bridge over the Russian River to reach the Willow Creek area of Sonoma Coast State Beach. In 2005, existing small state park facilities in the watershed of Willow Creek will be augmented dramatically by nearly 3,400 acres of redwood forest, grasslands, and stream corridors. Additional recreational facilities and trails are planned, but until they are available, public access to the newly acquired lands is available on a permit basis through the nonprofit LandPaths; call: 707-480-3760.

Two existing state park campgrounds remain open in the Willow Creek area. The only state park campsites on the Russian River are located at the Willow Creek unit in a willow thicket. Eleven sites available; no reservations; closed during the rainy season. There are picnic tables, fire rings, and toilets, but campers must carry in water; the campsites are within a quarter mile of the parking area.

Mouth of the Russian River

Willow Creek area of Sonoma Coast State Beach

In summer, a broad rocky river beach adjoins the campground, and swimming and fishing are popular. Steelhead trout and coho salmon are found in both the Russian River and Willow Creek, which joins the river near the campground. The adjacent woods and grasslands provide nesting areas for spotted owls and osprey, which feed along the creek, and sometimes river otters can be seen.

In contrast to the Willow Creek environmental campsites, the Pomo Canyon sites are located on a slope above the valley floor, in a redwood grove where sword ferns, redwood sorrel, and trillium grow. There are 20 campsites; campers must carry gear from the parking area, a distance of less than a quarter mile. One campsite near the parking area is wheelchair accessible. Picnic tables, fire rings, and toilets; running water nearby. No reservations. For information, call: 707-875-3483.

RED HILL: *E. of Hwy. One, 3.5 mi. S. of Jenner.* This 900-acre state park addition links Sonoma Coast State Beach with the watershed of Willow Creek to the east. A three-mile-long trail starts at Hwy. One at the entrance to Shell Beach and leads up and over the coastal hills to Pomo Canyon Campground. The trail passes through grasslands, over seasonal streams, and into forests, with expansive views of the lower Russian River and the ocean from the 1,200-foot summit of Red Hill.

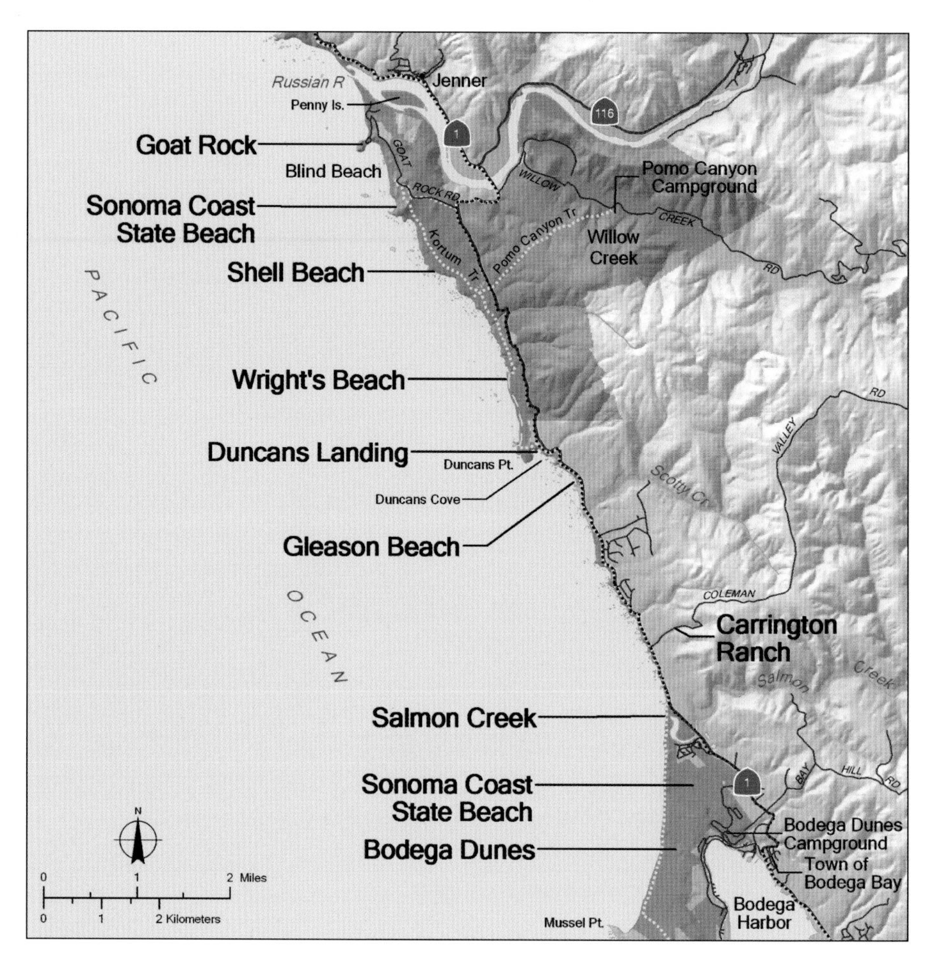

Russian R
Jenner
Penny Is.
116
1
Goat Rock
Blind Beach
GOAT
ROCK RD
WILLOW
CREEK
RD
Pomo Canyon Campground
Sonoma Coast State Beach
Kortum Tr
Pomo Canyon Tr
Willow Creek
Shell Beach
PACIFIC
OCEAN
Wright's Beach
Duncans Landing
Duncans Pt.
Duncans Cove
Gleason Beach
Scotty Cr
VALLEY
RD
COLEMAN
Carrington Ranch
Salmon Creek
Salmon Creek
Sonoma Coast State Beach
Bodega Dunes
BAY
HILL
RD
Bodega Dunes Campground
Town of Bodega Bay
Bodega Harbor
Mussel Pt.
N
0
1
2 Miles
0
1
2 Kilometers

Shell Beach

Sonoma Coast State Beach

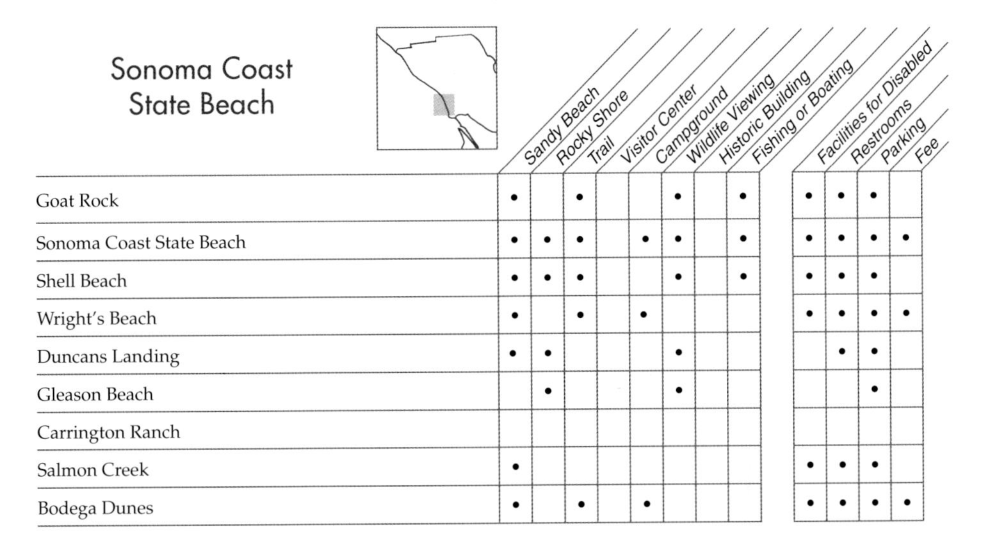

	Sandy Beach	Rocky Shore	Trail	Visitor Center	Campground	Wildlife Viewing	Historic Building	Fishing or Boating	Facilities for Disabled	Restrooms	Parking	Fee
Goat Rock	•		•			•		•	•	•	•	
Sonoma Coast State Beach	•	•	•		•	•		•	•	•	•	•
Shell Beach	•	•	•			•		•	•	•	•	
Wright's Beach	•		•		•				•	•	•	•
Duncans Landing	•	•				•				•	•	
Gleason Beach		•				•					•	
Carrington Ranch												
Salmon Creek	•								•	•	•	
Bodega Dunes	•		•		•				•	•	•	•

GOAT ROCK: *Goat Rock Rd., off Hwy. One.* The Goat Rock access road leads to beaches near the mouth of the Russian River, separated by Goat Rock. Along the access road that winds along the high bluff is a parking pull-out with a fine elevated vista of the large arched rock located offshore. The Kortum Trail leads south along the ocean's edge to Shell Beach and Wright's Beach.

Blind Beach can be reached by a steep trail down the hill, or at sea level from the Goat Rock parking area. Surfers ride the waves off the long curving sandy beach. Next to Goat Rock on the north side, the beach is made of coarser sand and pebbles; fishing for perch, small sharks, and crabs is popular here. Farther north, a parking lot in the dunes provides access to the beach extending north to the river's mouth; a sandbar often forms, and the lower river becomes a lagoon. The sandspit between ocean and river is wide, composed of fine sand, and scattered with driftwood. Kayakers cruise the river, and osprey may be seen diving for ocean fish and then commuting to a snag on Penny Island in the river to consume the catch. Harbor seals regularly gather on the north end of the strand; stay well back to avoid disturbing the seals and their pups. Restrooms at the northern parking lot have running water. No dogs allowed on the beach.

SONOMA COAST STATE BEACH: *Hwy. One, Russian River to Bodega Head.* This popular state park stretches along a dozen miles of coast, beginning north of Jenner and extending to Bodega Head, with numerous points of beach access. Some beaches are sandy, some are rocky; and they vary in accessibility. Picnic sites are scattered along the scenic cliffs, and campgrounds are located at Wright's Beach, Bodega Dunes, and near Willow Creek. On sea stacks near Gleason Beach, seabird colonies can be observed relatively close to Hwy. One, and a population of harbor seals regularly hauls out on the sandbar at the mouth of the Russian River. The surf is rough here, and the water is not suitable for swimming. Unpredictable waves can reach up onto beaches or rocks, and visitors who are fishing or beachcombing should observe warning signs and keep an eye on the ocean. Call: 707-875-3483.

SHELL BEACH: *Hwy. One, 3.5 mi. S. of Jenner.* The Shell Beach parking lot west of Hwy. One is a crossroads, providing access to Shell Beach at the base of the high bluff, as well as a link via the Kortum Trail north to Goat Rock (2.3 miles) or south to Wright's Beach (1.5 miles). From the parking area, hikers also head east on the Pomo Canyon Trail over Red Hill to Willow Creek and the Pomo Canyon campground.

Shell Beach is at the bottom of a steep trail and rough staircase bordered by wildflowers including yarrow, lupine, and Indian paintbrush. The beach is sandy and strewn with driftwood and bits of kelp. Rocks onshore and off support tidepools and sea birds, including cormorants; surf scoters may be seen in the waves. On top of the high bluff at the north end of the beach, a small picnic area overlooks the rocky coast and sea stacks.

WRIGHT'S BEACH: *Hwy. One, 6 mi. N. of Bodega Bay.* A picnic area at this state park unit is located below the coastal terrace, near sea level; open 8 AM to sunset. There are 27 developed campsites, some sheltered by trees. Running water, but no shower facilities. This wide sandy beach is the southern terminus of the Kortum Trail, which leads through a meadow along the bluff edge to Blind Beach near Goat Rock.

DUNCANS LANDING: *Hwy. One, 5 mi. N. of Bodega Bay.* The coastal terrace overlooks a small cove and some of California's most unpredictable and dangerous surf. Picnic tables on the bluff offer a view of the waves, as well as of wildflowers in spring and of the coastal hills. A steep trail leads to the small sandy beach at the bottom; portions of the area are closed for safety reasons.

In spite of rough waters, the site was a doghole lumber port in the 19th century. From 1862 to 1877 a horse-drawn railroad brought lumber from the Russian River sawmill of Alexander and Samuel Duncan to the landing, where schooners were loaded for shipment to San Francisco. The schooners were anchored in the cove on the south side of Duncan's Point; iron rings used to secure the ships can still be seen.

GLEASON BEACH: *Hwy. One, 4.5 mi. N. of Bodega Bay.* Pull-outs at Duncans Cove and Gleason Beach, units of Sonoma Coast State Beach, offer particularly close-up viewing of seabirds roosting or nesting on offshore rocks and sea stacks.

Surfing on the Sonoma coast

Salmon Creek Beach

CARRINGTON RANCH: *Coleman Valley Rd., .3 mi. E. of Hwy. One.* This new 344-acre unit of Sonoma Coast State Beach offers dramatic ocean vistas from steeply rising slopes north of Salmon Creek. Day use facilities are planned, including trails, interpretive exhibits, and restrooms.

SALMON CREEK: *Hwy. One, 2.5 mi. N. of Bodega Bay.* Salmon Creek is an important small coastal stream with a variety of wetland habitats, including saltwater and brackish marshes near the creek mouth and freshwater marshes upstream. Those near Hwy. One attract several species of waterfowl during migration; tundra swans, not commonly seen in most coastal areas, are found here almost every winter. A barrier beach forms across the stream mouth in summer, creating a freshwater lagoon. A small parking area and restrooms are located at the end of Bean Ave. near the wide sandy beach, which extends two miles south to Mussel Point. To protect the habitat of the threatened snowy plover, a bird species that nests on the beach, dogs and fires are not allowed on Salmon Creek Beach.

BODEGA DUNES: *Hwy. One, .5 mile N. of Bodega Bay.* This unit of Sonoma Coast State Beach includes both a day-use area located on the beach near the mouth of Salmon Creek and a separate campground among the dunes. The entrance road to both areas winds through rolling dunes, which have been planted with non-native, European dune grass as a stabilization measure. Dogs and fires are prohibited on the beach.

Bodega Dunes Campground offers 98 campsites, hot showers, trailer sanitation station, campfire center, picnic tables, horseback riding, and hiking trails. Campsites are wind-sheltered by trees and dunes. Leashed dogs are allowed in the Bodega Dunes Campground. From the campground, pedestrians can walk to the edge of nearby Bodega Harbor.

Coastal Erosion

THE CALIFORNIA COASTLINE is the product of the competing natural forces of geologic uplift and coastal erosion. Along most of the California coast, Earth forces cause uplift of the land, creating bluffs and sea cliffs. Over geologic time, sea levels have both risen and fallen, bringing erosive waves to bear on this rising coast. The result is flat terraces, sea cliffs, sea caves, arches, sea stacks, and other unique coastal landforms. The land wears away through gradual rockfall, as well as by more dramatic landslides and the collapse of sea caves. These processes are hazards to life and property along the coast, but they are an integral part of what makes the California coast one of the most spectacular meetings of land and sea in the world. The sea cliffs and beaches that comprise Sonoma Coast State Beach are an ideal place to observe some of these landforms.

From a vantage point on one of the rocky points along Sonoma Coast State Beach, such as Duncans Landing, one can see that the land rises from the coast in a series of stairstep-like terraces. These terraces formed during periods of rising sea level, as waves cut an offshore platform and a sea cliff that marched landward through time as waves attacked the cliff at its base. A similar terrace is being cut offshore today. During periods when sea level fell, the shoreline moved seaward, and because the land continued to rise, the terrace was lifted above the waves. As sea level has risen and fallen through geologic time, a whole flight of terraces has formed, with the highest terraces being the oldest. These terraces are covered with sediments representing ancient beach deposits, now lifted high above the sea.

Gleason Beach

Erosion does not act uniformly along the coast. Waves are bent, or refracted, toward shallow water, so they tend to hit promontories with more force than coves and embayments. Accordingly, headlands tend to erode at higher rates than coves. On the other hand, harder rocks, or rocks with few fractures, naturally resist the erosive forces better than weaker rocks, and so many promontories actually owe their existence to the presence of stronger rocks.

Erosion is concentrated in zones of weakness in the rocks making up the shoreline. For example, fractures in the rock may be enlarged by the pounding surf and may grow into sea caves through time. The partial collapse of a sea cave may form an arch, such as at Arch Rock offshore southwest of Goat Rock. Further sea cave collapse, or the collapse of a narrow point, may leave an isolated rock spire offshore, known as a sea stack. Several sea stacks and larger offshore rocks, such as Goat Rock, are remarkably flat-topped. These surfaces are isolated remnants of the terraces seen onshore. Conversely, many of the flat terrace surfaces on land are broken by a towering rock pillar—an ancient sea stack now lifted above the waves. Several of these are visible south of Duncans Landing.

Just as coastal erosion contributed to this dramatic landscape, so too erosion threatens the works of humans placed too close to the hazard. The small community of Gleasons Beach is a prime example. Here ongoing erosion threatens some two dozen homes. The sea cliff is collapsing, in a series of coalescing landslides. Several homes already have been lost, and owners of others are constructing seawalls and retaining walls in an attempt to hold the bluffs together.

While intended to protect life and property, seawalls, revetments, and other defensive shoreline structures have detrimental impacts. They can be visually intrusive, they occupy space on the beach that otherwise could be used for beachcombing and strolling (and as haul-out space for marine mammals), and they can prevent access to the beach by covering footpaths with rubble mounds and concrete walls. Most problematic, though, is the almost inevitable destruction of beaches. During a period of sea level rise, such as the present, waves impacting the base of a sea cliff cause the cliff naturally to recede landward, taking the "back" of the beach along with it. At the same time, the "front" of the beach (nearest the ocean) is submerged beneath the advancing sea. Although it does so by fits and starts, the whole cliff-beach complex gradually moves landward. All else being equal, the beach will stay roughly the same width during this migration. But if a seawall armors the sea cliff and, by doing exactly what it was designed to do, prevents the landward migration of the sea cliff and the back of a beach, then there is no place for the advancing beach front to go. Ultimately, the front and back of the beach are submerged, and waves crash against the seawall year-round, at high tide and low.

Coastal erosion is a natural process, but there are steps that can be taken to minimize its effects. Returning the river-borne flow of sand to the coast, now largely trapped by dams, will build healthy beaches that help protect cliffs from the waves over the long run. Erosion is greatest when runoff is allowed to flow over cliff edges and when groundwater levels rise as a result of too much irrigation. And coastal visitors can help by avoiding shortcuts down steep, coastal bluffs and over dunes. Remember your personal safety when in this eroding environment—avoid walking close to a crumbling cliff edge, and lay out your beach towel away from the foot of a sea cliff. The question is not whether rockfall will occur, but when.

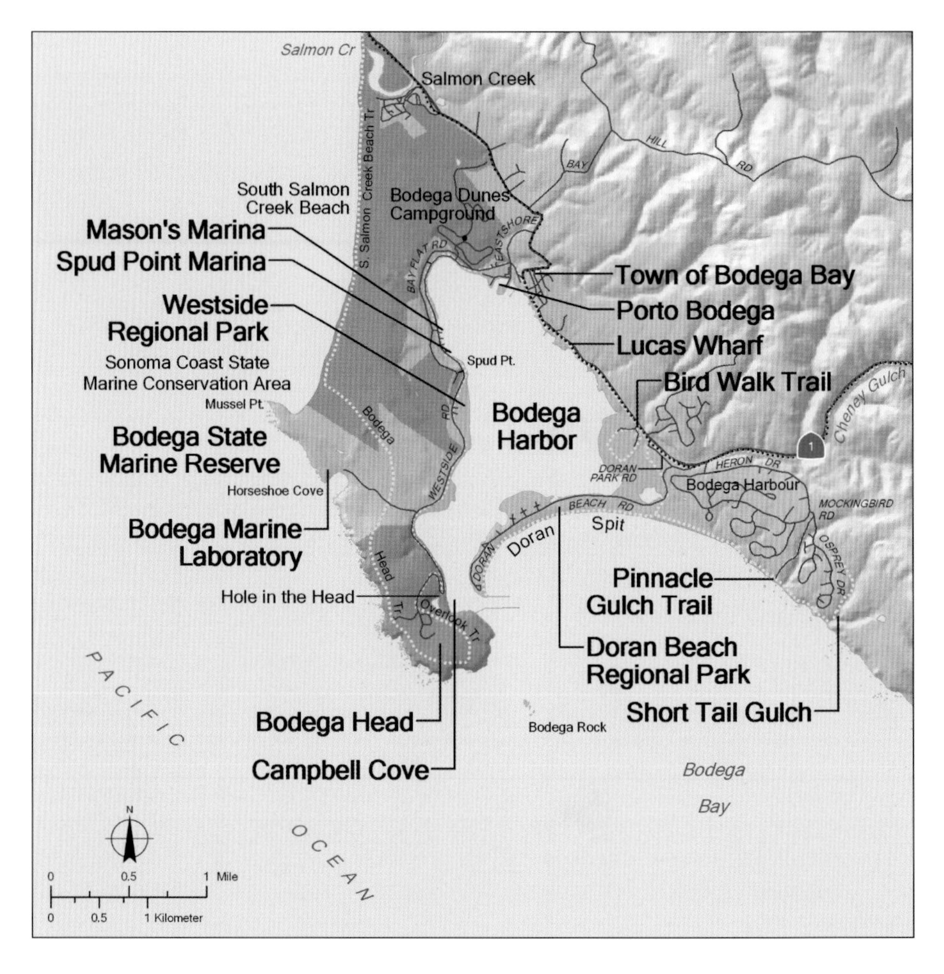

Doran Beach Regional Park

Bodega Bay

	Sandy Beach	Rocky Shore	Trail	Visitor Center	Campground	Wildlife Viewing	Historic Building	Fishing or Boating	Facilities for Disabled	Restrooms	Parking	Fee
Bodega Bay and Harbor	•	•				•		•				
Town of Bodega Bay						•	•	•		•	•	
Mason's Marina								•		•	•	
Spud Point Marina								•	•	•	•	
Westside Regional Park					•			•	•	•	•	•
Bodega Marine Laboratory and Marine Reserve				•					•	•	•	
Campbell Cove	•					•		•	•	•	•	
Bodega Head	•	•	•			•			•	•	•	
Porto Bodega					•			•	•	•	•	•
Lucas Wharf								•	•	•	•	
Bird Walk Trail		•				•			•	•	•	•
Doran Beach Regional Park	•				•	•		•	•	•	•	•
Pinnacle Gulch Trail	•		•							•	•	•
Short Tail Gulch	•		•									

BODEGA BAY AND HARBOR: *West of Hwy. One.* Bodega Bay is the crescent of ocean stretching from Bodega Head to Tomales Point in Marin County; Bodega Harbor is the sheltered water body north of Doran Spit. A panoramic view of the bay and harbor is available from the east side of Bodega Head.

Bodega Harbor is home to a large commercial and recreational fishing fleet, and its shallow waters and surrounding marshes draw shorebirds and waterfowl. Novices as well as experienced birders will find much to see here, both in sheer numbers of birds and in diversity of species (over 100 species can be sighted in the Bodega Bay area on any day during the peak month of December). Loons, grebes, and ducks occupy the open waters of the harbor, while shorebirds and gulls forage in the mudflats at low tide. Freshwater wetlands provide habitat for long-tailed marsh wrens, red-winged blackbirds, sora, and Virginia rails. Bird Walk Trail, Doran Beach Regional Park, and other harbor overlooks offer great wildlife viewing.

TOWN OF BODEGA BAY: *Hwy. One adjacent to Bodega Harbor.* The town of Bodega Bay originated as a small fishing port in the 1870s. In the early 1900s, families from the inland areas of the county began to build summer homes at Bodega Bay, and with the opening of Hwy. One in the 1920s, the town became a tourist destination. The dredging of Bodega Harbor in 1943 stimulated the growth of Bodega Bay's fishing industry. In the early 1960s, Alfred Hitchcock's classic horror movie, *The Birds*, was filmed partly in Bodega Bay. The birds' attack at the schoolhouse was shot in the nearby town of Bodega, but spliced with footage of the old Tides Wharf to create a single setting, a hill overlooking the bay.

Today the town retains much of its old-time character as a low-key resort and fishing center. The Fishermen's Festival, with a parade and blessing of the fleet, occurs annually, generally in April; call: 707-875-3866. The Tides Wharf at 800 Hwy. One offers a close-up view of fishing boats at work, as well as a fish market, restaurant, and other facilities. A golf course with sweeping ocean views is available to the public within the Bodega Harbour residential community; call: 707-875-3538. Kayaking trips on the protected waters of Bodega Harbor are popular, and Bay Flat/Westside Rd. along the harbor's edge invites sightseeing on bicycles, which can be rented locally. Digging for gaper, littleneck, and Washington clams, which inhabit gravel areas in the intertidal zone, is popular at low tide; a fishing license is required.

North of the town is the Children's Bell Tower, a collection of Italian-made bells that ring spontaneously in the breeze. The tower is a memorial to young Nicholas Green, a Bodega Bay resident who was killed by highway robbers while traveling with his family through Italy in 1994. Seven Italians received organ donations from seven-year-old Nicholas, and he and his family became widely known symbols of hope. The memorial is behind the Community Center, west of Hwy. One.

MASON'S MARINA: *1820 Westside Rd., Bodega Bay.* Private marina with berths, docks, fuel, marine supplies, and snack shop. For information, call: 707-875-3811.

SPUD POINT MARINA: *1818 Westside Rd., Bodega Bay.* The 244-slip marina is shared by commercial and recreational boats. Twenty-four-hour fuel dock, commercial flake-ice machine, hoist, laundromat, and showers. A breakwater surrounding the marina serves as a public fishing pier, where a fishing license is not necessary although catch limits apply.

WESTSIDE REGIONAL PARK: *Off Westside Rd., Bodega Bay.* Day-use boating area with launch ramp; 48 campsites; fees are charged. Picnic tables and wheelchair-accessible rest-

Lucas Wharf, Bodega Bay

rooms are available. For camping reservations, call: 707-565-2267. For information, call: 707-875-3540.

BODEGA MARINE LABORATORY AND MARINE RESERVE: *Off Westside Rd., Bodega Bay.* The Marine Reserve, a part of the University of California's Natural Reserve System, includes 362 acres of rocky intertidal habitat, sandy beaches, saltmarsh, dunes, and coastal prairie, all managed for teaching and research. The laboratory is open to the public Fridays 2 PM to 4 PM; guided tour includes an overview of marine research laboratory and aquariums. Call: 707-875-2211. The Department of Fish and Game manages the adjoining Bodega State Marine Reserve, extending 1,000 feet offshore, where all marine plants and animals are protected. From Mussel Point to north of Salmon Creek is the Sonoma Coast State Marine Conservation Area, where marine plants and certain invertebrates may not be taken; check Department of Fish and Game regulations.

CAMPBELL COVE: *End of Westside Rd., Bodega Bay.* A sheltered beach is located just inside the mouth of Bodega Harbor; fishing boats pass nearby on their way to sea. Opposite the parking lot is a freshwater pond known as the Hole in the Head. Once the beginnings of a nuclear power plant defeated by community opposition in the 1960s, the excavated pool is now a haven for wildlife. From the wooden boardwalk, look for rails and swamp sparrows in winter months, or black-crowned night herons nesting in spring or summer.

BODEGA HEAD: *End of Westside Rd., Bodega Bay.* Bodega Head, like Tomales Point to the south, owes its prominence to the resistant granitic bedrock making up the peninsula. The northernmost sliver of granite exposed on the California coast, the rock making up Bodega Head has traveled northward some 350 miles over the last 15 to 20 million years along the San Andreas Fault, which lies along the peninsula's eastern margin. The rock has eroded into a series of spires, arches, and caves along fractures and zones of weakness. These features are best seen by taking the Bodega Head trail around the tip of the point, which also offers great views of Bodega Rock and Tomales Point. East of the San Andreas Fault lie the softer shales and sandstones of the Franciscan Complex. These rocks are easily worn by wind and water relative to the granite of Bodega Head, creating the lowlands around Bodega Harbor. Into this depression are blown sands eroded from the bluffs, or carried to the coast by rivers and streams, such as nearby Salmon Creek. This sand has formed an extensive field of dunes between Bodega Harbor and South Salmon Creek Beach; the dunes can best be seen by hiking the network of trails southwest of Bodega Dunes Campground.

To reach Bodega Head, take Westside Rd. to the end, then continue uphill and take either the right fork to the ocean parking lot or the left fork to the eastern parking lot overlooking the narrow entrance channel of Bodega Harbor. At the ocean parking lot's south end, note the natural rock bridge below and,

Fisherman with black rockfish

in spring, look for pigeon guillemots and pelagic cormorants nesting in niches on the sheer cliffs. Or take the half-mile trail north to the Horseshoe Cove overlook to reach the highest promontory on the peninsula, where Bodega Bay's most impressive view takes in massive dune fields to the north, the shallow waters of Bodega Harbor to the east, the Point Reyes Peninsula to the southeast, and a vast swath of open ocean. Weathered rock outcroppings covered with lichens are scattered about, and the windswept hills are dotted in season with wildflowers including buttercups, lupines, and seaside daisies.

From the eastern parking lot on the head, follow the trail past the restrooms one-half mile to a dramatic view of wave-washed Bodega Rock offshore, where field glasses reveal sea lions hauled out on the sand and seabirds such as pelagic and Brandt's cormorants. The trail continues along the cliffs past a memorial to fishermen lost at sea, terminating at the ocean parking lot. Great blue herons may be seen most anywhere on Bodega Head, and in the spring, brilliantly colored goldfinches dart among the brush. Whales can be seen offshore from November to early May. Restrooms are located at both the west and east parking lots on Bodega Head.

PORTO BODEGA: *1500 Bay Flat Rd., Bodega Bay.* Ocean trips are available to fish for salmon, halibut, rockcod, lingcod, or crabs or to look for whales; call: 707-875-3344. Facilities at the marina include 77-berth boat dock and launch, 57-space RV park with hookups, restaurant, and tackle shop. For information, call: 707-875-2354.

Pinnacle Gulch Trail

Osprey

LUCAS WHARF: *599 Hwy. One, Bodega Bay.* Public fishing pier, with good views over Bodega Harbor. Fresh fish market, restaurant, and snack shop.

BIRD WALK TRAIL: *W. of Hwy. One opposite N. Harbour Way, Bodega Bay.* Old levees built to hold silt dredged from Bodega Harbor now provide a superb elevated viewing platform for strolling and wildlife observation. From the parking area, walk to the far end of the levee to gain an unobstructed view of the harbor and marshes. Birds also use the freshwater ponds within the levees. Songbirds, shorebirds, and waterfowl can be sighted in great abundance. Near the parking area are sheltered picnic tables and a barbecue grill. Parking fee.

DORAN BEACH REGIONAL PARK: *Doran Park Rd., Bodega Bay.* The two-mile-long curving sandspit offers both bay and harbor recreation. The sandy ocean beach is broad, and modest breakers are good for beginning surfers. Kite flyers and wind surfers take advantage of prevailing breezes. On the other side of the sandspit are a boat launch facility and fine views of the harbor. The park offers 138 campsites, showers, picnic tables, restrooms, trailer sanitation station, fishing off the rock jetty at the end of the sandspit, and fish cleaning station; fees apply. Dogs must be leashed at all times. For camping reservations, call: 707-565-2267. For information, call: 707-875-3540.

PINNACLE GULCH TRAIL: *Mockingbird Rd., Bodega Harbour subdivision.* Take So. Harbour Way off Hwy. One, then left on Heron Dr. to Mockingbird Road. Restrooms with running water; fee for parking. A half-mile-long trail, steep in places, begins across the street and descends a narrow canyon to a quarter-mile-long sandy beach. Neighboring coves are accessible on either side.

SHORT TAIL GULCH: *Mockingbird Rd., Bodega Harbour subdivision.* Follow Mockingbird Rd. past the Pinnacle Gulch Trail to Osprey Dr., then turn left to trailhead. A trail newly opened in 2004 leads through a wooded gulch to the shore, including some 200 steps. At the sandy beach, an option is to hike northwest to the Pinnacle Gulch Trail, making a loop back to Osprey Drive. Street parking is available.

Waves and Tides

THE OCEAN is always in motion. From a ship in open water, the ocean surface seems like random choppiness. From a high promontory such as Trinidad Head or Table Bluff County Park in Humboldt County, this surface chop slowly transforms into more orderly patterns and landward trending wave trains. At the beach you notice the breaking waves, wave run-up, and changing tide levels. All these conditions are waves.

Waves are complex phenomena of energy transmission. Surface waves in the ocean have three separate sources—wind, tides, and seismic events—and oceanographers refer to these waves collectively as gravity waves. Waves have several basic components—the *crest*, the *trough* or low part, *wave length* or horizontal distance between one crest and another, *wave height* or vertical distance between trough and crest, and *wave period* or the time it takes for a crest to travel one wave length.

Most of the deep-water chop and waves that break along the shore are wind waves. As storms and wind develop over the open ocean there is a transfer of energy from air to water. The energy transfer and the size of the waves increase with wind strength, the time the wind blows, and the distance over which it blows (the force, duration, and fetch). The height of individual waves is limited to a steepness of approximately 1:7 (the ratio of height to length). Waves will collapse or break as they exceed this steepness limit, creating the chop and whitecaps that appear throughout the ocean. If winds continue, waves eventually grow into fully developed seas, where the waves have absorbed as much energy as they can. Long-duration storms with winds of 30 to 40 knots can generate waves with average heights of 15 to 30 feet, and extreme heights approaching 50 to 60 feet. In rare situations, waves can approach 100 feet in height. Such a storm event in December 1914 may have generated the waves that extinguished the light at the Trinidad lighthouse, 196 feet above sea level.

Waves are not confined to areas of high winds. Once formed, they can travel thousands of miles. Waves transport energy; the water itself moves very little. If you watch floating birds, they bob up and down, but, like the water, they stay pretty much in the same location. As waves travel, they interact with other waves; crests combine with troughs from another wave train to create flat areas; crests combine with other crests to create extremely high water levels. Most likely "rogue" waves, the dread of sailors, form as the rare combination of several extreme wave crests.

Waves convey energy very efficiently through the deep ocean. When the depth is about half the length of the wave, water depth begins to influence the waves. Waves are said to "feel" the bottom and they begin to bend and align to conform to the bottom contours. When waves get into water depths about 1.3 times the wave height, they be-

The tossing waves, the foam, the ships in the distance,
The wild unrest, the snowy, curling caps—
That inbound urge and urge of waves,
Seeking the shores forever.

—Walt Whitman, *From Montauk Point*

come unstable and break along the shore. Very large waves can break on offshore features, re-form, and break again closer to shore. Wave energy increases with the square of the height. A four-foot-high wave contains 33 foot-tons of energy per foot of crest; a 12-foot-high wave contains 295 foot-tons. A three-fold increase in wave height has a nine-fold increase in energy. Waves can be more powerful, and dangerous, than they appear, threatening both structures built along the coast and people playing on the beach. When the waves seem higher than usual, exercise extra caution while visiting the shore.

Wind waves have periods ranging from about four seconds up to 20 or 22 seconds, and lengths from hundreds of feet to several thousand feet. By comparison, tide waves have a 12 hour 25 minute period and a wavelength that is half the circumference of the earth. They travel 700 to 800 miles per hour, with the crest being identified as high tide and the trough as low tide. Tide waves are generated by the gravitational pull exerted on the earth by the moon, the sun, and other planets. The moon has by far the most influence on the tides, and the tide wave period relates to the time it takes for the earth to revolve relative to the moon. Spring tides, the highest tide ranges of the month, occur when the moon and sun are aligned and their gravitational pulls are combined. Neap tides, the smallest tide ranges, occur when the moon and sun are in opposition. The distance between the earth and the moon also influences tides. The moon makes an elliptical orbit around the earth. When it is closest to the earth, at perigee, the high tides are higher. When the moon is farthest from the earth, at apogee, the tides are lower. At least twice a year, the spring and neap tides combine with the perigean or apogean tides to produce the highest and lowest tides of the year.

California has a mixed tide—each day there are two high and two low tides of different magnitudes. There is a pattern of higher high tide followed by a lower low tide, a lower high tide and a higher low tide. The level of the tide is stated in relation to the average of the daily lower low water levels, for instance +3.5 MLLW (mean lower low water). In general, the tide range is higher in the northern part of the state than in the central or southern region. In San Francisco the average range from higher high to lower low tide is 5.84 feet; in Eureka it is 6.86 feet.

Tsunami waves, often inappropriately called tidal waves, are generated by impulses—large seismic events, volcanic eruptions, or submarine landslides. These waves, while quite spectacular, occur only rarely. The wavelengths can be 100 to 200 miles, with velocities up to 500 miles per hour and up to 10 minutes between crests. These high-energy waves are barely noticeable on the open ocean, but at the shoreline, they can exceed 100 feet in height. In 1960, an earthquake in Chile generated a tsunami wave that raised water levels in Crescent City by over five feet. Four years later, an earthquake in Alaska generated waves that killed 12 people, caused property damage in the millions, and inundated downtown Crescent City.

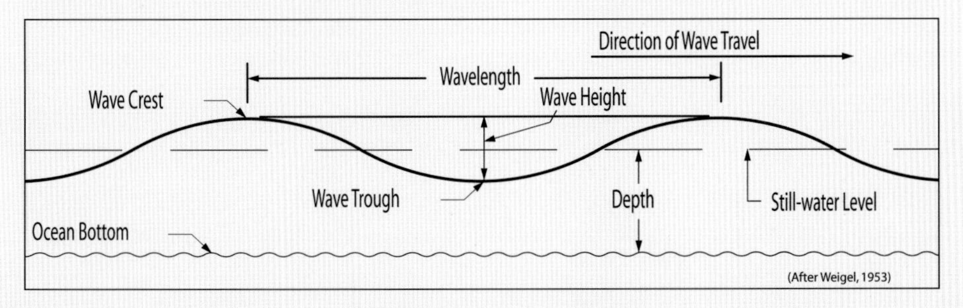

Cordell Bank and Gulf of the Farallones National Marine Sanctuaries

MANY SEA STACKS dot the waters off Northern California; most are near shore and few are large enough to be labeled "islands." Cordell Bank is a rocky feature some nine miles long and five miles wide, located 21 miles west of Point Reyes, that might be called an "underwater island." Like the Farallon Islands, Cordell Bank was originally part of the Sierra Nevada range, and its granite formations have inched northwestward over the past 30 million years, moving about an inch per century. During the last Ice Age, 15,000 years ago, the bank was a real island, when sea level was about 350 feet below its present level. That ancient exposure of rock to the force of breaking waves can be read even today in the form of underwater terraces on Cordell Bank cut by wave action, much like those visible on the Sonoma coast.

You cannot see Cordell Bank from land or from the water's surface, but the bank makes its presence known. Mariners passing over it notice the effects of currents on their vessels. The water over the bank is shallower than to the sides, and as waves refract in the shallower water, they change the direction of objects moving on the surface. Fishermen have long noticed the greater abundance of fish found over the bank, along with the large number of birds, whales, sea lions, and harbor seals.

The bank is densely packed with organisms that thrive on the nutrients brought by currents carrying the products of upwelling. The water is clear here, untouched by the sediments discharged by the Golden Gate and coastal rivers, and sunlight penetrates deeply. About 200 feet of water covers most of the bank, although pinnacles in some places have less than 130 feet of water over them. The tops of the bank's spires are crowded with a high diversity of sponges, anemones, hydrocorals, and barnacles. Fish are plentiful too. To the west, water depths increase dramatically; only a couple of miles away, the water is more than 3,000 feet deep.

The typical kelp forests found along much of the continent's edge are not found on Cordell Bank, and most of the plants and animals that grow on the bank are typically found, not surprisingly, in offshore rather than nearshore settings. But some of the

Juvenile rockfish swarm over the invertebrate-covered pinnacles of Cordell Bank

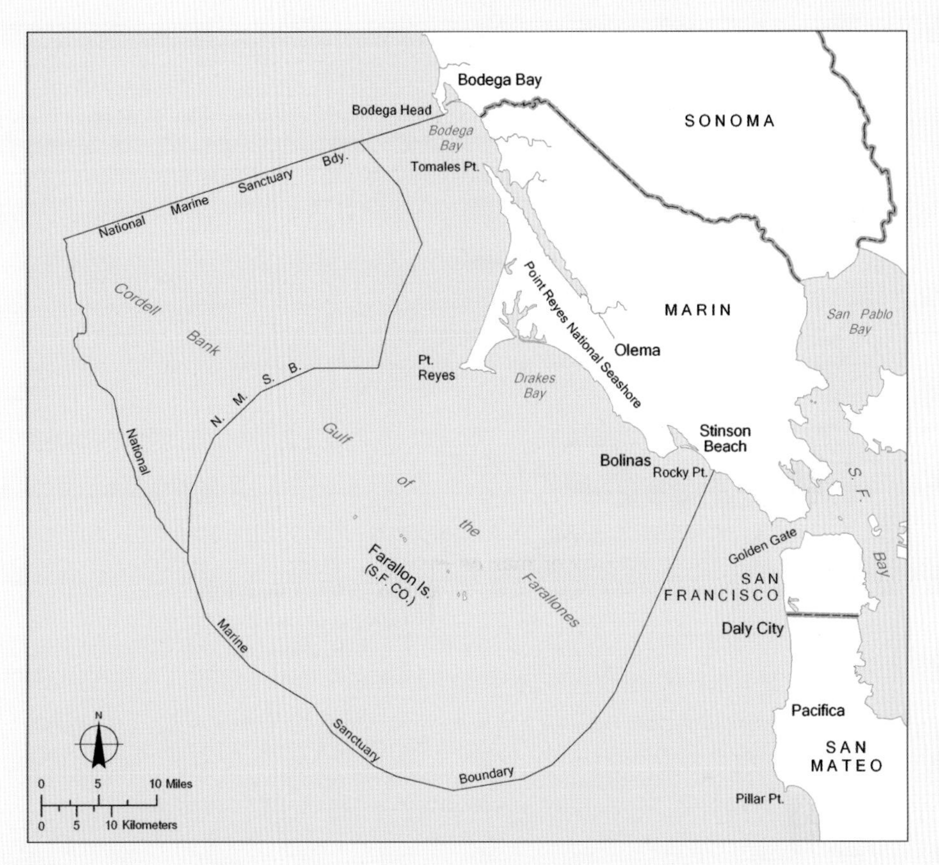

bank's species, such as certain turban snails and red-striped acorn barnacles, usually are found only in intertidal areas. These species have pelagic larvae that likely settled and survived on the bank.

The hard substrate of Cordell Bank is always submerged and does not experience the cycles of desiccation and immersion that characterize intertidal environments. The bank is also free of the violent action of crashing surf, and it is insulated from the strong actions of wind-driven waves on the surface. In its relatively bright and calm setting, fed by the biologically rich soup of nutrients brought by upwelling, the bank is an underwater habitat with few equals. The strong currents around the bank make it a great place for filter feeding invertebrates. Able to detect the abundant food sources at the bank, marine mammals and birds are drawn to it to forage.

Most of the organisms living attached to Cordell Bank are invertebrates that take advantage of the available substrate in a food-rich environment. Of the plants that occur here, red algae are most common, probably because green plants cannot absorb efficiently the blue and green light rays that reach the bank through more than 100 feet of water. To a diver, red organisms such as algae ordinarily appear muted or even black, because red light rays are absorbed by the water. On Cordell Bank, one common red organism appears brightly colored even underwater, due to fluorescence. Robert Schmieder, who wrote *Ecology of an Underwater Island* based on his many dives to the

Cordell Bank, found that the strawberry anemone emits red light rays while absorbing blue-green light, giving it a bright red hue even underwater.

The Cordell Bank National Marine Sanctuary was established in 1989 to protect the ecosystem of the bank and its living resources, including birds, mammals, and invertebrates. The bank is off-limits to oil and gas exploration and other development. Dumping is prohibited, as is the taking of plants or benthic organisms from the top of the bank. Ship traffic passes freely overhead, although the area has been closed to fishing since 2003 because it falls within the Rockfish Conservation Area as defined by the Pacific Fisheries Management Council.

The neighboring Gulf of the Farallones National Marine Sanctuary extends from nearshore waters such as Tomales Bay and Bolinas Lagoon up to 40 miles offshore, surrounding the Farallon Islands, which harbor the largest concentration of breeding seabirds in the contiguous United States. The southern tip of the Marin County coast is touched by a corner of another federally designated sanctuary, the Monterey Bay National Marine Sanctuary, which is centered on the resources of the massive submarine canyon off Monterey Bay. Each sanctuary has regulations to help protect the habitats and species that characterize them. For example, the Gulf of the Farallones Sanctuary prohibits disturbing seabirds or marine mammals by flying motorized aircraft at altitudes of less than 1,000 feet over waters within one nautical mile of the Farallon Islands or Bolinas Lagoon. Within marine sanctuaries, as elsewhere in California, the State Department of Fish and Game enforces fishing restrictions and rules that apply to the taking of plants and invertebrate species in intertidal areas.

Visitors can learn about the resources of both the Cordell Bank and Gulf of the Farallones Sanctuaries at the visitor center maintained by the Farallones Marine Sanctuary Association at the Old Coast Guard Station at the west end of Crissy Field in the Presidio in San Francisco. Visitor center hours are Wednesday–Sunday, 10 AM–4 PM.

Boat trips are available to Cordell Bank and the Gulf of the Farallones, where visitors can view whales, other marine mammals, and seabirds that congregate there. September and October are the best months to visit, when strong northwest winds die down, the ocean's surface can be flat, and visibility can be good. During summer and fall months, blue and humpback whales may be seen.

The sanctuary staff monitors the underwater environments using a two-person submersible. One collaborative research project involves tracking the movements of the black-footed albatross, a long-distance seabird that visits the bank area and flies regularly across the northeastern Pacific Ocean. During spring, the black-footed albatross nests on uninhabited islands 1,000 miles northwest of Hawaii. Parent birds fly on feeding journeys of 22 to 30 days to California's North Coast, while their young wait on the nest for their long-distance take-out meal. Some of the albatrosses have been tagged, utilizing global positioning technology. You can learn more about this project and link to maps showing their movement at the Cordell Bank Sanctuary's website at www.cordellbank.noaa.gov. Learn more about the Gulf of the Farallones National Marine Sanctuary at www.farallones.org and at www.farallones.noaa.gov.

Page opposite: Stinson Beach, Marin County

Marin County

To St. Helena
116
12
SONOMA
116
101
12
1
VALLEY
FORD
RD
Estero Americano
Estero de San Antonio
BODEGA
Dillon Beach
Tomales
TOMALES
PETALUMA RD
AVE
Petaluma
Tomales Pt.
D ST
116
Tomales
Laguna Lake
121
Marshall
RED
HILL
RD
Tomales Bay SP
Bay
37
Abbotts Lagoon
MARIN
NOVATO
BLVD
Novato
Nicasio Res
BLVD
Inverness
POINT REYES
PETALUMA RD
Point Reyes
National Seashore
DRAKE
NICASIO
Pt. Reyes Station
Nicasio
LUCAS
1
SIR
FRANCIS
Olema
VALLEY
RD
VALLEY
RD
Golden Gate
NRA
DRAKE
BLVD
101
SIR
FRANCIS
Drakes
Bay
Pt. Reyes
Kent Lake
Fairfax
San
Rafael
San Anselmo
Alpine Lake
Point Reyes
National Seashore
Kentfield
580
Larkspur
PACIFIC
Mill
Valley
PANORAMIC
HWY
Bolinas
Bolinas Pt.
Stinson
Beach
Mount
Tamalpais
SP
Duxbury Pt.
Tiburon
Angel Is.
Bolinas
Bay
Muir Beach
Golden Gate
NRA
Rodeo Lagoon
Pt. Bonita
OCEAN
1
N
N. Farallon Is.
SAN
FRANCIS
0
5
10
Miles
0
5
10 Kilometers
Farallon Is.
Farallon NWR
To Half
Moon Bay

Marin County

ALTHOUGH CLOSE to population centers of the Bay Area, the Marin County coast is largely rural. Parks and agricultural and open space lands dominate; small villages and scattered ranch houses dot the landscape. Once-upon-a-time plans for a city in the Marin headlands and freeways through the coastal hills never came to pass, and now west Marin County boasts three national parks and monuments, three state parks, and numerous county parks and open space preserves, along with conservation lands managed by non-profit entities. Beaches, bays, rocky shore, and forests invite visitors to explore the coast.

Outdoor activities span the seasons in Marin. Whale watching is popular from fall through spring. From spring into summer, coastal visitors enjoy viewing wildflowers on Point Reyes and hiking trails throughout the coastal area. Autumn brings the southerly migration of hawks, falcons, eagles, and other birds of prey, and Marin offers exceptional opportunities to view them. Many raptors prefer to fly over land rather than the sea, and as they head south, they tend to funnel over the Marin Headlands. The Golden Gate Raptor Observatory counts and monitors the birds from year to year at a location off Conzelman Road in the Golden Gate National Recreation Area known as Hawk Hill. Visitors are welcome at the site, where occasional talks and banding demonstrations are offered from September through November; call: 415-331-0730.

Annual events include the Western Weekend parade and barbecue in Point Reyes Station in June and the Fourth of July parade in Bolinas. Yacht races on Tomales Bay, summer Shakespeare on the beach at Stinson, and the Labor Day sand castle contest at Drakes Beach are recurring activities. For information on local events and visitor attractions, contact the West Marin Chamber of Commerce in Point Reyes Station at 415-663-9232 or the Marin County Visitors Bureau at 415-925-2060. The West Marin Stagecoach provides limited weekday transit service from San Anselmo along Sir Francis Drake Blvd. to Olema, Point Reyes Station, and Inverness, and from Marin City at Hwy. 101 to Muir Beach, Stinson Beach, and Bolinas; call: 415-526-3239. The San Francisco Municipal Railway provides bus service to the Marin Headlands on Sundays and holidays; call: 415-673-6864.

Surfboard rentals and equipment:
Live Water Surf Shop, Stinson Beach, 415-868-0333.
Marin Surfsports on Hwy. One off Hwy. 101, Mill Valley, 415-381-9283.
Point Reyes Surf Shop, Point Reyes Station, 415-663-1072.

Kayaking equipment and tours:
Blue Waters Kayaking, offering rentals, tours, and camping trips to remote beaches in the Point Reyes National Seashore, 415-663-1743; or 19225 Shoreline Hwy., Marshall, 415-663-1754; or 12938 Sir Francis Drake Blvd., Inverness, 415-669-2600.
Eskape Sea Kayaking, offering tours and instruction, 707-583-2390.

Mountain bike rentals:
Cycle Point Reyes, Point Reyes Station, 415-663-9164.
Olema Ranch Campground, Olema, 415-663-8001.

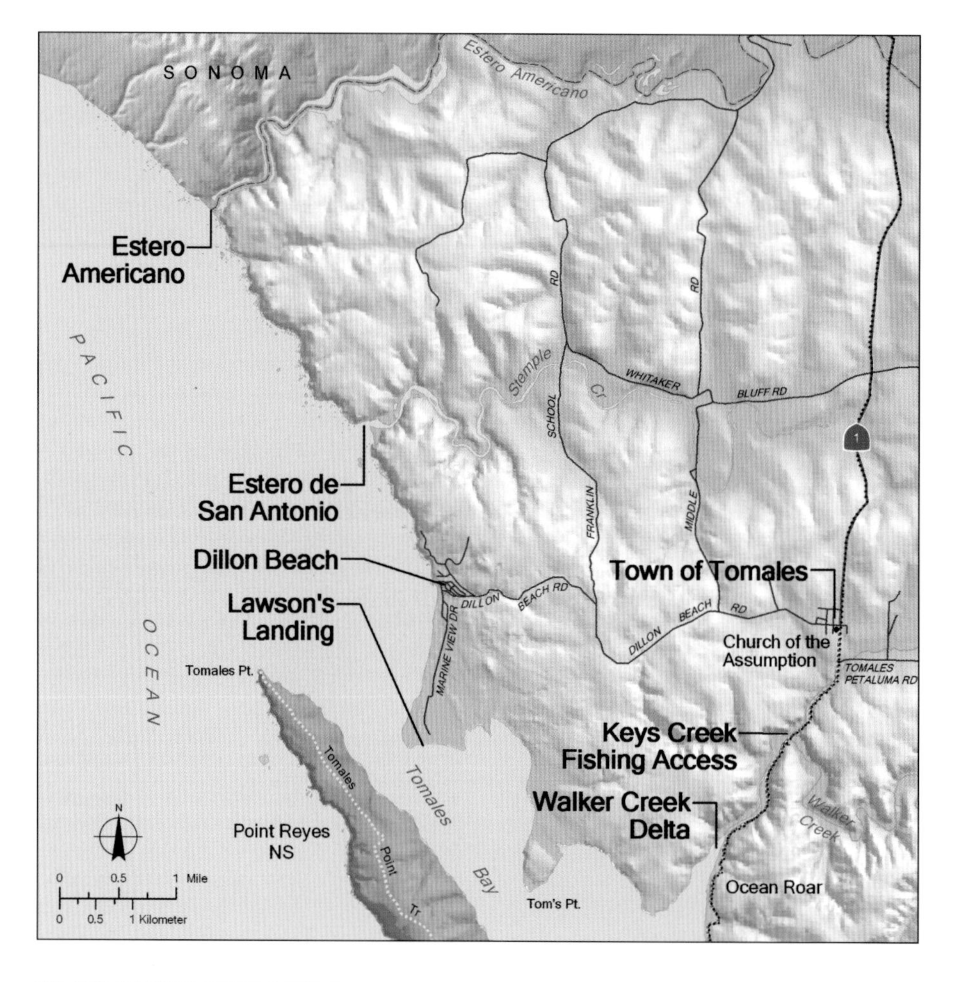

Dillon Beach

Tomales / Dillon Beach

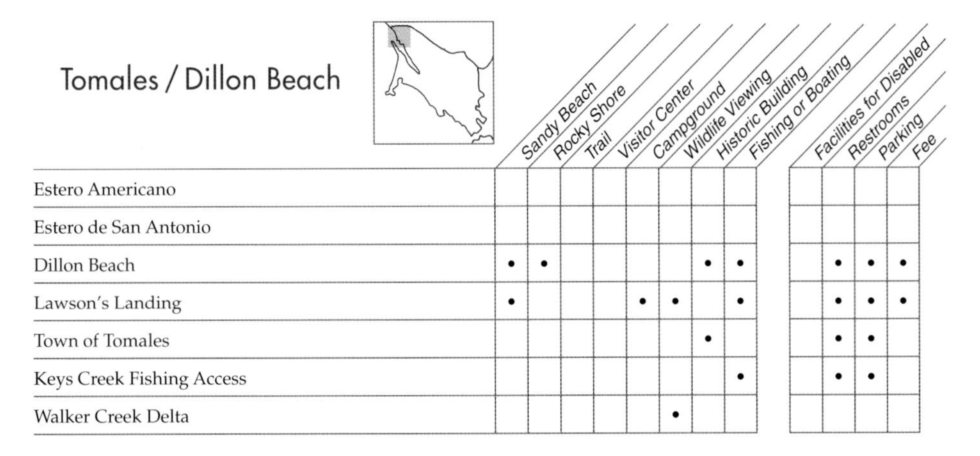

	Sandy Beach	Rocky Shore	Trail	Visitor Center	Campground	Wildlife Viewing	Historic Building	Fishing or Boating	Facilities for Disabled	Restrooms	Parking	Fee
Estero Americano												
Estero de San Antonio												
Dillon Beach	•	•					•	•		•	•	•
Lawson's Landing	•				•	•		•		•	•	•
Town of Tomales							•			•	•	
Keys Creek Fishing Access								•		•	•	
Walker Creek Delta						•						

ESTERO AMERICANO: *Off Hwy. One, 9 mi. N.W. of Tomales.* Americano Creek winds through rolling coastal hills, forming the fjord-like Estero Americano. The creek lies in a "drowned valley," the term for an area gradually flooded by a relative rise in sea level. The change in sea level came about either through an actual rise of ocean waters as glaciers melted after the last Pleistocene Ice Age, or through the earth's subsidence, or through a combination of both factors. The estero is a shallow body of mixed salt and fresh water; the saltwater influence extends up to three miles inland. In summer, reduced freshwater flow allows a sandbar to build, blocking the mouth of the estero and creating heightened salinity behind the bar. The long, narrow configuration of the estero is unusual among California estuaries.

The estero includes more than one square mile of open water. Pacific herring deposit their eggs on the eelgrass growing in the channel near the mouth of the estero; other fish found in the estero include staghorn sculpin, shiner surfperch, starry flounder, and smelt. The sandy bottom serves as a nursery for juvenile Dungeness crabs. Owls and hawks inhabit the surrounding uplands, and tundra swans, uncommon along the coast,

Estero Americano

Estero de San Antonio

have been observed along the upper reaches of the estero. The estero is surrounded by privately owned agricultural land, and although public roads cross Americano Creek several miles inland, there is no direct public access to the estero itself.

ESTERO DE SAN ANTONIO: *Off Hwy. One, 5 mi. N.W. of Tomales.* The Estero de San Antonio, lying in the drowned valley of Stemple Creek, is a long, narrow estuary similar in form to the Estero Americano. Wading birds feed and rest here. Waterfowl such as pintails, American wigeons, canvasbacks, and ruddy ducks can be found here, but generally in smaller numbers than at the neighboring Estero Americano, which has greater open water area. Western pond turtles commonly bask in the sun along the margins of the Estero de San Antonio. There is no public access to the estero, which is surrounded by private land, but the upper end of the estero may be seen from Whitaker Bluff Road.

DILLON BEACH: *End of Dillon Beach Rd., 4.1 mi. W. of Tomales.* Rows of modest cottages dating from the 1920s line the narrow streets of this resort community. Day use of the privately owned sandy beach is available for a fee. Clinging to the rocks at the north end of the beach is a rich assortment of intertidal species, such as sea palms, sea anemones, California mussels, goose barnacles, and ochre sea stars. Picnicking, crabbing, clamming, and fishing are popular; dogs allowed on beach. Call: 707-878-2094.

LAWSON'S LANDING: *5 mi. W. of Tomales, off Hwy. One, S.W. of Dillon Beach.* Take the road south from Dillon Beach through sand dunes, some of which are 150 feet high, to this privately operated resort on a wide sandy beach at the mouth of Tomales Bay. Beach sagewort plants may be seen among the dunes, harbor seals haul out on the sandy shore, and tule elk may be seen on Tomales Point across the narrow bay. Picnic tables, fire rings, and trailer and tent camping are available in meadows surrounded by sand dunes; call ahead in wet weather, when the campground may be closed. RV dump stations available. Dogs allowed on leash. Car day-use and camping fees apply.

Boating facilities include self-launch, launching service by tractor, boat rentals, fuel, and repairs. Popular area for gaper clamming. Pier fishing for perch, smelt, and crabs during daylight hours, usually February through October. Salmon and halibut fishing are generally best from mid-June through August. The boathouse offers bait and tackle, propane, firewood, and fishing licenses; open limited hours during December and January. Call: 707-878-2443.

TOWN OF TOMALES: *Hwy. One, 4.5 mi. S. of the Sonoma County line.* The village of Tomales was once an ocean port and one of Marin County's most important towns. The first European settler was an Irishman named John Keys, who, beginning in 1850, built a house and store. Other businesses followed, along with a line of warehouses, which served the ships that carried butter, hogs, beef, and potatoes down Keys Creek to Tomales Bay and thence to San Francisco. Within 20 years the creek bed had silted in, and ships could sail no closer to Tomales than Ocean Roar at the mouth of Keys Creek. When the North Pacific Coast Railroad reached Tomales in 1875, linking the area to the rail-ferry wharf in Sausalito, transportation of farm produce was greatly improved. A number of 19th century structures remain near the intersection of Main St. and Dillon Beach Road. The Church of the Assumption, located just south of the present business section, was built in 1860 and restored after the 1906 earthquake.

KEYS CREEK FISHING ACCESS: *Hwy. One, 1 mi. S.W. of Tomales.* This small public fishing access area has picnic tables and restrooms. A trail leads from the gravel parking lot down to the bank of Keys Creek, where coho salmon and steelhead are caught from late fall until spring. Call: 415-045-1455.

WALKER CREEK DELTA: *2 mi. S. of Tomales, W. of Hwy. One.* Walker Creek with its tributary, Keys Creek, is the second largest of the streams feeding Tomales Bay. Although its flow is reduced by siltation, the stream still supports runs of coho salmon and steelhead, and efforts have been made by land managers to improve the habitat for fish. The Walker Creek delta includes over 100 acres of marsh and mudflats where salt marsh plants predominate. Shorebirds use the marsh, including whimbrels, short-billed dowitchers, and occasional long-billed curlews. The delta is managed by Audubon Canyon Ranch, a land preservation and education organization, and access to the marsh is reserved primarily for educational and scientific purposes and is by appointment only; for information, call: 415-663-8203.

Keys Creek Fishing Access

The San Andreas fault zone, visible in the linear form of Tomales Bay (looking southeast)

San Andreas Fault

THE SAN ANDREAS FAULT, perhaps the most famous fault in the world, slices through California from the Gulf of California to Cape Mendocino. North of San Francisco, the fault runs along the coast, intermittently passing on and offshore, between Bolinas Lagoon and Shelter Cove. The fault separates rocks of the Pacific Plate on the west from rocks of the North American Plate to the east. The Pacific Plate is sliding northward along the fault at the rate of about two inches per year, although most of the motion occurs suddenly during major earthquakes that occur at intervals of about 400 years. For example, the Pacific Plate lurched 24 feet to the north in the Olema area during the 1906 San Francisco Earthquake, offsetting fences, opening cracks in the ground, and shaking the earth sufficiently to level much of San Francisco.

The fault has left its mark on coastal geography. The rocks near the fault are highly fractured and sheared as a result of hundreds of such earthquakes, repeated over geologic time since the fault first became active about 30 million years ago. These fractured rocks are eroded more easily than the more coherent rocks away from the fault, and for this reason the trace of the fault is often marked by long, linear valleys that cut across the land. Tomales Bay is one such valley, in this case flooded by the sea.

The Point Reyes area is an excellent place to observe the San Andreas Fault. The fault trace is readily visible not just at Tomales Bay, but also as the long valley running between Point Reyes Station and Stinson Beach, along Highway One. The Earthquake Trail, a half-mile long paved loop that starts at the Bear Valley headquarters of Point Reyes National Seashore, lets you observe the fault up close. The trail's many interpretive panels describe the geology of the area and explain how the fault has shaped the landscape.

Offset fence at Olema after 1906 earthquake

San Francisco after 1906 earthquake and fire

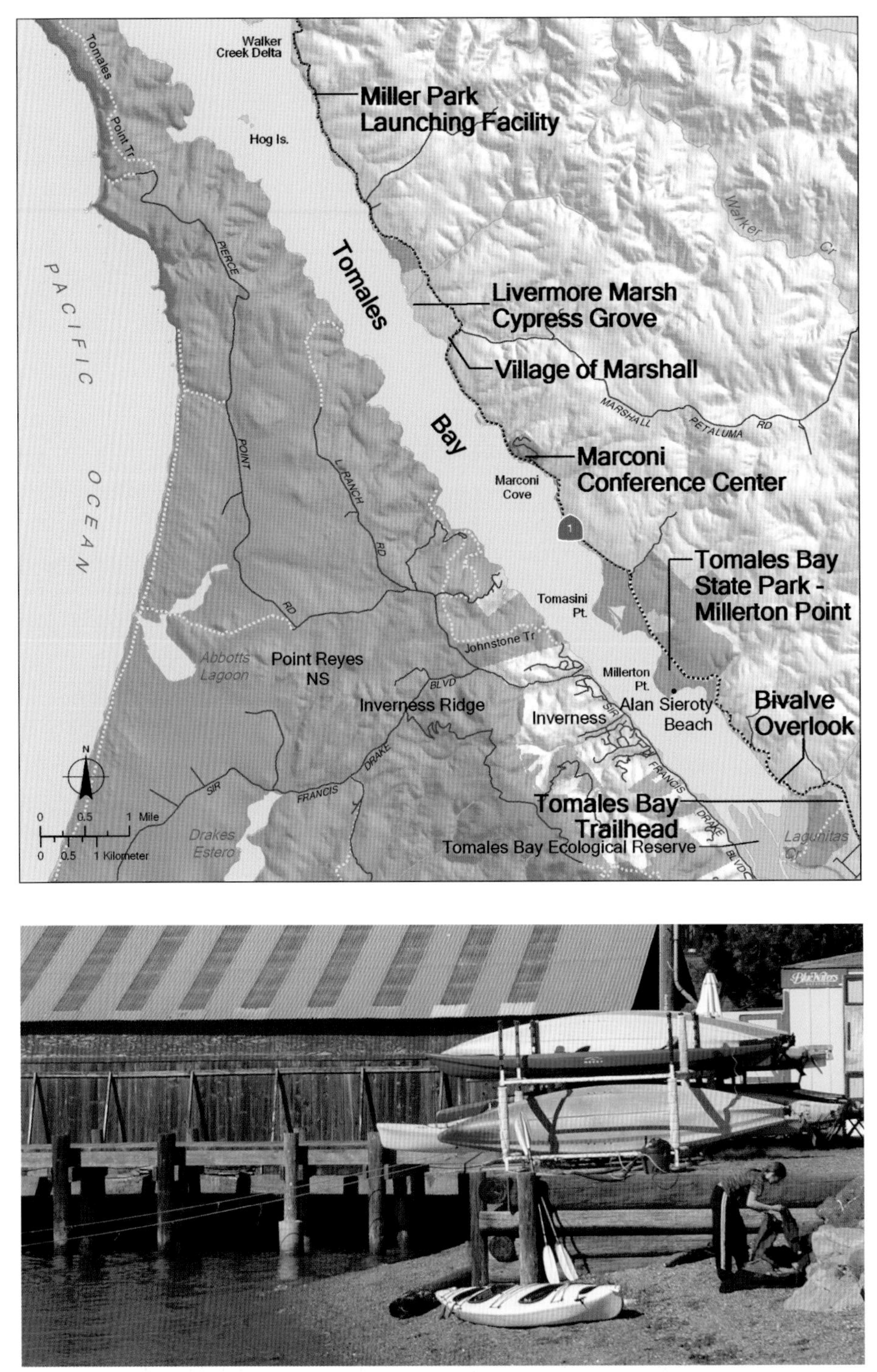

Village of Marshall

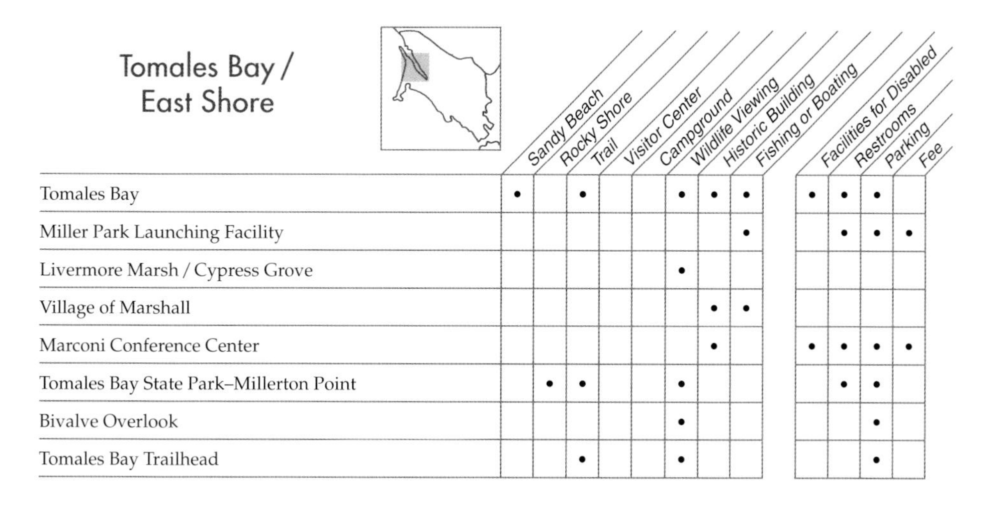

Tomales Bay / East Shore

	Sandy Beach	Rocky Shore	Trail	Visitor Center	Campground	Wildlife Viewing	Historic Building	Fishing or Boating	Facilities for Disabled	Restrooms	Parking	Fee
Tomales Bay	•		•			•	•	•	•	•	•	
Miller Park Launching Facility								•		•	•	•
Livermore Marsh / Cypress Grove						•						
Village of Marshall							•	•				
Marconi Conference Center							•		•	•	•	•
Tomales Bay State Park–Millerton Point		•	•			•				•	•	
Bivalve Overlook						•					•	
Tomales Bay Trailhead			•			•					•	

TOMALES BAY: *N.W. of Point Reyes Station.* When Spaniard Sebastián Vizcaíno sailed homeward from Cape Mendocino on a voyage of exploration in 1603, he mistook the mouth of Tomales Bay for the outlet of a great river and named it Rio Grande de San Sebastián. It was not until 1793 that another Spaniard, Captain Juan Matute, scouted the interior of the bay, where he was greeted by peaceful Miwok Indians who lived in shoreline settlements.

The bay is 13 miles long, one mile wide, and very shallow; the south end is less than ten feet deep, and wide expanses of mudflats are exposed at low tide. The bay is fed not by a great river, but rather by small streams, of which Walker/Keys Creek, Lagunitas Creek, and Olema Creek are the largest. The amount of fresh water entering the bay is modest, and the bay is not a true estuary, where substantial mixing of fresh water and salt water takes place. Localized estuarine conditions do exist at scattered points along the shoreline, however, and both salt and freshwater marshes are found here.

Nearly 100 species of water-associated birds have been identified at Tomales Bay. Shorebirds commonly seen include marbled godwits, black turnstones, and willets. Waterfowl seen include surf scoters, buffleheads, and ruddy ducks. Great blue herons and snowy egrets wade in the shallow waters of the bay in search of fish.

Tomales Bay

Marin County ranks second in California behind Humboldt Bay in the production of oysters, clams, and mussels. In Tomales Bay, seed oysters are imported and attached to large oyster shells suspended from racks or rafts, where they grow to harvestable size. Rows of racks are visible at low tide at several points along the east shore of the bay, and oysters are offered for sale along Highway One. In the extensive eelgrass beds in the shallow waters near the mouth of Tomales Bay, Pacific brant feed and Pacific herring deposit their eggs. The herring have supported an important commercial fishery since Gold Rush days in California. At first, the fish themselves were sold on the domestic market, but now the principal target of the fishery is the herring roe, which is exported to Japan as a delicacy. Herring boats cast their nets in the bay during the annual season beginning each January; the fishing season is brief in order to avoid depleting the herring population.

MILLER PARK LAUNCHING FACILITY: *Hwy. One, 3.6 mi. N. of Marshall.* This county park has a pier, jetty, paved parking, and concrete boat launch ramp. Planned improvements include wheelchair-accessible restrooms. The park offers fishing in Tomales Bay or picnicking in a grove of trees, overlooking Hog Island and Inverness Ridge, across the water. Fee for parking.

LIVERMORE MARSH / CYPRESS GROVE: *Marshall-Petaluma Rd., Hwy. One intersection.* The marsh is managed by Audubon Canyon Ranch, a land preservation and education organization. Access is reserved primarily for educational and scientific purposes and is by appointment only; for information, call: 415-663-8203.

VILLAGE OF MARSHALL: *Hwy. One, 10 mi. N. of Point Reyes Station.* In the 19th century, this was an important transfer point between ocean-going schooners and the North Pacific Coast Railroad, which ran along the east shore of Tomales Bay. Today Hwy. One occupies the old railroad right-of-way at Marshall. Commercial and recreational fishing continues in the area; boat repair and service facilities are located just south of town.

MARCONI CONFERENCE CENTER: *E. of Hwy. One, 1 mi. S. of Marshall.* A Miwok Indian village once occupied the cove at Marconi. The 28-room Marconi Hotel was built in 1913 by the Marconi Wireless Company in the style of an Italianate villa to house workers at what was the first transpacific radio transmission facility. Only a year before, the prompt rescue of survivors of the wreck of the *Titanic* in the North Atlantic Ocean had been made possible by Marconi's application of the new technology of wireless communication to marine transportation. Radio equipment at the Marconi site included a mile-long antenna held aloft by 270-foot-high towers; both international and ship-to-shore communications were conducted at the site. In 1929, Radio Corporation of America moved parts of the radio operation to a site on the Point Reyes Peninsula that was better suited to the new shortwave radio technology, and radio operations at Marconi ceased in 1939.

On the blufftop above the cove at Marconi, the site of the old Marconi Hotel is now a state park, used as a conference center. Future restoration of the hotel is planned, which will allow expansion of the conference facilities. Use for meetings and conferences is available for groups, by reservation only. Overnight accommodations are also offered, on a space-available basis, to travelers not attending a conference; reservations taken no more than three days in advance. Call: 415-663-9020.

TOMALES BAY STATE PARK–MILLERTON POINT: *Hwy. One, 5.4 mi. N. of Pt. Reyes Station.* This unit of Tomales Bay State Park juts into Tomales Bay opposite the village of Inverness. Walk to the point for views of the bay as well as across the water to forested Inverness Ridge. A small salt marsh borders the beach, known as Alan Sieroty Beach, near an osprey platform built to encourage nesting. Near the gravel parking area are picnic tables, grills, and restrooms that are wheelchair accessible with assistance. Dogs on leash allowed only on the fire road running on high ground from the parking lot. Tomasini Point, which is undeveloped state park property, is located one mile to the north of Millerton Point. An opening in the fence at Tomasini Point allows pedestrian access to a trail running down to the shore of the bay; no dogs allowed. For information, call: 415-669-1140.

BIVALVE OVERLOOK: *Hwy. One, 3.2 mi. N. of Pt. Reyes Station.* A fine view of the southern end of Tomales Bay is available from a gravel pull-out along Hwy. One. The site is known as "Bivalve" from the days when it was a stop on the North Pacific Coast Railroad, which operated along the shore of Tomales Bay. The railroad's original embankment can be seen cutting diagonally across the marsh.

TOMALES BAY TRAILHEAD: *Edge of Tomales Bay Ecological Reserve, 2.2 mi. N. of Pt. Reyes Station.* A trail leads from Hwy. One to the southern tip of Tomales Bay at the edge of the Tomales Bay Ecological Reserve. Small gravel parking lot.

Bivalve Overlook

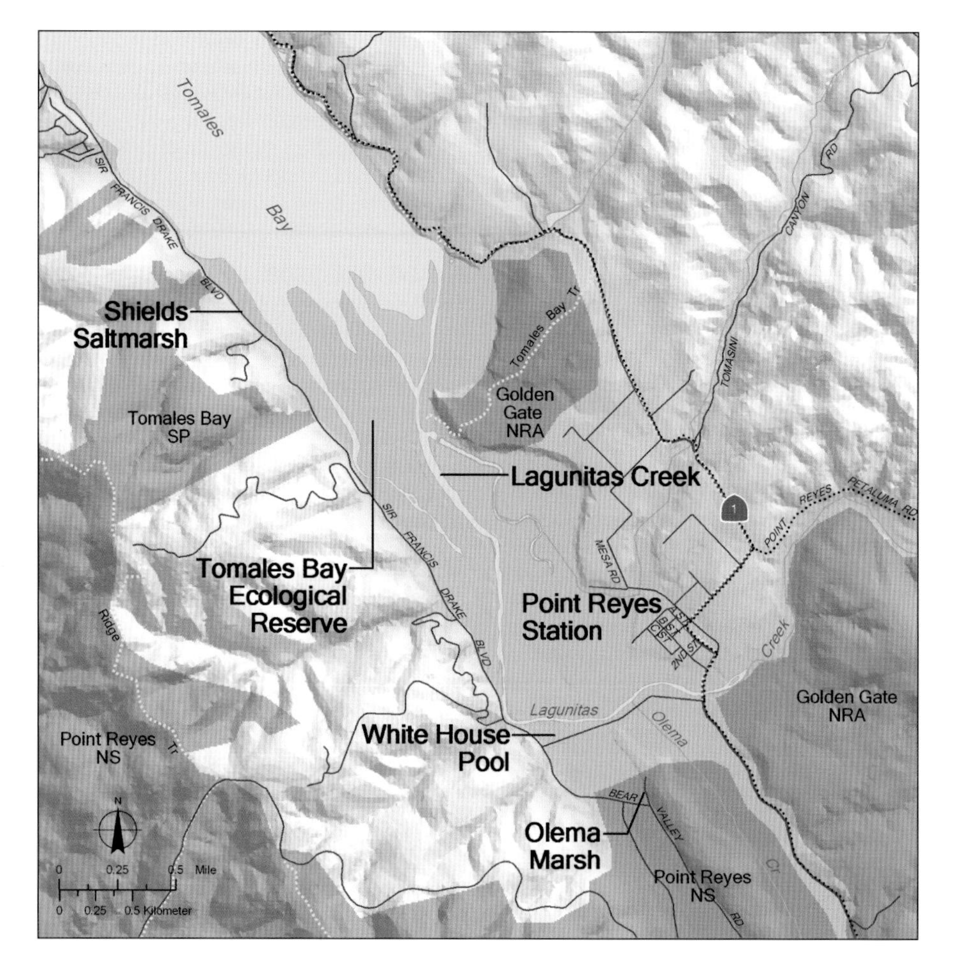

California buckeye tree in bloom

Point Reyes Station

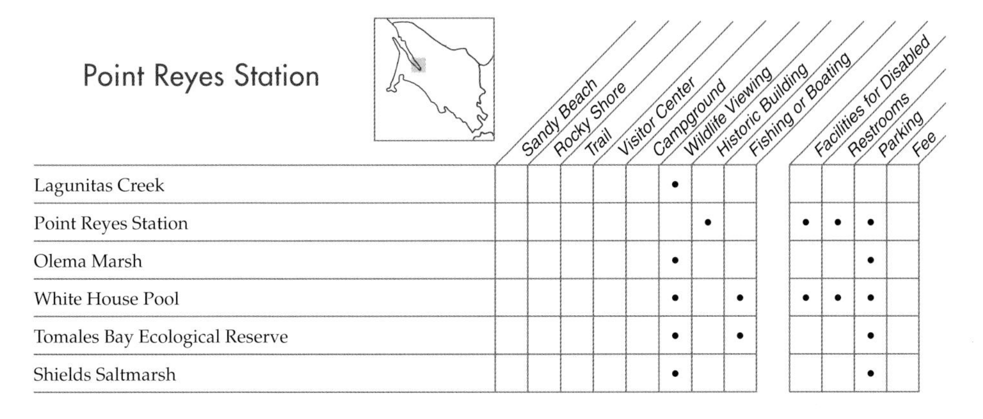

	Sandy Beach	Rocky Shore	Trail	Visitor Center	Campground	Wildlife Viewing	Historic Building	Fishing or Boating	Facilities for Disabled	Restrooms	Parking	Fee
Lagunitas Creek						•						
Point Reyes Station							•		•	•	•	
Olema Marsh						•					•	
White House Pool						•		•	•	•	•	
Tomales Bay Ecological Reserve						•		•			•	
Shields Saltmarsh						•					•	

LAGUNITAS CREEK: *Hwy. One at Pt. Reyes Station.* Lagunitas Creek, also called Paper Mill Creek, drains the northern slopes of Mount Tamalpais and flows through redwood forest and riparian woodland before entering the south end of Tomales Bay. Lagunitas Creek is the largest of the streams feeding Tomales Bay. Its significance for salmon and steelhead trout far outweighs its modest scale; some ten percent of the coho salmon in northern California streams swim up Lagunitas Creek. Fishery experts estimate that the annual number of salmon dropped below 100 fish by the 1980s, due to placement of dams that block fish passage and loss of gravelly bottom habitat. Restoration efforts have brought a dramatic increase in returning salmon, although the run is still much diminished from a century ago, when perhaps 5,000 fish traveled upstream annually to deposit eggs on the gravelly bottom of the creek. Lagunitas Creek is also habitat for the endangered California freshwater shrimp, which live only in a few creeks in Marin, Sonoma, and Napa Counties.

POINT REYES STATION: *Hwy. One at Pt. Reyes–Petaluma Rd.* Point Reyes Station was originally a stop on the North Pacific Coast Railroad, which linked Sausalito in southern Marin County to northern Marin and Sonoma Counties. In 1875 the line reached the town of Tomales, and by 1877 it extended north to the lumber mills on the Russian River. Southeast of Point Reyes Station, the tracks (originally narrow-gauge) ran through dense redwood forest along Lagunitas Creek; north of town the tracks ran mostly on a trestle along the eastern shore of Tomales Bay. The railroad hauled lumber, farm produce, commuters, and weekend sightseers on trips that were not without their mishaps; more than one locomotive left the tracks and landed in the creek or upside down in the bay mud.

Point Reyes Station

Lagunitas Creek

The townsite of Point Reyes Station was laid out in 1883, and the town grew up around the railroad station. Many early immigrants came from northern Italy or Italian-speaking Switzerland, and the names of local geographical features and dairies reflect this heritage. Buildings on the south side of "A" Street date from around the turn of the century, including the red brick Mission Revival–style Grandi Building, now in disrepair, at the corner of 2nd Street. After train service to Point Reyes Station ceased in 1933, commercial structures were built on the north side of "A" Street on the former railroad right-of-way; the old depot now serves as the post office. Other buildings in town are named for their original functions: the Old Creamery Building, the Livery Stable, the Hay Barn. Today, Point Reyes Station is the commercial center of west Marin, with a bank, grocery store, and hardware store, as well as restaurants, galleries, and gift shops.

OLEMA MARSH: *Sir Francis Drake Blvd. and Bear Valley Rd. intersection, S.E. corner.* This 40-acre wetland, the largest of the freshwater marshes around the edge of Tomales Bay, is formed by Olema Creek upstream from its junction with Lagunitas Creek. The marsh is bordered by dense willow thickets, and public entry is not permitted, but many birds may be seen from the roadside along Sir Francis Drake Blvd. or from a small parking area off Bear Valley Rd. Over 150 species of birds have been observed, including the yellow warbler, the Virginia rail, and the sora, an uncommon relative of the rail. Kingfishers and red-winged blackbirds are common among the cattails, and during fall and winter migratory water birds including American coots use the marsh. Western pond turtles are found in abundance. The marsh is part of the Point Reyes National Seashore; no facilities.

WHITE HOUSE POOL: *Sir Francis Drake Blvd. and Bear Valley Rd. intersection, N.E. corner.* White House Pool is a public fishing access point near the junction of Lagunitas and Olema Creeks. Good birding opportunities among the willow thickets along the creeks.

TOMALES BAY ECOLOGICAL RESERVE: *Southern portion of Tomales Bay.* A large wetland and marsh complex is found at the shallow southern end of Tomales Bay where Lagunitas Creek enters the bay. Some 500 acres of wetlands are included in the Ecological Reserve managed by the Department of Fish and Game. Waterfowl hunting and fishing are permitted in the reserve with a valid hunting or fishing license; for more information, call: 707-944-5500. The land area is closed from March 1 through June 30.

Some of the historic marsh adjacent to Tomales Bay has been diked off and used as grazing land as part of the historic Giacomini Ranch. Since acquisition of the ranch in 2000 by the National Park Service, planning has proceeded for eventual habitat restoration of much of the diked area. Meanwhile, salt-tolerant plants such as pickleweed and saltgrass grow in areas that are subject to tidal influence, while freshwater wetlands exist near seeps and small streams. Altogether, the wetlands at the southern end of Tomales Bay exhibit a tremendous diversity of birds as well as other uncommon and endangered creatures such as the California red-legged frog and the tidewater goby.

SHIELDS SALTMARSH: *Sir Francis Drake Blvd., 3 mi. N. of Pt. Reyes Station.* Views of the marsh are available from an overlook with a bench and interpretive displays. Bring field glasses to spot distant birds or wildlife across the grassy plain. Limited parking in roadside pull-out. The area is managed by Audubon Canyon Ranch, an organization devoted to preservation, research, and education with wildlife reserves near Bolinas Lagoon and elsewhere; for information, call: 415-663-8203.

Tomales Bay Ecological Reserve

Bishop Pine Forest

THE BISHOP PINE is a "closed–cone" pine, which means that the cones only open after fire or on hot days. Otherwise, they are closed tight, and the seeds are not released. Because of this characteristic, stands of bishop pine forest are characteristically even-aged, originating after fire. You can see how bishop pines have regenerated in large numbers at Point Reyes, following the Mount Vision fire of 1995. A fire-free period of 80 or more years might cause the trees to succumb to disease and die without reproducing. Bishop pine forest can be explored on the northern end of Inverness Ridge in the Point Reyes National Seashore, in Tomales Bay State Park, and at Salt Point State Park.

Bishop pine

The **bishop pine** *(Pinus muricata)* can be identified by looking at the number of needles in each fascicle, or bundle. This is California's only pine with just two needles per fascicle. The species thrives in a maritime climate. Bishop pine forest generally occupies headlands and low hills from near sea level to an elevation of approximately 1,300 feet, usually within seven to eight miles of the ocean. Fog and fog drip during the dry summer provide critical moisture that supplements winter precipitation.

Pacific rhododendron

The beautiful native **Pacific rhododendron** *(Rhododendron macrophyllum)* reaches a height of 10 to 15 feet and has wonderful rosy pink flowers and large, leathery evergreen leaves. The rhododendron grows in areas with high rainfall on somewhat acid soils (that is, with low soil pH), often beneath an overstory of conifers near the ocean. To see the grand rhododendron, visit Salt Point State Park, Sea Ranch trails, or Kruse Rhododendron State Reserve. Be sure to go in April and May for the peak bloom.

Bolete

B**oletes** are mushrooms with tubes, instead of gills, on the underside of the cap. The most famous member of the family is the *Boletus edulis*, often called the King of Mushrooms for its impressive size and flavor. These mushrooms have a mycorrhizal association with the roots of bishop pines, meaning that the boletes aid the trees by helping them to absorb nutrients, while receiving sugars from the trees' roots. If you go out to the woods after a fall rainstorm, look for the king bolete hiding in the pine needle duff.

Salal *(Gaultheria shallon)* is a low shrub that grows along the coast in the shade of the woods. Look for the broad, glossy, bright green leaves, which have a sharp point at the tip. Salal is a lovely groundcover, growing not only in bishop pine forest, but also in redwood and Douglas-fir forest. Loose clusters of small, white or pinkish urn-shaped blossoms appear between March and June. The flowers are followed by black edible berries that birds love. The dark, juicy berries were a plentiful and important fruit for Native Americans living along the coast. The berries were eaten fresh and also dried into cakes for later enjoyment.

Salal

The **pygmy nuthatch** *(Sitta pygmaea)* is the smallest of the North American nuthatches at 3.75 to 4.5 inches in length. It is a small, social, and noisy little bird and is one of only two nuthatch species in the world known to have helpers at the nest; offspring from previous years help their parents raise the young. Like all nuthatches, they work their way up and down a tree trunk searching for grubs and insects in the bark. These birds often gather together in small flocks and cuddle together at night for warmth. Listen for their constant twittering when you enter the bishop pine forest, and search for the pygmy nuthatches moving straight up or down the trunk of the tree.

Pygmy nuthatch

Chickadees personify inquisitiveness. They investigate every hole, crack, and cranny in the forest, and all the while they are chatting with each other. Chickadees are aptly named after their distinctive call: "tseck-a-dee-dee." **Chestnut-backed chickadees** *(Poecile rufescens)* are easily recognizable by their beautiful chestnut cape. These little acrobats are fairly common in coastal bishop pine forests, but sometimes they are invisible due to the density of the greenery. Their cheery call may be the only indication you will have of their presence.

Chestnut-backed chickadee

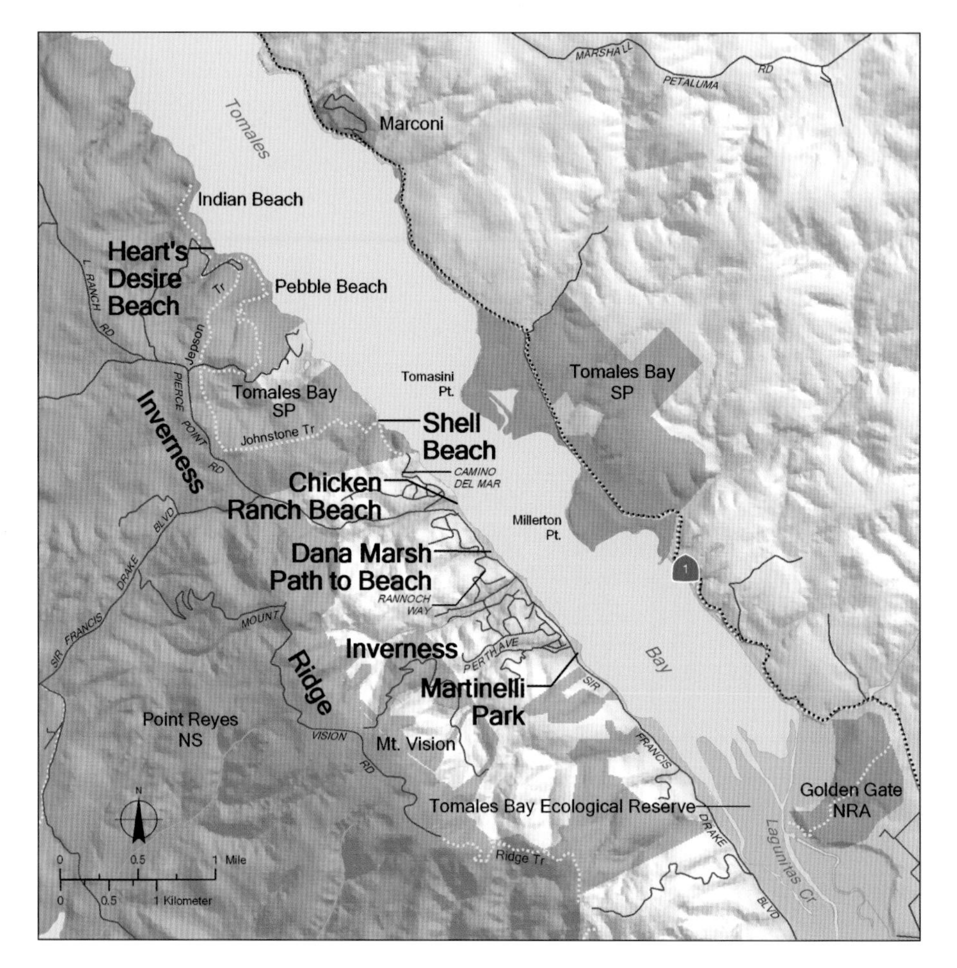

Tomales Bay near Inverness

Tomales Bay / West Shore

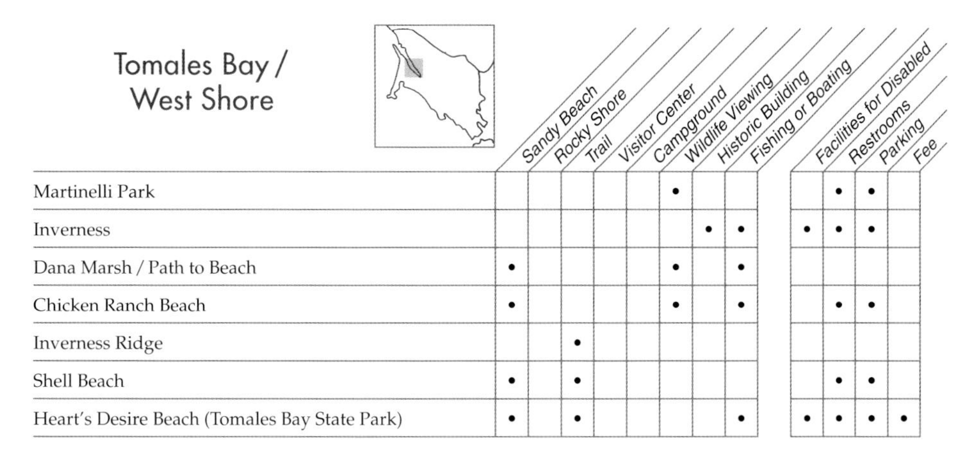

	Sandy Beach	Rocky Shore	Trail	Visitor Center	Campground	Wildlife Viewing	Historic Building	Fishing or Boating	Facilities for Disabled	Restrooms	Parking	Fee
Martinelli Park						•				•	•	
Inverness							•	•	•	•	•	
Dana Marsh / Path to Beach	•					•		•				
Chicken Ranch Beach	•					•		•		•	•	
Inverness Ridge			•									
Shell Beach	•		•							•	•	
Heart's Desire Beach (Tomales Bay State Park)	•		•					•	•	•	•	•

MARTINELLI PARK: *Sir Francis Drake Blvd., Inverness.* A small park adjacent to Tomales Bay with fine views of the water and forested ridge.

INVERNESS: *Sir Francis Drake Blvd., 3.1 mi. N. of Bear Valley Road.* Originally a summer resort on the wooded slopes of Inverness Ridge overlooking the waters of Tomales Bay, Inverness is now both a year-round and seasonal community. The village offers several points of public access to the shore of the bay, along with overnight accommodations and a handful of eating establishments.

DANA MARSH / PATH TO BEACH: *Sir Francis Drake Blvd., .25 mi. N. of Inverness.* Dana Marsh is a small restored wetland adjacent to Tomales Bay. There is an unpaved path to a small bay beach, accessible from Sir Francis Drake Blvd. at the foot of Rannoch Way. Adjacent private property; do not trespass.

CHICKEN RANCH BEACH: *Sir Francis Drake Blvd., .5 mi. N. of Inverness.* This small sandy beach on Tomales Bay is reached via a bridge over a creek. There is limited street parking; chemical toilet, not wheelchair accessible.

INVERNESS RIDGE: *W. of Tomales Bay.* The steep, forested Inverness Ridge that runs parallel to the drowned rift valley of the San Andreas Fault beneath Tomales Bay presents a sharp contrast to the rolling grassy hills on the east side of the bay. The highest point on Inverness Ridge is Mount Wittenberg, elevation 1,407 feet, located southwest of Point Reyes Station. A dense canopy of Douglas-fir covers the southern end of Inverness Ridge, while bishop pine grows on the northern end.

Inverness

Forest near Shell Beach

Near the point where Sir Francis Drake Blvd. crosses the crest of the ridge is a dwarf bishop pine forest where severely acidic soils cause stunting; the trees here, however, are not as dwarfed as those in the Mendocino pygmy forest. Found in association with the bishop pine forest on Inverness Ridge is the rare Mount Vision ceanothus; this is the only known location for this species of plant. The forest on Inverness Ridge harbors the southernmost colony in the coastal range of the native mountain beaver, an elusive, cat-sized, nocturnal rodent. Ospreys nest in snags on the ridge, and blue Steller's jays are abundant. Pygmy nuthatches may be seen among the trees, and king bolete mushrooms may be found here. Much of the bay side of Inverness Ridge is privately owned; Point Reyes National Seashore and units of Tomales Bay State Park compose the rest.

An undeveloped unit of Tomales Bay State Park is located on the ridge above Inverness. Starting from the top of Perth Ave., there are hiking trails through the forest that connect to Mount Vision in the Point Reyes National Seashore. Check at the Heart's Desire park headquarters for more information, or call: 415-669-1140.

SHELL BEACH: *End of Camino del Mar, off Sir Francis Drake Blvd., 1 mi. N. of Inverness.* This small unit of Tomales Bay State Park includes a sandy curve of bay beach at the base of a hillside cloaked with dense bishop pine forest. The quarter-mile-long trail from the parking lot winds downhill through trees festooned with moss and surrounded by ferns and huckleberry. Bright orange lichens cover the north-facing boulders along the beach, and heart cockles inhabit the sand. Day use only; dogs are not allowed on the beach. From Shell Beach, the 4.3-mile-long Johnstone Trail leads north through the forest to the Heart's Desire Beach unit of Tomales Bay State Park.

HEART'S DESIRE BEACH: *Pierce Point Rd., 2 mi. N. of Inverness.* This is the main facility of Tomales Bay State Park. Heart's Desire Beach is popular for picnicking, clamming, and boating. The beach is sandy and gently sloping, and the shallow bay waters invite swimming. Nearby are picnic tables and restrooms with running water. A separate larger picnic area including several group sites and offering fine filtered views of Tomales Bay through the pine trees is located at the vista point, a few hundred yards south of Heart's Desire. Dogs on leash are allowed in the vista point picnic area, but not on trails or at the beach. No camping allowed.

Several short to moderate hiking trails start at Heart's Desire. A half-mile-long nature trail starts at the north side of the cove and leads north through bishop pine and oak forest to Indian Beach. Along the path are interpretive signs explaining the uses of plants such as coffeeberry, bracken fern, and poison oak by the Coast Miwok people who once lived in small villages along this shoreline. On Indian Beach are replicas of Miwok houses. From the south side of Heart's Desire Beach, the Johnstone Trail leads 4.3 miles to Shell Beach, along the way passing small secluded Pebble Beach and the Jepson Trail, which winds through a particularly fine stand of virgin bishop pine. The forest along the trails is fragrant with California bay and pine, and huckleberry and thimbleberry crowd the path. Animals that inhabit the area include foxes, raccoons, badgers, weasels, squirrels, rabbits, deer, bobcats, and gopher and garter snakes. Birds include chestnut-backed chickadees, goldfinches, and woodpeckers, and monarch butterflies may be found here.

Heart's Desire Beach

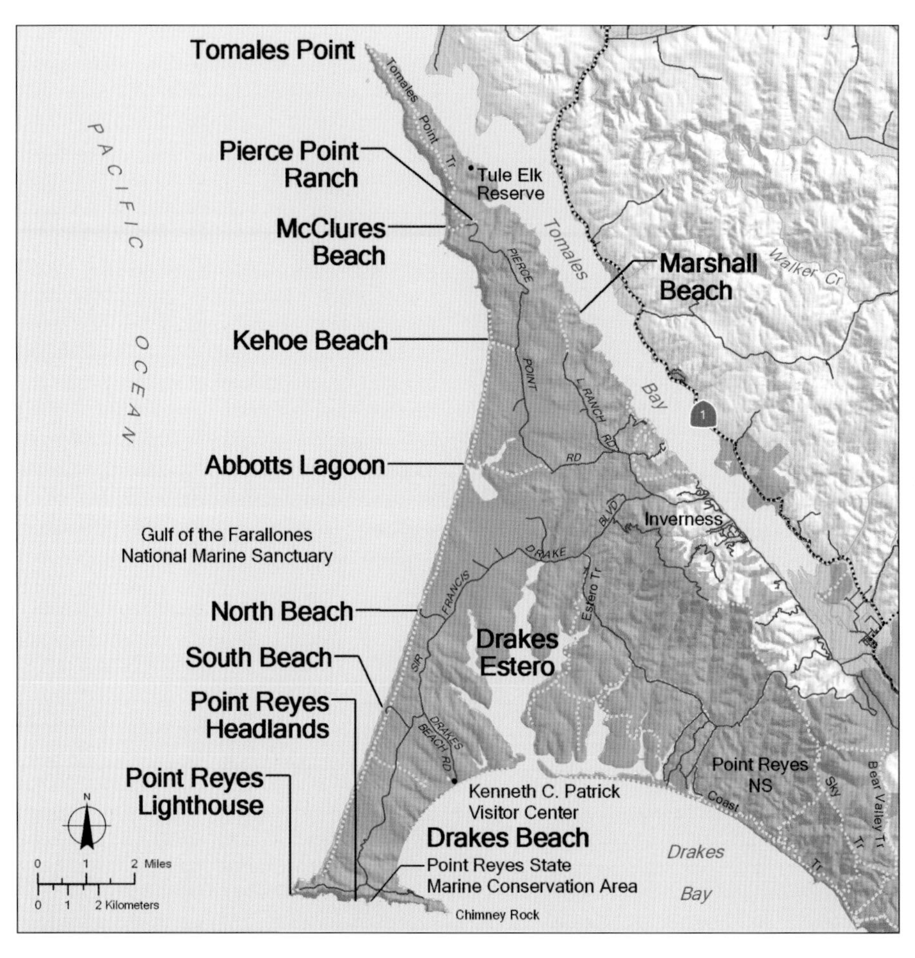

View south from McClures Beach

Tomales Point to Point Reyes Lighthouse

	Sandy Beach	Rocky Shore	Trail	Visitor Center	Campground	Wildlife Viewing	Historic Building	Fishing or Boating	Facilities for Disabled	Restrooms	Parking	Fee
Marshall Beach	•		•							•	•	
Tomales Point / Pierce Point Ranch		•	•	•		•	•		•	•	•	
McClures Beach	•	•	•			•			•	•	•	
Kehoe Beach	•		•			•			•	•	•	
Abbotts Lagoon	•		•			•		•	•	•	•	
Drakes Estero			•			•			•	•	•	
North Beach and South Beach	•								•	•	•	
Point Reyes Headlands		•	•			•	•		•	•	•	
Point Reyes Lighthouse		•	•	•		•	•		•	•	•	
Drakes Beach	•			•					•	•	•	

MARSHALL BEACH: *"L" Ranch Rd., off Pierce Point Rd., 1.3 mi. N. of Sir Francis Drake Blvd. junction.* A steep trail leads to this Tomales Bay beach, which is part of Point Reyes National Seashore. Small dirt parking lot; restrooms available.

TOMALES POINT / PIERCE POINT RANCH: *End of Pierce Point Rd., 7 mi. N. of Inverness.* Pierce Point Rd. winds north through the hills from Sir Francis Drake Blvd., past historic dairy ranches established in the 1860s. The road climbs in elevation toward the end, offering water views that alternate between Tomales Bay and the ocean. At the end of the road is a cluster of buildings from the historic Pierce Point Ranch, preserved by the National Park Service but no longer in operation as a dairy. In the 19th century, each cow in the Pierce Ranch's herd typically produced one pound of butter per day, for shipment via schooner to San Francisco and other points in California. Visitor facilities at Pierce Ranch include a self-guided trail with

Pierce Point Ranch

View south, Tomales Point to Point Reyes Headland

interpretive signs about the history of the ranch and the buildings.

The northern four miles of Tomales Point, including the Pierce Point Ranch buildings, are within the National Seashore's Tule Elk Reserve. The reserve was created in 1978 to restore to the elk some of the habitat originally theirs before the introduction of dairy farming on the peninsula. Some 3,000 acres are available for the elk to roam; they may be seen year-round. Tule elk share the Point Reyes Peninsula with native black-tailed deer, along with introduced fallow deer and axis deer. A three-mile-long trail leads from the parking area at Pierce Ranch through grasslands inhabited by badgers and California voles to the north end of Tomales Point, with striking views of Tomales and Bodega Bays and the ocean; no beach access.

MCCLURES BEACH: *End of Pierce Point Rd., 9 mi. N. of Inverness.* Exposed to the full force of storms and waves rolling in off the open ocean, this is one of the most dramatic sections of the Marin coast; it is also extremely unsafe for swimming. Backed by steep bluffs, the beach is a broad sandy crescent; portions are covered with rocks smoothed by the surf. A smaller sandy beach to the south is reachable through a cleft in the rocks, at extreme low tides.

Cormorants, brown pelicans, and common murres may often be seen on the granite sea stacks off the south end of the beach, and black oystercatchers may be spotted on the offshore rocks. Surf scoters ride the ocean waves. Red-tailed hawks sometimes hang nearly motionless, riding the wind, above the high bluff backing the beach. Abundant intertidal life, including giant green anemones and ochre sea stars, is found among the rocks on the beach. A steep half-mile-long trail used by hikers and equestrians leads from the parking lot to the beach; dogs are not permitted. In summer, look for sticky monkeyflowers blooming along the trail.

KEHOE BEACH: *Trail off Pierce Point Rd., S. of McClures Beach.* A half-mile-long trail to the beach runs beside Kehoe Marsh, a freshwater pond that supports migratory waterfowl and year-round bird residents. Spring wildflowers include California poppies, baby blue eyes, wild hollyhock, phacelia, and cream cups. Most of the trail is wheelchair negotiable. Restrooms available; parking along road shoulder. Kehoe Beach is nesting habitat for western snowy plovers, a threatened bird species. From about mid-March to early September, the beaches remain open to visitors, although breeding areas may be roped off and dogs are prohibited from near Kehoe Beach south to near the Point Reyes Beach North parking lot; leashed dogs allowed at other times.

ABBOTTS LAGOON: *Off Pierce Point Rd., 3.4 mi. N.W. of Sir Francis Drake Blvd.* An easy 1.5-mile-long trail leads from the parking lot to the lagoon, where hikers may continue to the sand dunes and the beach. The upper part of the lagoon nearest the road is freshwater, and shorebirds that are uncommon elsewhere, such as Baird's sandpipers and semipalmated sandpipers, are found in fall and winter around the lagoon. Western grebes and pied-billed grebes are frequently sighted; Caspian terns may be seen in the summer. Look also for common yellowthroats. Closer to the ocean, the lagoon is brackish, since the sandbar at the mouth is breached occasionally by winter storm waves and high tides. Canoeing is permitted in the lagoon.

DRAKES ESTERO: *Estero Trail off Sir Francis Drake Blvd., 4 mi. W. of Inverness.* This is the largest of the saltwater lagoons along the Marin coast. The shallow estero receives freshwater flow only from minor streams, and broad mudflats are exposed at low tide. A mile-long trail leads downhill from the Estero Trailhead to the water's edge. Giant geoduck clams live in the mudflats, along with abundant phoronids, or "stringworms." The rocky intertidal area of the estero is inhabited by limpets, sea anemones, ochre sea stars, and several varieties of crabs. The estero harbors rays and leopard sharks and also serves as a nursery for fish such as lingcod. The largest harbor seal breeding colony at Point Reyes can be seen at Drakes Estero. Birding is best during the fall migration and winter layover periods; look for shorebirds, hawks, and osprey.

Drakes Estero

Chimney Rock, Point Reyes Headlands

NORTH BEACH AND SOUTH BEACH: *Off Sir Francis Drake Blvd.* This 12-mile-long beach is sandy, windy, and often wild. The surf here is undiminished by offshore reefs or islands, and the horizon seems extra-large. The cold waters do not invite swimming, nor are the currents safe for doing so. Huge breakers pound this windward shore; when strolling, look out for hazardous sneaker waves.

Two main access areas are served by parking lots. In the wide dune field behind the beach are stands of endangered Point Reyes lupine, along with colorful, non-native ice plant. Dogs on leash and campfires allowed on the beach. North Beach supports nesting snowy plovers during spring and summer, when dogs are prohibited from near the North Beach parking area to near Kehoe Beach; leashed dogs allowed at other times.

POINT REYES HEADLANDS: *End of Sir Francis Drake Blvd., 15 mi. S.W. of Inverness.* The headlands extending from the Point Reyes Lighthouse east to Chimney Rock form a massive granite promontory up to 600 feet high. The submarine topography is similarly steep: depths of 150 feet and more are found close to shore. Along the shoreline at the base of the headlands are sea stacks, smooth granite rocks, wave-carved sea caves, and coarse-sand pocket beaches. The variety of surfaces makes the area attractive to many different kinds of clinging plants and invertebrates. The subtidal zone is home to giant green anemones, rose anemones, and red sea urchins. Red abalone fasten to undersea rocks, where they feed on kelp.

Censuses have shown that the headlands area is the principal location in the county for breeding birds. Particularly numerous are common murres, thousands of which nest in the summer on the rocks almost directly below the lighthouse, and Brandt's and pelagic cormorants. California sea lions occupy offshore rocks, and there is a breeding colony of Steller sea lions in one of the coves. Since 1981, an elephant seal breeding colony near Chimney Rock has grown to some 2,000 animals; an overlook east of the lighthouse provides views of the animals as they come ashore during the winter breeding period, summer molting period, and at other times. In spring, look for blue-eyed grass and sky lupine blooming in the coastal prairie. The headlands from the lighthouse east to Chimney Rock are part of the Point Reyes State Marine Conservation Area, where no plants or animals may be taken. The waters surrounding the Point Reyes Peninsula are also part of the Gulf of the Farallones National Marine Sanctuary.

POINT REYES LIGHTHOUSE: *End of Sir Francis Drake Blvd., 15 mi. S.W. of Inverness.* The tip of the Point Reyes Peninsula, thrusting seaward some 20 miles from the main shoreline and subject to persistent winds and fogs, has been the site of numerous shipwrecks. The earliest recorded was in 1595 when the *San Agustin,* a Spanish galleon loaded with cargo from the Philippines and captained by Sebastián Rodríguez Cermeño, went aground in what is now called Drakes Bay. In 1870 the Point Reyes Lighthouse was built on the bluff face 294 feet above sea level. An automated light and foghorn were installed in 1975, although the original lighthouse and Fresnel lens remain. The point receives up to 2,700 hours of fog annually, but it gets relatively little rainfall, less than 20 inches per year.

Gray whales pass near Point Reyes on their annual winter migration south to Baja California, where calving takes place in warm lagoons. Point Reyes is a favorite whale-watching spot between January and early May; prime viewing months are mid-January (southbound animals) and mid-March (northbound), with the last cows and calves heading north from April to early May.

Parking at the lighthouse is limited; during peak whale-watching months, the National Park Service operates a shuttle from the Drakes Beach parking area, weekends and holidays only. From the parking area, a half-mile-long paved path leads to the Point Reyes Light Visitor Center, view platform, and 300-step stairway leading down to the lighthouse. Special parking and drop-off areas and arrangements for visitors with limited mobility are available. Interpretive walks are led through the lighthouse; Visitor Center open Thursday through Monday, 10 AM–4:30 PM; stairs closed when wind and weather conditions warrant. Call: 415-669-1534.

DRAKES BEACH: *Drakes Beach Rd., off Sir Francis Drake Blvd.* A popular broad sandy beach with chalk white cliffs; swimming and beach fires permitted. Moon jellies and by-the-wind-sailors sometimes wash ashore here. The Kenneth C. Patrick Visitor Center has interpretive displays and ranger-led walks. Open daily except Wednesday and Thursday during summer months, and open weekends and holidays all year, 10 AM–5 PM. Facilities include snack bar, picnic tables, and restrooms. Call: 415-669-1250.

Near the visitor center is a monument to English explorer Francis Drake, thought by some historians to have landed near here in 1579. The white cliffs and summer fogs match the descriptions brought back to England by Drake's expedition.

Point Reyes Lighthouse

Point Reyes from Double Point, Marin County

© 2004, Tom Killion

Beaches

BEACHES form at the dynamic junction between water, air, and land. While beaches are part of everyone's idea of the California coast, less than one-third of the shoreline between San Francisco and the Oregon border supports a beach. Beaches are in a constant process of accretion and erosion as they respond to changes in wave energy from mild summer conditions to stormy, high-energy winter conditions. Summer beaches tend to be wide and gently sloped, while in winter they become narrower and more steeply sloped. Large volumes of sand can be carried from the dry summer beach to submerged winter berms and then back to the beach.

Beach material begins as eroded continental material—sand, gravel, cobbles, and boulders—most of it washed to sea by streams and rivers. The Eel River has the highest level of sediment delivery of all rivers in California, annually carrying approximately 3.7 million cubic yards of sediment to the coast, or more than half that contributed by all northern California rivers combined. Beaches near the mouths of rivers can fluctuate greatly from year to year, widening after flood events deliver large amounts of material, and then narrowing gradually as waves and currents rearrange the sand and distribute it along the coast.

Beaches are classified according to sand size, foreshore slope, and the size of onshore berms. Steeper beaches generally have larger-grained sands, while more gently sloping beaches have a finer-grain composition. Beaches along the North Coast can be made of glass or sand, cobbles or boulders. Many Sonoma County beaches have a noticeable greenish color, due to the serpentine rock of the area, while in both Humboldt and Marin Counties there are beaches named Agate Beach, for the pieces of agate and jade that can be found mixed with the sand.

Limantour Beach, Marin County

The rivers and streams of the North Coast also carry logs from watersheds to the sea, adding to the visual character of the beaches. As waves run up onto the beach face and bend around large pieces of driftwood, they often leave fan-shaped formations of heavier rocks and pebbles immediately landward of the driftwood, in its lee. Many North Coast beaches have driftwood; Pelican Beach in Del Norte County is one of the more regularly photographed driftwood beaches.

Beaches are very harsh environments for living organisms. Plants and animals that inhabit the beach are adapted to changing conditions such as submersion, exposure to sand and wind, and changes in the form and size of the beach itself. Even though beaches appear relatively barren, they harbor a unique community of creatures. Beach fauna are generally low in species diversity, but high in abundance. Most of these animals live buried in the sand and are extremely mobile, making them difficult to study. Some species you may observe on the beach are not beach-dwellers at all, but instead are marine plants or animals, such as kelp or jellyfishes, that are washed up on shore after a significant storm event.

Razor clam

Burrowed in the wet sand lives a bivalve mullosk known as the **razor clam** *(Siliqua patula)*. It has a long, thin, brittle shell shaped much like an old-fashioned straight razor. Each year, a razor clam produces a distinct growth ring, making its age easy to determine. Razor clams in California have an average life span of nine years. They are ideally suited for digging, since their smooth shells offer little resistance to the surrounding sand. Razor clams have a neck (called a siphon) sticking out one end of their shell and a foot sticking out the other. They eat plankton and minute plants filtered through their siphon from the surrounding seawater. This species is harvested extensively, both commercially and recreationally.

Sand crab

A fully grown **sand crab** *(Emerita analoga)* is only about one inch long and looks like a tiny lobster. It burrows just beneath the surface of the wet sand. The sand crab does everything backwards; it swims and walks backwards, it even digs into the sand backwards and enters tail first. To feed, sand crabs face up the slope of the beach and extend their feathery antennae to trap plankton and detritus from the wave backwash as it flows back down the slope. Sand crabs are important organisms in the sandy beach ecosystem, providing food for shorebirds, including plovers, sandpipers, sanderlings, and willets.

The **marbled godwit** *(Limosa fedoa)* is a long-legged, long-necked cinnamon-colored shorebird with a slightly recurved (turned-up) bill. The distinctive bill is pink at the base and black on the tip. A marbled godwit finds its food by touch, walking along the shore with its sensitive bill thrust forward, feeling for crustaceans and other invertebrates. You may see one stop suddenly and plunge its bill into the sand to find dinner. This species breeds in the northern Great Plains and winters on the coast of California. Look for them in groups along the shore between August and April.

Marbled godwit

The adult **western gull** *(Larus occidentalis)* is one of the largest and darkest of the Pacific gulls. It has pink legs and a bright red dot at the tip of the yellow bill. Western gulls are fairly aggressive birds and are known to challenge other birds for food, forcing them to regurgitate their catch. Sometimes they follow marine mammals and feed on squid and fish that are forced to the surface by the mammals. Their diet includes clams, crabs, urchins, carrion (dead, decaying flesh), and, unfortunately, garbage.

Western gull

The **willet** *(Catoptrophorus semipalmatus)*, a large member of the sandpiper family, looks fairly nondescript from the ground, but once it takes flight it is unmistakable. It has a striking black-and-white wing pattern and a piercing raucous call that distinguishes it from similar species. It is named for its loud territorial call, which sounds like "*pill will willet, pill will willet!*" Willets gather in loose congregations at the water's edge. If any of the birds in the group are startled and take flight, the rest will immediately follow. The heavy bill and the bluish-gray legs of the willet help to distinguish it from the similar-looking species of shorebird, the greater yellowlegs.

Willet

Western snowy plover

The **western snowy plover** *(Charadrius alexandrinus nivosus)* is a small shorebird. With a white belly, a pale brown back, and distinct dark patches on either side of the neck, it blends in well with the beach and can be difficult to see. Snowy plovers breed on beaches between early March and September. The nests, which are started by the male during courtship, are simple depressions in the sand, located in open flat areas. Plovers forage near the water's edge searching for invertebrates such as crabs, flies, and beetles. Recreational activities, free-running dogs, beach cleaning, construction of breakwaters and jetties, dune stabilization, and off-road vehicles adversely affect snowy plover breeding sites. At present, the snowy plover is on the endangered species list, due primarily to predation and human disturbances to habitat.

Moon jelly

The **moon jelly** *(Aurelia labiata)* is named for its translucent moon-like umbrella-shaped body. This jellyfish has short, fine fringe-like tentacles that help direct food into its mouth. The jelly feeds mostly by trapping microscopic plankton in a film of mucus that flows over the surface of the bell (the large dome-like gelatinous body) and is picked off as it reaches the edges by the thick mouth tentacles underneath. Hanging from the center of the bell, the moon jelly has lacy appendages covered with stinging cells. Although moon jellies may sting their prey, they pose little threat to humans. Look carefully at the next moon jelly that you find washed upon the shore. Sometimes a purplish clover pattern is visible inside the jelly's bell; these are the moon jelly's reproductive organs.

By-the-wind-sailors

By-the-wind-sailors *(Velella velella)* are small blue pelagic jellyfish, about three inches across, that generally occur floating on the surface of offshore waters. They have a sail that is angled at 45 degrees to their main body axis, so that animals drift with the wind. Right-handed sailors occur off California, where prevailing northerly winds usually hold the population offshore. However, occasionally prolonged westerly or southerly winds can drive the little jellyfish up on beaches in huge numbers.

Sand dollars *(Dendraster excentricus)* usually live in subtidal areas beyond the surf, to depths of up to 90 feet. They form dense congregations, sometimes exceeding 625 animals per square yard of sandy ocean floor. They situate themselves half-buried in the sand, at an angle facing into the current. Food particles are trapped by their spines and conveyed to their mouths. Most people only see the sand dollars when the exoskeleton, or shell, washes up on the beach. Look for the beautiful five-petaled flower on the upper surface of the white circular shell. The species name, *excentricus,* refers to the fact that the flower is off-center.

Sand dollars

Piles of dead and decaying seaweed washed ashore, called beach wrack, are inhabited by a wide variety of insects and other arthropod species. **California beach hoppers** *(Megalorchestia* spp.*)* are among the most abundant macrofauna (animals that are large enough to be seen with the naked eye) found in beach wrack. These tiny creatures stay hidden under the sand and wrack during the day, both to reduce desiccation and to limit predation. After dark, they emerge and forage on the kelp. To people living along the coast, the stink of beach wrack is a temporary problem, soon solved by beach hoppers. Besides serving as "custodians" of the sandy shore, they are also an important source of protein for shorebirds.

California beach hopper

Beach wrack, Bonham Beach, Mendocino County

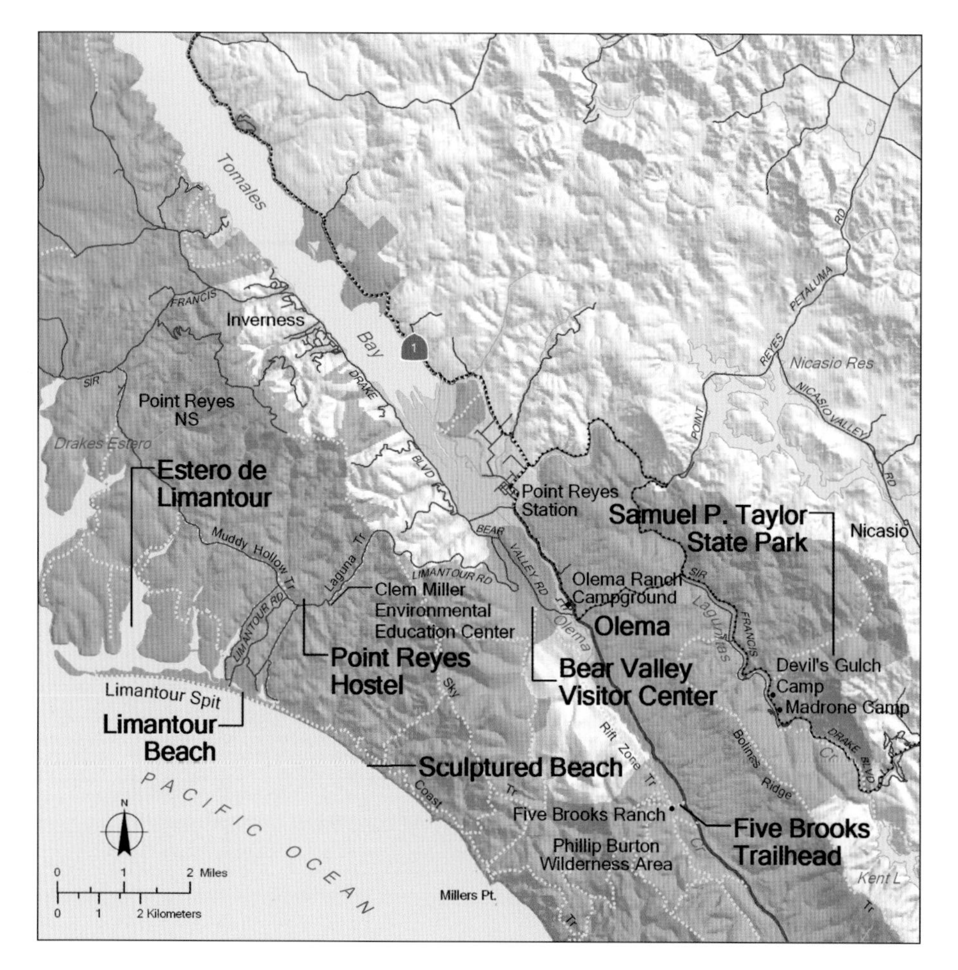

Limantour Spit

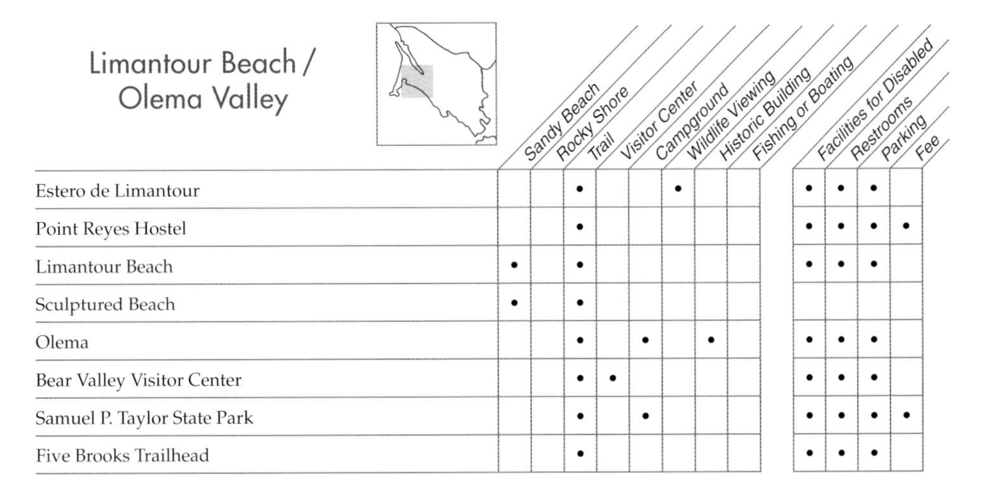

Limantour Beach / Olema Valley

	Sandy Beach	Rocky Shore	Trail	Visitor Center	Campground	Wildlife Viewing	Historic Building	Fishing or Boating	Facilities for Disabled	Restrooms	Parking	Fee
Estero de Limantour			•			•			•	•	•	
Point Reyes Hostel			•						•	•	•	•
Limantour Beach	•		•						•	•	•	
Sculptured Beach	•		•									
Olema			•		•		•		•	•	•	
Bear Valley Visitor Center			•	•					•	•	•	
Samuel P. Taylor State Park			•		•				•	•	•	•
Five Brooks Trailhead			•						•	•	•	

ESTERO DE LIMANTOUR: *End of Limantour Rd.* Extending along the north side of the two-mile-long Limantour Spit is the sheltered Estero de Limantour. Pickleweed grows in the salt marsh on the margin of the lagoon, and eelgrass grows in deeper water. Water birds such as white pelicans and brant feed and rest in the estero, and harbor seals haul out on shore. Numerous invertebrates are found in the mudflats, including blue mud shrimp and Lewis's moon snails. The 500-acre estero is a state marine conservation area, where viewing of wildlife is encouraged, but State Department of Fish and Game regulations prohibit the removal of plants or animals without a permit.

POINT REYES HOSTEL: *Off Limantour Rd., 7 mi. from Bear Valley Rd.* The hostel has 44 beds, including bunkhouse space for groups, wood stoves, and a full kitchen. A wheelchair ramp leads from the parking lot to the main house. Overnight fees are charged. Advance reservations are strongly recommended year-round; available by mail only. For information, write P.O. Box 247, Pt. Reyes Station 94596, or call 415-663-8811 from 7:30–9:30 AM or 4:30–9:30 PM only.

LIMANTOUR BEACH: *End of Limantour Rd.* A long sandy crescent backed by grassy dunes stretches east from the mouth of Drakes Estero. Due to the sheltering effect of the Point

Estero de Limantour

Olema

Point Reyes National Seashore was created to protect wild lands, natural ecosystems, and open space close to the urban centers of the San Francisco Bay Area. Recreational activities in the National Seashore include beachcombing, hiking, kayaking, bicycling, and wildlife viewing. Some 80 miles of unspoiled coastline are preserved in the park, and there are beaches, grasslands, forests, and estuaries to explore.

About half of the 71,000-acre National Seashore has been designated the Phillip Burton Wilderness Area, composed of a patchwork of tracts interspersed among grazing and other private lands. A network of foot trails leads through the wilderness area; there are no structures or pavement. Horses are allowed on wilderness trails; no bicycles or motorized vehicles permitted.

Camping in the National Seashore is permitted at four walk-in camps by reservation only; Wildcat walk-in camp is a group site. Camping permits are available at the visitor center or by phone reservation from 9 AM to 2 PM daily; call: 415-663-8054. There is no car camping available in the park. National Seashore information is available at the Bear Valley Visitor Center, or call: 415-464-5100 or 415-464-5137.

The Point Reyes National Seashore Education Program offers classes and field seminars on environmental education and natural history, and a summer youth camp. The Clem Miller Environmental Education Center, an overnight facility available to groups for educational purposes, features hiking trails and interpretive programs. For information or reservations for field programs or the Clem Miller Center, write: Point Reyes National Seashore Association, Point Reyes Station 94596, or call: 415-663-1224.

Reyes Headlands, wave action is generally calmer here than on the west-facing beaches of the Point Reyes Peninsula. A wheelchair-accessible trail runs along the lagoon southeast of the main access road (watch for sign). The beach can also be reached via the Coast Trail from Palomarin at the southeast end of the national seashore.

SCULPTURED BEACH: *6 mi. from Bear Valley via trail.* A sandy beach backed by colorful eroded bluffs is located three miles south of Limantour Beach, accessible by trail from the Bear Valley Visitor Center or from Palomarin. Sculptured Beach can also be reached from points north or south along the shore at very low tide. For National Seashore information, including recorded weather, trail, and tide conditions, call: 415-464-5100.

OLEMA: *Hwy. One at Sir Francis Drake Blvd.* Olema retains the look of the 19th century stagecoach stop that it once was. The building occupied by the Olema Inn at the corner of Hwy. One and Sir Francis Drake Blvd. dates from 1876, and a handful of other historic buildings give the village its old-time flavor. Olema never grew large; when the North Pacific Coast Railroad was constructed through West Marin in the 1870s, Olema was bypassed in favor of nearby Point Reyes Station.

Tent and RV camping are available at Olema Ranch Campground, which has 203 sites, some with full or partial hook-ups. RVs up to 50 feet long can be accommodated; dump station available. The campground offers hot showers, a general store and laundromat, play area and recreation hall, mountain bike rentals, wireless Internet access, and a post office. Group camping facilities available. Dogs allowed on leash. Reservations recommended during summer months; call: 415-663-8001.

BEAR VALLEY VISITOR CENTER: *Bear Valley Rd., .5 mi. N.W. of Olema.* The visitor center, housed in a modern high-ceilinged wooden

Bear Valley Visitor Center

structure, features a bookstore and exhibits about park attractions. Open 9 AM–5 PM weekdays, 8 AM–5 PM weekends and holidays. Nearby facilities include a Coast Miwok Indian village, a wheelchair-accessible Earthquake Trail, and a Morgan horse farm. Trails from the visitor center span the park, leading to beaches, lagoons, and forests.

SAMUEL P. TAYLOR STATE PARK: *Sir Francis Drake Blvd., 5.2 mi. E. of Olema.* 2,600 acres of wooded canyons along Lagunitas (or Paper Mill) Creek. The park is the former site of the first paper mill on the Pacific Coast, built in 1856. The park features picnic areas, trails, and campgrounds. There are 60 developed family sites, an area set aside for hike or bike campsites, and about 30 enroute sites. Five group camping areas are available, two at Madrone and three primitive sites (one for equestrians) at Devil's Gulch, located one mile north of the other camping areas. Group campers must check in at the main park entrance. For camping reservations, call: 1-800-444-7275. Picnic areas, campground restrooms, and six family campsites are wheelchair accessible; there are four miles of paved trails. Streams in the park are closed to fishing and boating year-round; there is a children's swimming area in the creek. Paved and unpaved cycling paths available. Call: 415-488-9897.

FIVE BROOKS TRAILHEAD: *Hwy. One, 3 mi. S. of Olema.* A network of trails begins here for the southern part of the Point Reyes National Seashore. Many trails lead over wooded Inverness Ridge to beaches in the Phillip Burton Wilderness Area of the National Seashore. Trails also lead to the Coast Trail and to walk-in campsites. Check with the Bear Valley Visitor Center for trail conditions; trail maps available at all park visitor centers. Parking and restrooms at the trailhead; the main trail is partially wheelchair accessible with assistance. For park and trail information, call: 415-464-5100 or 415-464-5137. Five Brooks Ranch offers horseback riding lessons and trail rides; for information, call: 415-663-1570.

Dairy land at Tomales Point

Farms and ranches dominate the North Coast landscape. Half of Marin County's land is devoted to agriculture, and the county produces much of the Bay Area's milk. In 1862, Marin County already provided one-quarter of the state's butter, which, when salted, was less perishable than milk for shipment to market in the days before refrigeration. In addition to butter and cheese, Marin County farms produced potatoes, beets, peas, wheat, and barley, which were loaded onto ships at Bolinas or Tomales or onto the North Pacific Coast Railroad for transport to San Francisco's growing population. Hay grown in Marin helped fuel the state's horse-drawn transportation facilities.

Raising potatoes on slopes in northwestern Marin County resulted in sediment clogging coastal streams. Look at Keys Creek and try to imagine even a small ocean-going schooner making its way upstream to the village of Tomales. Today, farming practices have changed, and the raising of cows, calves, and sheep, along with dairy operations, are predominant in Marin County. Specialty products, including free-range and grass-fed animals, cheese, yogurt, and ice cream, are on the increase.

Although there are now fewer dairy operations in Marin County than in the 1950s, those that remain are larger and produce more milk per cow. Most operations remain family-owned, and nearly all farmers want to continue farming, despite often slim economic returns. The preservation of agriculture is encouraged by North Coast counties, which have local ordinances that discourage creation of residential subdivisions in farming areas. Some 35,000 acres of land on 53 family farms and ranches in Marin County have been permanently protected for agriculture through conservation easements held by the Marin Agricultural Land Trust. Established in 1980, the trust was the nation's first private nonprofit set up specifically to preserve farmland. Grazing co-exists with recreation and resource protection in the Point Reyes National Seashore and Golden Gate National Recreation Area, which include 32,000 acres of land leased to farmers.

About half the farms in Marin have partial or total organic operations. The Straus Family Creamery on the east shore of Tomales Bay was the first organic dairy in the western United States. A new grass-fed livestock certification program in Marin and Sonoma Counties recognizes participating farms where animals feed on natural forage and are given no antibiotics or supplemental hormones. Forage production in the generally moist coastal environment is up to six times higher than in the arid interior of California.

The Grown in Marin marketing program, as well as local farmers markets throughout the North Coast, allow visitors to sample the outstanding products of this evolving industry, which forms an enduring part of the character and economy of the California coast.

Farm stays and tours are available in many North Coast locations; check local tourism offices, University of California Cooperative Extension offices, chambers of commerce, or farm bureaus for more information.

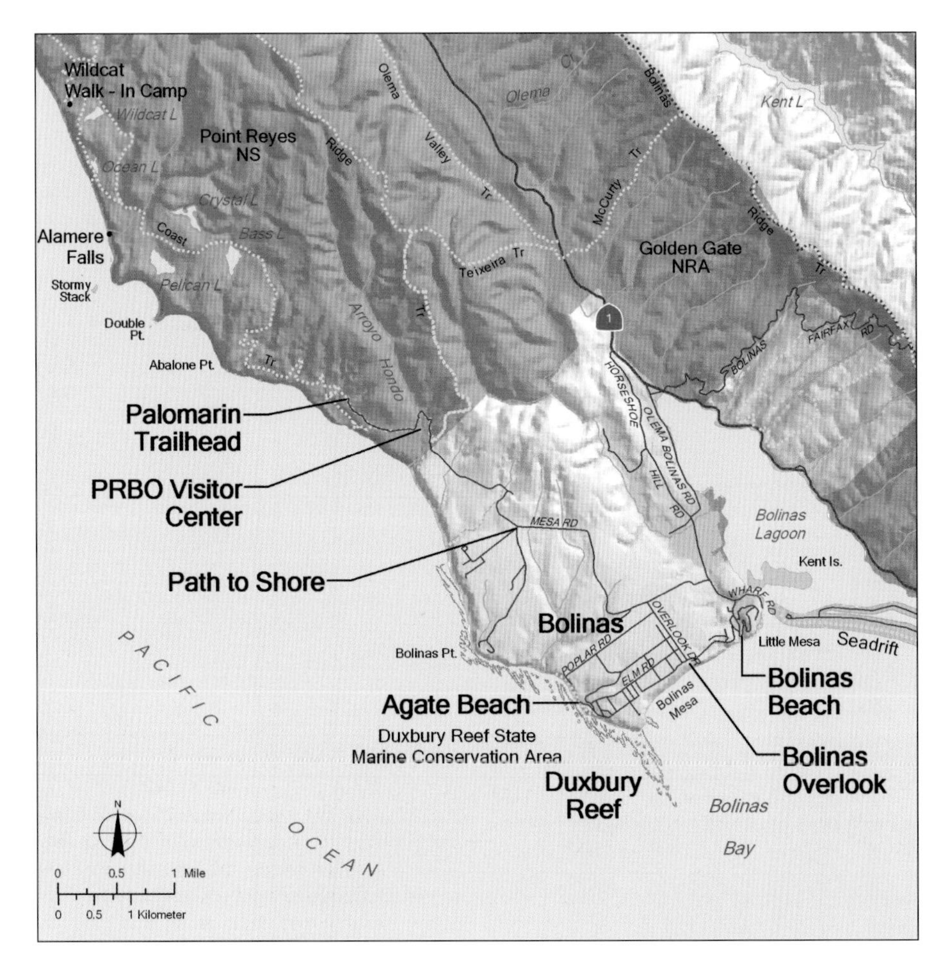

View toward Palomarin Trailhead

Bolinas

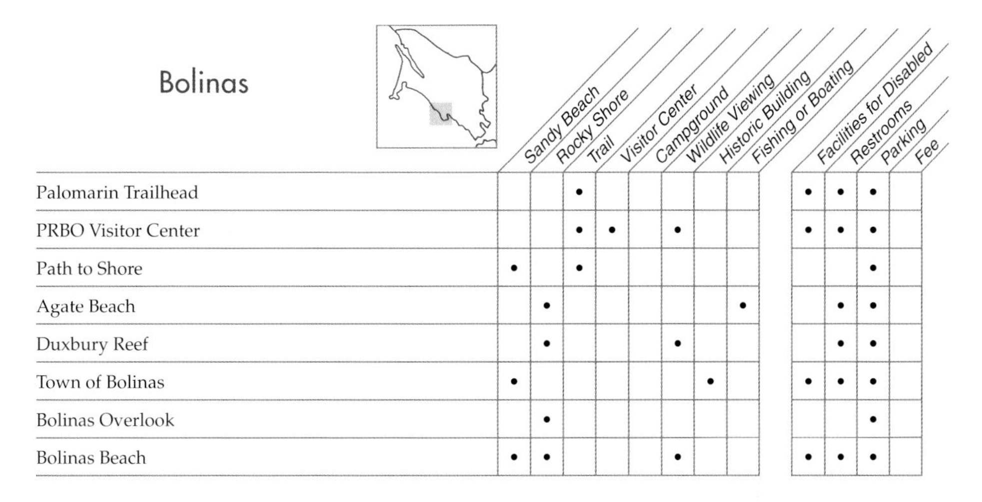

	Sandy Beach	Rocky Shore	Trail	Visitor Center	Campground	Wildlife Viewing	Historic Building	Fishing or Boating	Facilities for Disabled	Restrooms	Parking	Fee
Palomarin Trailhead			•						•	•	•	
PRBO Visitor Center			•	•		•			•	•	•	
Path to Shore	•		•								•	
Agate Beach		•						•		•	•	
Duxbury Reef		•				•				•	•	
Town of Bolinas	•						•		•	•	•	
Bolinas Overlook		•									•	
Bolinas Beach	•	•				•			•	•	•	

PALOMARIN TRAILHEAD: *End of Mesa Rd., 5 mi. N.W. of Bolinas.* The Palomarin Trailhead provides access to the southern end of the Point Reyes National Seashore. A network of trails leads along the bluffs and to lakes, beaches, and walk-in camps. At Double Point, two Monterey shale outcroppings enclose a small bay, where acorn barnacles and California mussels cling to the reef. Purple and giant red sea urchins lie in the crevices, and red rock crabs are common. Rocks north and south of Double Point, particularly Stormy Stack off North Point, are important feeding and roosting sites for Brandt's cormorants, common murres, grebes, and California brown pelicans. On top of the bluff is Pelican Lake, drained by a stream flowing down to the beach. Four nearby freshwater lakes, named Bass, Crystal, Ocean, and Wildcat, contain small marshy areas and provide habitat for waterfowl. Alamere Falls cascades down the bluff north of Double Point. The Coast Trail continues north along a roadless stretch of shoreline to Sculptured Beach and beyond.

PRBO VISITOR CENTER: *Mesa Rd., 4.5 mi. N.W. of Bolinas.* Formerly known as the Point Reyes Bird Observatory, PRBO Conservation Science carries out research and education focused on protecting birds and other wildlife. PRBO research activities have helped lead to creation of the three national marine sanctuaries that touch Marin County's coast: the Gulf of the Farallones, Cordell Bank, and Monterey Bay National Marine Sanctuaries. Visitors are welcome to view some of the research activities, including the bird-banding process in which birds are safely captured in a fine net, examined, and released. Banding takes place in the morning, Tuesday through Sunday, from May 1 through Thanksgiving, and on Wednesdays and weekends the remainder of the year, weather permitting. Drop-in visitors in groups of 10 or fewer are welcome to view the mist-netting process; for information on scheduling a school or community group tour, or to check conditions, call: 415-868-0655.

A small visitor center with displays of the observatory's research projects is open from early morning to 5 PM; a half-mile-long self-guided nature trail leads through coastal scrub habitat and the small canyon of Arroyo Hondo. Monthly bird walks are offered year-round by PRBO, to locations throughout Marin County, including Bolinas Lagoon, Muir Beach, Point Reyes, and Limantour Estero. For information, call: 415-868-1221.

PATH TO SHORE: *451 Mesa Rd., N.W. of Bolinas.* A half-mile-long trail leads from Mesa Rd. to the shoreline; part of Point Reyes National Seashore. Call: 415-464-5100. A second unimproved path to the shoreline is located a quarter-mile north, near mile marker 2.68 on Mesa Road.

AGATE BEACH: *End of Elm Rd., Bolinas.* This county park includes a northwest-facing beach at the edge of Duxbury Reef. The park is small, but it provides access to nearly two miles of shoreline at low tide. Rocks around the beach that are splashed by the waves hold associations of barnacles, limpets, and periwinkles. The nearshore waters north and south of Agate Beach are part of the Duxbury Reef State Marine Conservation Area, where the taking of aquatic plants and certain animals is banned; check Department of Fish and Game regulations, or call: 707-875-4260.

DUXBURY REEF: *S.W. perimeter of Bolinas Mesa.* Duxbury Reef is California's largest exposed shale reef. The reef lies at the base of the headlands known as the Bolinas Mesa, formed of Monterey shale. At low tide, a mile-long stretch of reef is exposed. The shape of the reef is continuously changing, as the waves erode the edges of the reef and the adjacent cliffs. The relatively soft shale is habitat for an unusual assemblage of rock-boring clams and worms.

To view the reef up close, park at Agate Beach at the end of Elm Road. Shallow gravel tidepools contain a type of acorn worm that is apparently unique to the Duxbury Reef, and a small burrowing anemone is found only here and at Puget Sound in Washington State. The intertidal area includes gooseneck barnacles, ochre sea stars, and huge

Duxbury Reef

Bolinas Overlook

beds of California mussels. The entire reef is a marine reserve, where regulations prohibit removal of aquatic plants and limit the taking of animals, including fish; check Dept. of Fish and Game rules, or call: 707-875-4260.

TOWN OF BOLINAS: *End of Bolinas-Olema Rd., off Hwy. One.* When San Francisco grew explosively after the discovery of gold in 1848, Bolinas shared the boom in a modest way. Redwoods on the hills surrounding Bolinas were cut to build the wooden row houses of San Francisco, and smaller trees were cut for household fuel. From sawmills near the shore, lumber was loaded on flat-bottomed lighters in Bolinas Lagoon and transferred to ships anchored offshore. In the U.S. Census of 1850, the first to include California, two-thirds of Marin County's few hundred residents lived in the Bolinas area. Some 19th century buildings remain, particularly near the old downtown along Wharf Road. Part of Smiley's Bar dates from 1852, the Bigson House building dates from 1875, and St. Mary Magdalene Catholic Church from 1878. Hundreds of thousands of monarch butterflies overwinter in a grove of eucalyptus trees in Bolinas.

BOLINAS OVERLOOK: *End of Overlook Dr., Bolinas.* A spectacular view of the southern portion of Duxbury Reef, the ocean, and San Francisco in the distance is available from a small pull-out at the end of Overlook Drive.

BOLINAS BEACH: *Ends of Brighton and Wharf Aves., Bolinas.* Bolinas Beach curves around the headland known as the Little Mesa, and the beach can be reached from two street ends. At the end of Brighton Ave., a gated concrete ramp leads to a sand and pebble beach popular with surfers. At high tide, the beach is narrow. A small county park with restrooms and tennis courts is located on Brighton Ave. near the beach. The second entrance to Bolinas Beach is from the end of Wharf Ave., which has very limited parking and no turn-around at the end. The mouth of Bolinas Lagoon, which is only a narrow channel at low tide, separates Bolinas Beach from the Seadrift sandspit. Brown pelicans, gulls, and waterfowl can be seen at the entrance to the lagoon.

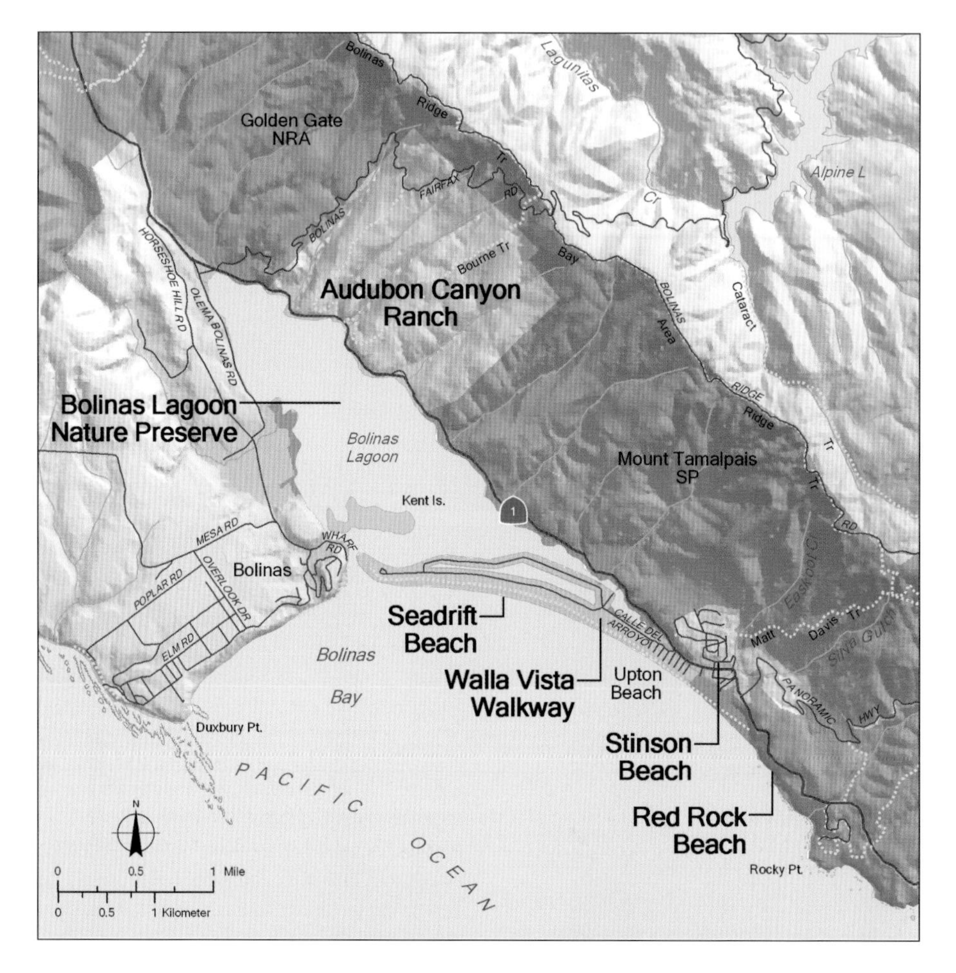

Harbor seals at Bolinas Lagoon

Stinson Beach

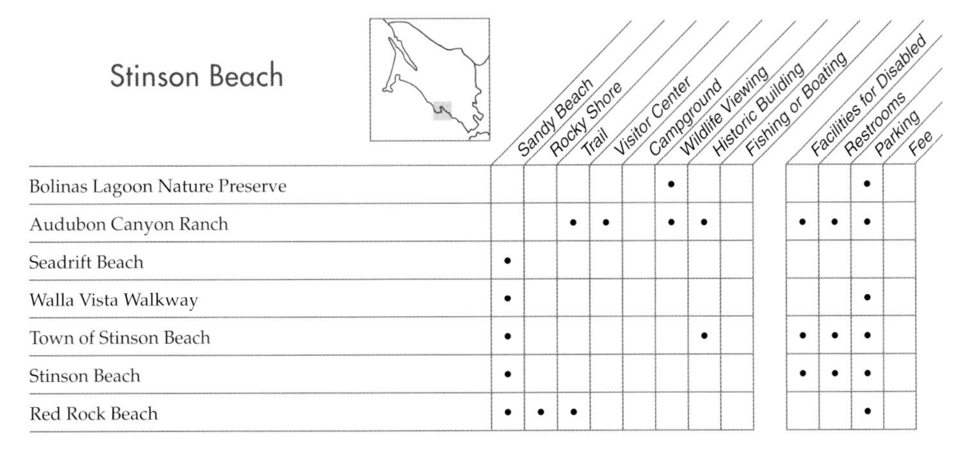

	Sandy Beach	Rocky Shore	Trail	Visitor Center	Campground	Wildlife Viewing	Historic Building	Fishing or Boating	Facilities for Disabled	Restrooms	Parking	Fee
Bolinas Lagoon Nature Preserve						•					•	
Audubon Canyon Ranch			•	•		•	•		•	•	•	
Seadrift Beach	•											
Walla Vista Walkway	•										•	
Town of Stinson Beach	•						•		•	•	•	
Stinson Beach	•								•	•	•	
Red Rock Beach	•	•	•								•	

BOLINAS LAGOON NATURE PRESERVE: *Along Hwy. One and Olema-Bolinas Rd.* Bolinas Lagoon is a county-owned preserve that includes over 1,200 acres of sheltered water, salt marsh, and mudflat. The lagoon is a major stopover point for migrating shorebirds and other waterfowl, including sandpipers, plovers, geese, and ducks. As many as 35,000 birds have been counted, representing over 60 species. Herons and egrets are resident year-round and are abundant on the mudflats. Surf scoters, ruddy ducks, greater scaups, avocets, and pintails are also commonly seen along with black-necked stilts. Uncommon, but occasionally present, are Virginia rails.

Bolinas Lagoon is home to a large colony of harbor seals, which can be seen at low tide hauled out on the mudflats, often far out in the lagoon. Bring field glasses for a closer look; do not attempt to approach these shy creatures. The lagoon also serves as a nursery for fish such as starry flounder, cabezon, and several varieties of perch. Pickleweed grows along the margin of the lagoon, and western pygmy blue butterflies may be seen in the area. The shallowness of the lagoon has resulted from siltation due, in part, to historic logging on the surrounding hills. The preserve offers no visitor facilities other than excellent vistas from numerous roadside pull-outs along Highway One; view wildlife respectfully from a distance. Call: 415-499-6387.

AUDUBON CANYON RANCH: *Hwy. One, 3 mi. N. of Stinson Beach.* Inland of Hwy. One near Bolinas Lagoon are several heavily wooded canyons separated by grass-covered ridges. High in the redwood trees in one of the canyons is a large nesting colony of great blue herons and great egrets. Audubon Canyon Ranch, a non-profit environmental preservation, education, and research organization, operates the Bolinas Lagoon Preserve. A half-mile-long trail leads to an overlook on the hill where fixed telescopes allow close-up views of all the activities of a nesting colony of great blue herons, including courtship, mating, incubation, feeding, and fledgling flights. Two three-mile-long loop trails and a self-guided nature trail into other canyons are also available.

Facilities at Audubon Canyon Ranch include an exhibit hall, nature bookshop, and picnic area adjacent to the 1875 home of Captain Peter Bourne. In a neighboring canyon, the ranch operates a wildlife education center in another ranch house built in the 1870s. The ranch is open to the public during the nesting season on weekends and holidays, 10 AM–4 PM, from the second weekend in March through the second weekend in July. Courting displays of the birds take place early in the season, while eggs and chicks are visible in April and May. Visits to the preserve are self-guided. Blinds for viewing nesting birds, picnic tables, the display hall, and the bookstore are wheelchair accessible. School and other groups may visit Tuesday through Friday by appointment. Dogs not allowed, other than service animals. No fee for entry, but donations are encouraged. Call: 415-868-9244.

Some 40 years ago, Bolinas Lagoon faced a fate far different from the serene beauty enjoyed by today's visitors. In the early 1960s, a freeway was slated for construction at sea level along the east side of the lagoon, paralleled on the ridge above by a "parkway" that would have sliced through the forested slopes. Bolinas Lagoon was proposed to be dredged for an enormous marina, to be sited next to a heliport and a hotel built on filled land in the lagoon (located atop the San Andreas Fault). Inevitably, the highways would have brought urban development on a massive scale to the quiet valleys of West Marin, with a catastrophic loss of wildlife habitat and natural values. Alarmed citizens campaigned against the development plans and raised funds to acquire key parcels of land in and around the lagoon. In the end, the freeway plans were dropped, the Bolinas Harbor District was disbanded, and open space was protected. Bolinas Lagoon is now a county nature preserve, and Audubon Canyon Ranch operates an adjacent preserve centered on a magnificent heron and egret rookery. The Golden Gate National Recreation Area and Mount Tamalpais State Park encompass neighboring slopes and are managed for protection of wildlife and scenic beauty and for public recreation.

SEADRIFT BEACH: *N.W. of Stinson Beach.* The mile-and-a-half-long Seadrift sandspit curves upcoast from Stinson Beach, where the generally south-facing shore and nearby mountains create a relatively warm beach environment. Although the roads in the Seadrift subdivision are for residents only, the beach from the water inland to 60 feet from the private seawall is open to the public.

WALLA VISTA WALKWAY: *Calle Del Arroyo, off Hwy. One.* A public pedestrian accessway leads from Calle del Arroyo, outside the Seadrift Subdivision gate, along the west side of Walla Vista, a short lane leading to the beach. Turn right to Seadrift Beach, or turn left to the Marin County–maintained Upton Beach; for policy regarding dogs, call: 415-499-6387. Limited road shoulder parking on Calle del Arroyo, on the north side only.

TOWN OF STINSON BEACH: *Hwy. One, .3 mi. N. of Panoramic Hwy.* Early settler Alfred D. Easkoot promoted Stinson Beach as a destination for campers and swimmers; his 1875 house, now restored, is located on Highway One just west of the village center. Around 1900, a summer tent colony called Willow Camp appeared in the dunes near the beach. Cabins and houses began to be built after a subdivision was created by Nathan H. Stinson in 1906. Early day visitors to Stinson Beach arrived by stagecoach or buggy via a road around the north side of Mount Tamalpais; a 1911 plan to extend the Mount Tamalpais and Muir Woods Railway from Mill Valley was never carried out.

Introduced eucalyptus and Monterey pine trees in the village are winter resting areas for colonies of monarch butterflies. Easkoot Creek, bordered by dense riparian vegetation, flows through the center of town and feeds Bolinas Lagoon. The former creek mouth was located on the beach, emptying directly into the bay. Efforts are under way to restore the creek's habitat value for steelhead trout. Galleries, eating places, and a few small inns are located in the village.

STINSON BEACH: *W. of Hwy. One, N. of Panoramic Hwy.* Probably California's most popular beach north of San Francisco, Stinson Beach offers sunbathing, volleyball, surfing, and other recreational pursuits. Air and water temperatures are often milder along this broad, sandy stretch than on other northern California beaches because it faces south and is sheltered by coastal mountains. Lifeguards are on duty from late May to mid-September.

The main part of the beach, parking area, and support facilities lie within the Golden Gate National Recreation Area (GGNRA). There are picnic tables and barbecue grills set among willows and cypress trees, a snack bar (open summer only), restrooms, and showers. The vehicle entrance gate is open from 9 AM to about sunset; check current closing time upon entry. Volleyball equipment and a beach wheelchair can be borrowed from the main lifeguard tower. No pets allowed on the beach. Call: 415-868-1922. Upton Beach, extending two-thirds of a mile upcoast from the Golden Gate National Recreation Area beach, is maintained by Marin County; for information, call: 415-499-6387.

RED ROCK BEACH: *Hwy. One, 1 mi. S. of Stinson Beach.* From an unpaved parking lot at milepost 11.45, a trail leads down the steep, red chert bluff to a remote, clothing-optional beach within Mount Tamalpais State Park. Springs seeping from the cliff face support wildflowers, and giant boulders are scattered on the beach.

Stinson Beach

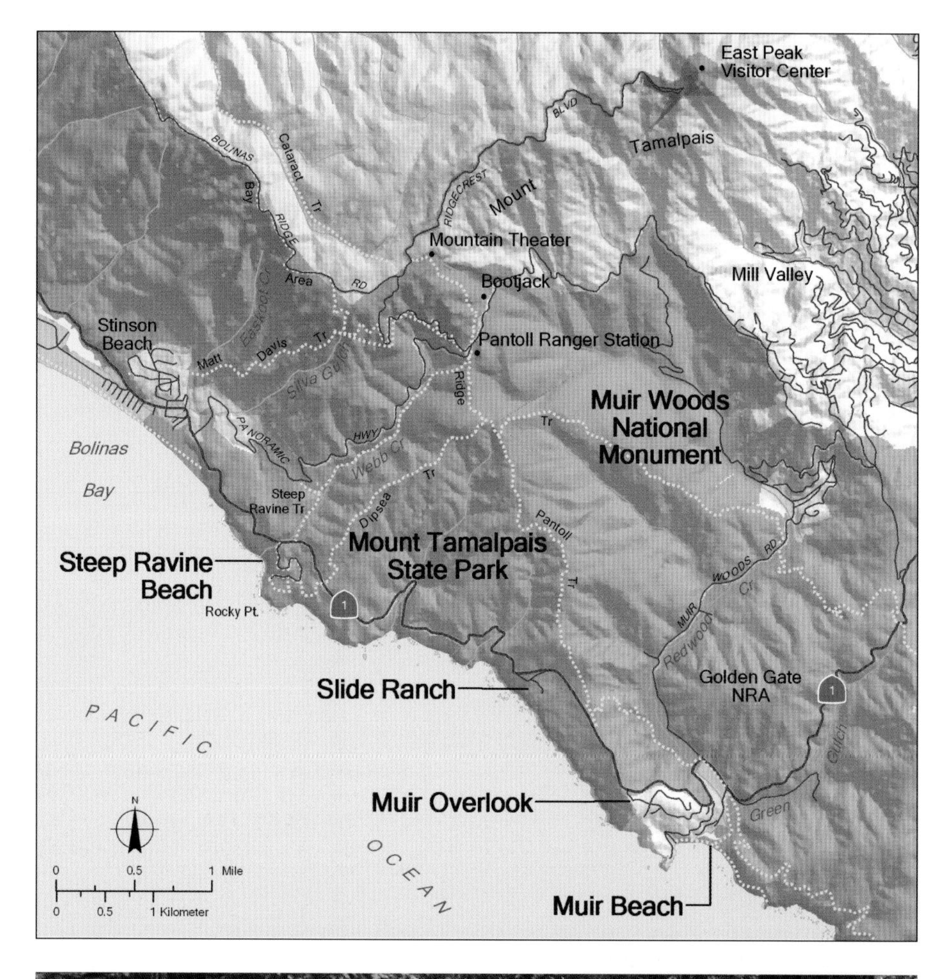

East Peak Visitor Center
Tamalpais
Mount
BLVD
RIDGECREST
BOLINAS
Bay
RIDGE
Cataract
Tr
Mountain Theater
Bootjack
Mill Valley
Area
RD
Stinson Beach
Matt
Easkoot Cr
Davis
Tr
Silva Gulch
Pantoll Ranger Station
Ridge
Muir Woods National Monument
PANORAMIC
HWY
Webb Cr
Tr
Bolinas
Bay
Steep Ravine Tr
Dipsea
Pantoll
Tr
Mount Tamalpais State Park
Steep Ravine Beach
Rocky Pt.
MUIR
WOODS
RD
Cr
Redwood
Golden Gate NRA
Slide Ranch
PACIFIC
OCEAN
Gulch
Green
Muir Overlook
Muir Beach
N
0
0.5
1 Mile
0
0.5
1 Kilometer

Muir Woods

Muir Beach Area

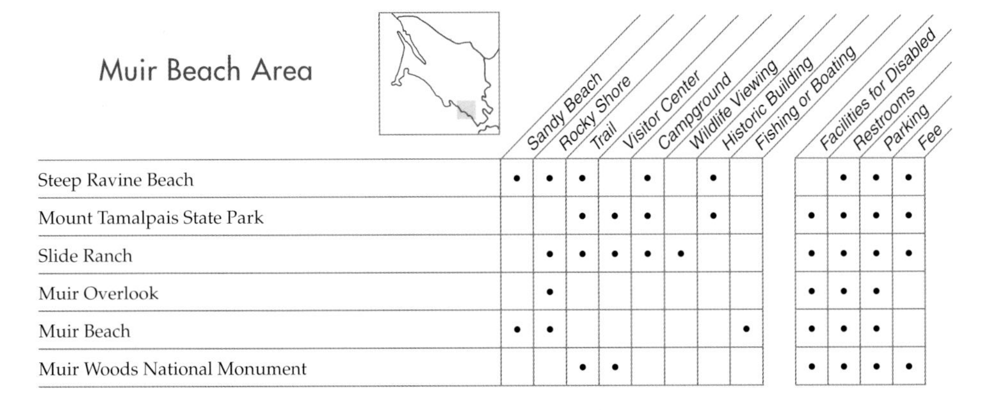

	Sandy Beach	Rocky Shore	Trail	Visitor Center	Campground	Wildlife Viewing	Historic Building	Fishing or Boating	Facilities for Disabled	Restrooms	Parking	Fee
Steep Ravine Beach	•	•	•		•		•			•	•	•
Mount Tamalpais State Park			•	•	•		•		•	•	•	•
Slide Ranch		•	•	•	•	•			•	•	•	•
Muir Overlook		•							•	•	•	
Muir Beach	•	•						•	•	•	•	
Muir Woods National Monument			•	•					•	•	•	•

STEEP RAVINE BEACH: *Hwy. One, 2 mi. S. of Stinson Beach.* Nine rustic cabins overlook the sea from a marine terrace within Mount Tamalpais State Park. Cabins have wood stoves, picnic tables and benches, sleeping platforms, and outdoor barbecues, but no running water or electricity. Primitive toilets, water faucets, and firewood are nearby. Seven primitive campsites, each with table, firepit, food locker, and space for a tent, are located a few hundred yards from the parking area. No showers available. A maximum of one vehicle and five people are allowed per cabin or campsite; no pets.

Reservations are required for all Steep Ravine sites, which tend to book up well in advance; call: 1-800-444-7275. Campers with reservations may drive to a parking area a quarter-mile down the road leading to the beach; beach day users may park in pullouts near the locked gate on Hwy. One and walk down the road.

MOUNT TAMALPAIS STATE PARK: *Off Panoramic Hwy.* More than 6,000 acres of parkland in Mount Tamalpais State Park adjoin recreational land holdings of the Golden Gate National Recreation Area, and other entities, linked by trails and scenic roads. From the top of 2,571-foot Mount Tamalpais, once reached by a winding railroad line and now accessible by road or trail, visitors have a panoramic view of the Farallon Islands offshore, the towers of San Francisco, and parts of the Coastal Range. Astronomy evenings with a talk by an astronomer and volunteer-provided telescopes are held at the Mountain Theater each Saturday nearest the first quarter moon, from March through October; for recorded information, call after 4 PM: 415-455-5370.

Some 250 miles of trails link the beaches with the top of Mount Tamalpais and the town of Mill Valley. The Steep Ravine and Cataract Trails follow streams with waterfalls. Trail maps are available at the visitor center on the summit of the mountain's east peak and at local bookstores. The Mount Tamalpais Interpretive Association offers volunteer-led hikes featuring a variety of park resources and attractions; for recorded information on hikes, call: 415-258-2410. Horses are allowed on fire roads, paved roads, and on posted equestrian trails; bicycles are allowed on fire roads and paved roads only. Dogs are not permitted on trails or in undeveloped areas of the state park.

Pantoll Campground on Panoramic Hwy. has 16 campsites with tables, rock barbecues, food lockers, and space for a tent; running water, restrooms, and parking are nearby; no reservations. There is also an enroute campground at Pantoll in the lower parking lot, where self-contained RVs can be accommodated on a one-night basis between 6 PM and 9 AM only. Panoramic Hwy. is narrow and winding, and motor homes over 35 feet in length are not recommended. Alice Eastwood Camp, located off Panoramic Hwy., has two group campsites, and Frank Valley Group Horse Camp, on Muir Woods Rd., has equestrian facilities. For reservations, call: 1-800-444-7275; for information on campgrounds, call Pantoll Ranger Station: 415-388-2070. Fees apply for camping and

Dorothea Lange and her camera seared the faces of 1930s-era migrant farm families into the nation's consciousness. During World War II she photographed shipbuilders in Richmond, California, and interned Japanese-Americans, and she traveled with husband Paul Taylor to Asia, South America, Egypt, and India, taking photos everywhere they went. Dorothea Lange died in 1965.

Steep Ravine 1960

Dorothea Lange's large circle of relations, stepchildren, and close family friends included photographer Imogen Cunningham; Cunningham's son, Rondal Partridge; and his daughter, Elizabeth. In *Dorothea Lange: A Visual Life*, a volume of images and essays, Elizabeth Partridge remembers: *Summers we often went to Steep Ravine. There, in Paul and Dorothea's little cabin perched on a cliff hanging over the Pacific Ocean, we shed most of our clothes, and all sense of time. We had nothing but the vastness of the ocean and tiny shell of a cabin, which cradled us all gently when the sun and wind wore us out. Here Dorothea walked slowly through the wet sand, the wind blowing in off the ocean, the ever-present Pentax camera hanging from her neck. Inside the cabin she hung her camera on a nail by the door, ready to take photographs indoors or out. Here the worries that tugged perpetually at adults didn't seem to bother them, and we kids ate and quarreled and slept happily together like a litter of puppies.*

Easter 1960 (Steep Ravine)

Copyright the Dorothea Lange Collection, Oakland Museum of California, City of Oakland. Gift of Paul S. Taylor.

for day-use parking within Mount Tamalpais State Park at Pantoll, Bootjack, and East Peak.

SLIDE RANCH: *Hwy One, 1 mi. N. of Muir Overlook.* Slide Ranch is a non-profit environmental education center located within the Golden Gate National Recreation Area, on a steep slope of Mount Tamalpais. Educational programs emphasize agriculture and sustainability. Wildlife, such as bobcats, red-tailed hawks, and lizards, use the property, along with domestic animals that include goats, sheep, and ducks. Hiking trails and tidepools on the site are used to teach about resource conservation.

Programs are offered to families and to community and school groups, with an emphasis on elementary school–age children. Family programs include camp-outs, toddler's days, and family farm days. Daytime and overnight programs for groups include exploration, campfires, and experience with farm chores. Pre-registration is required; fees charged, but scholarships may be available for those with limited income; for information, call: 415-381-6155.

MUIR OVERLOOK: *Off Hwy. One, 1 mi. N. of Muir Beach.* Spectacular views of steep serpentine cliffs are available from this bluff high above the sea. Offshore due south is the Potato Patch Shoal, where the sea at low tide is little more than 20 feet deep, causing turbulence and occasional freak waves that threaten boaters. Picnic tables, restrooms, and paved parking available.

MUIR BEACH: *Hwy. One at Muir Woods Rd.* Redwood Creek drains the southern slopes of Mount Tamalpais and flows through Frank Valley to the ocean, adjacent to the community of Muir Beach. The creek is a spawning stream for steelhead and coho salmon. Big-leaf maples, which turn golden in the fall, grow along the stream as do red alders.

Where Frank Valley and Green Gulch meet at the shoreline is a lagoon formed by Redwood Creek and a large beach with rocky areas and tidepools at both ends. Between mid-October and mid-March, monarch butterflies overwinter in a grove of Monterey pines near the entrance to the Golden Gate National Recreation Area parking lot. The sandy beach is popular for fishing, picnicking, and sunbathing.

MUIR WOODS NATIONAL MONUMENT: *Off Hwy. One, 3 mi. N. on Muir Woods Rd.* Muir Woods is a 554-acre redwood grove deep in Redwood Creek valley on the southern slopes of Mount Tamalpais. The park contains the only remaining old-growth redwood forest near San Francisco. Although its loop trails are very popular with visitors, Muir Woods connects to less-crowded trails that lead throughout Mount Tamalpais State Park and the Golden Gate National Recreation Area. The Pantoll Trail leads west to Stinson Beach on the coast, and the Dipsea Trail, which passes near the Muir Woods visitor center, connects Mill Valley with Steep Ravine Beach on the coast.

From the main entrance to Muir Woods, a wheelchair-accessible boardwalk leads along Redwood Creek into the woods; there are interpretive signs and exhibits including living examples of redwood-like tree species from around the world. Leopard lilies bloom in summer in the shaded forest; look for huckleberries in late summer. Facilities include a visitor center, a bookshop, a café, and restrooms; nature walks are available. The monument is open every day; call: 415-388-2595.

Muir Overlook

Bolinas Ridge to Duxbury Point, Marin County

© 2004, Tom Killion

Protecting Coastal Resources

SINCE 1849 California's official state seal has proclaimed, "Eureka," Greek for "I have found it." Those who chose this phrase may have had gold in mind, but the motto has turned out to apply equally well to the millions of newcomers who have found their way to this edge of the continent. More than 36 million residents now have a stake in the future of California's coast. Millions more visitors, from all parts of the globe, revel in the California coast's distinctive sense of place, unparalleled beauty, wildlife, and history. This guidebook points readers to a small fraction of the countless opportunities for recreation, solitude, thrills, and learning that can be experienced on the California coast.

The pressure of growing numbers of residents and visitors brings into focus the finite nature of California's coastal resources. If planned, financed, and constructed, additional recreation facilities may be added to future editions of this guide, but the fact remains that there will never be more miles of shoreline or more beaches. Protecting the coast and restoring lost resources, where possible, are paramount if Californians of the future are to experience the rich array of choices enjoyed by today's coastal visitors.

Some steps to protect and restore the coast's resources can be taken by any of us, acting individually (see p. 13). Other challenges facing the coast require concerted action by society as a whole. None of us alone can successfully address global climate change. As California winters become warmer and wetter in the future, as indicated by current trends, sea level along the shoreline may rise by up to 12 inches. Warming temperatures are likely to cause increased erosion of beaches and cliffs and loss of coastal wetlands in California, among other changes.

Preventing habitat loss for plants and creatures is another challenge that faces California, with its rapidly growing towns and suburbs. The consequences of the loss of a species like the majestic bald eagle, symbol of our nation's freedom, are easy to grasp. But lesser-known plants and animals also need to be protected from extinction. The biological diversity of living things in an ecosystem is an integral component of that ecosystem's health. The loss of a dominant, or "keystone," species, or the invasion of a non-native species, can affect how an ecosystem works. Preserving biodiversity is an important management goal for conservation lands, including parks and open space areas. Two species from California's North Coast, the marbled murrelet and the Myrtle's silverspot butterfly, although not easily observed by visitors, nevertheless contribute to the biological diversity of coastal ecosystems, and they may play an important role in ecosystem health.

The marbled murrelet *(Brachyramphus marmoratus)* is a small seabird that nests high in the canopy of coastal old-growth forests and feeds on fish in nearshore waters. This unusual bird is listed as a threatened species under the federal Endangered Species Act and an endangered species under California state law. During the summer nesting season, marbled murrelets fly daily

Marbled murrelet

from redwood groves to the ocean, to gather food for their young. When they are not nesting, the birds live at sea, spending their days feeding close to land and then moving farther offshore at night. The murrelets' reliance on old-growth forests for nesting and their use of coastal waters for feeding have brought the birds into conflict with human economic interests. This is especially true in northern California, where small, isolated populations are particularly vulnerable to local extinction from logging, accidental death in gill-nets, oil pollution, increases in predator populations, and declines in food supplies due to El Niño events. To try to catch a glimpse of this secretive species or maybe just to hear the plaintive call, *keer, keer, keer,* in the early morning fog, visit Humboldt Redwoods State Park, Jedediah Smith Redwoods State Park, or Prairie Creek Redwoods State Park, all of which are known to host small populations of the species.

Another endangered species that occurs along the coast is the Myrtle's silverspot butterfly *(Speyeria zerene myrtleae)*. Populations of Myrtle's silverspot are restricted to coastal dunes, coastal scrub, and coastal prairie. Formerly widespread on the San Francisco and Marin peninsulas, the butterfly is now known from only four populations in northern Marin County, including Point Reyes National Seashore. Adult Myrtle's silverspot butterflies live for only about three weeks during the late summer and early fall. During their short butterfly stage they feed on nectar, mate, and lay eggs. The eggs are laid on native violets, and they hatch in the fall. The Myrtle's silverspot spends the winter as a caterpillar, and in the spring the caterpillar feasts on the tender new growth of the violets. The caterpillar later metamorphoses into the beautiful adult butterfly. Invasive non-native plants are a serious threat to the Myrtle's silverspot caterpillar, which depends upon specific violet species for its forage. Where coastal scrub and dunes are blanketed by European beach grass *(Ammophila arenaria)* and ice plant *(Carpobrotus edulis)*, habitat for this striking butterfly is unavailable.

In California and throughout the world, biological invasions have become one of the most pressing environmental challenges. Invasive species of plants and animals can reduce native species diversity, change hydrology, and alter the morphology (structure) of habitats. European beach grass is an invasive coastal plant that can change dune morphology and availability of moisture significantly. Dunes that form with only

Myrtle's silverspot butterfly feeding on *Grindelia* flower, Point Reyes National Seashore

Ice plant and European beach grass at Bodega dunes, Sonoma County

native beach grasses and forbs have low slopes that run perpendicular to the beach. By contrast, European beach grass traps more sand than native grasses; consequently, dunes that form under European beach grass have steeper slopes and are aligned nearly parallel to the shoreline. This change in the form of dunes disrupts conditions that support native dune plant communities. The rich diversity of native species declines sharply under these conditions, and, without aggressive management measures, nearly pure stands of European beach grass form.

Extensive research has been conducted at the Lanphere Dunes unit of the Humboldt Bay National Wildlife Refuge to develop effective eradication techniques for unwanted European beach grass. With its extensive underground rhizome network, the grass is extremely tenacious, and its eradication has proven to be a continuing challenge to land managers. The arsenal of known removal techniques now includes manual, mechanical, and chemical alternatives, but refinements and other methods are being sought.

Another invasive coastal plant is hottentot fig, or ice plant, a succulent, mat-forming perennial plant from South Africa. This species was actively planted in the early 1900s with the goal of stabilizing dunes, and more recently it has been widely planted as a roadside ground cover. Ice plant has since spread from areas of cultivation into many native plant communities along the entire California coastline. This plant is extremely invasive and has the ability to effectively exclude native species from the areas that it occupies. Once ice plant becomes established, it may smother native plants by forming deep, dense mats, or it may alter soil moisture in content, stability, or pH (acid/alkaline balance), making it difficult or impossible for other species to survive. Research regarding the biology and ecology of ice plant, as well as methods of invasion control, has been conducted at the University of California's Bodega Marine Lab, located adjacent to Bodega Head in Sonoma Coast State Beach. Like its weedy brethren, European beach grass, ice plant is a very difficult invader to control.

Coastal ecosystems are threatened by aquatic invasives as well as those on land. The European green crab *(Carcinus maenas)* is the cause of great concern. This little crab was likely introduced by the release of ballast water from ships arriving from Atlantic ports. Green crabs were first discovered in San Francisco Bay and Bodega Bay in the late 1980s, and by the mid-1990s they had spread to bays and lagoons from Monterey to Eureka. European green crabs have the potential to damage Dungeness crab, oyster, and clam fisheries and to seriously affect many other species. This crab is a voracious feeder, and it will eat nearly anything it comes across, ranging from mussels to clams, and snails to worms. Chemical treatments are infeasible due to potential impacts to native plants and animals in the water, and controlling this invader remains difficult.

Toxic chemicals, pathogens, and nutrient loading are technical terms for another challenge facing California's coast: water pollution. Toxic substances such as minute particles of metals, polycyclic aromatic hydrocarbons (PAHs), polychlorinated biphenyls (PCBs), heavy metals, and pesticides such as DDT enter coastal waters through run-off during winter storms. As bottom-dwelling organisms such as clams and snails are exposed to these chemicals, they may in turn pose a risk to other organisms that eat them. Some toxins "bio-accumulate," meaning that the chemicals concentrate in the tissue of the organism that consumes them. The bald eagle was originally listed as endangered under the federal Endangered Species Act because many of the majestic birds were killed by their consumption of fish that had bio-accumulated DDT. Birds that survive on a diet of marine fish are at risk of developing concentrations of toxins, which may cause thinning of eggshells and loss of reproductive effort.

European green crab

Pathogens are disease-causing organisms, including viruses, bacteria, and parasites. These microorganisms are found in sewage, and when present in marine waters, they can pose a health threat to swimmers, surfers, divers, and seafood consumers. Fish and filter-feeding organisms such as shellfish concentrate pathogens in their tissues, and, just as toxins do, they may cause illness in organisms that consume them.

In some areas along the coast, excessive nutrients are added to marine waters through agricultural run-off and sewage. Nutrients stimulate the growth of algae, and as the algae die, they decay and rob the water of oxy-

Brown pelican

gen. The algae can become so dense that they prevent sunlight from penetrating the water. Fish and shellfish are deprived of oxygen, and underwater seagrasses are deprived of light. Animals that depend on seagrasses for food or shelter leave the area or may die. The excessive algae growth may also result in brown and red tides, which have been linked to fish kills.

Bald eagle

The continued health and biodiversity of coastal ecosystems depends on the protection and management of high-quality habitat on land and in water and on an abundance of native species. The health of the coast also depends on sensible land use decisions that take into account the need for all living creatures to maintain their foothold on the sometimes crowded shores of California. These challenges cannot be met by individuals alone, and neither can they be met by the plant and animal residents of the coast. Instead, awareness and action by all of us are necessary. Next time you go to the coast, enjoy your experience, but learn as much as you can about the places you visit and become aware of what needs to be done to protect these places for future generations to enjoy.

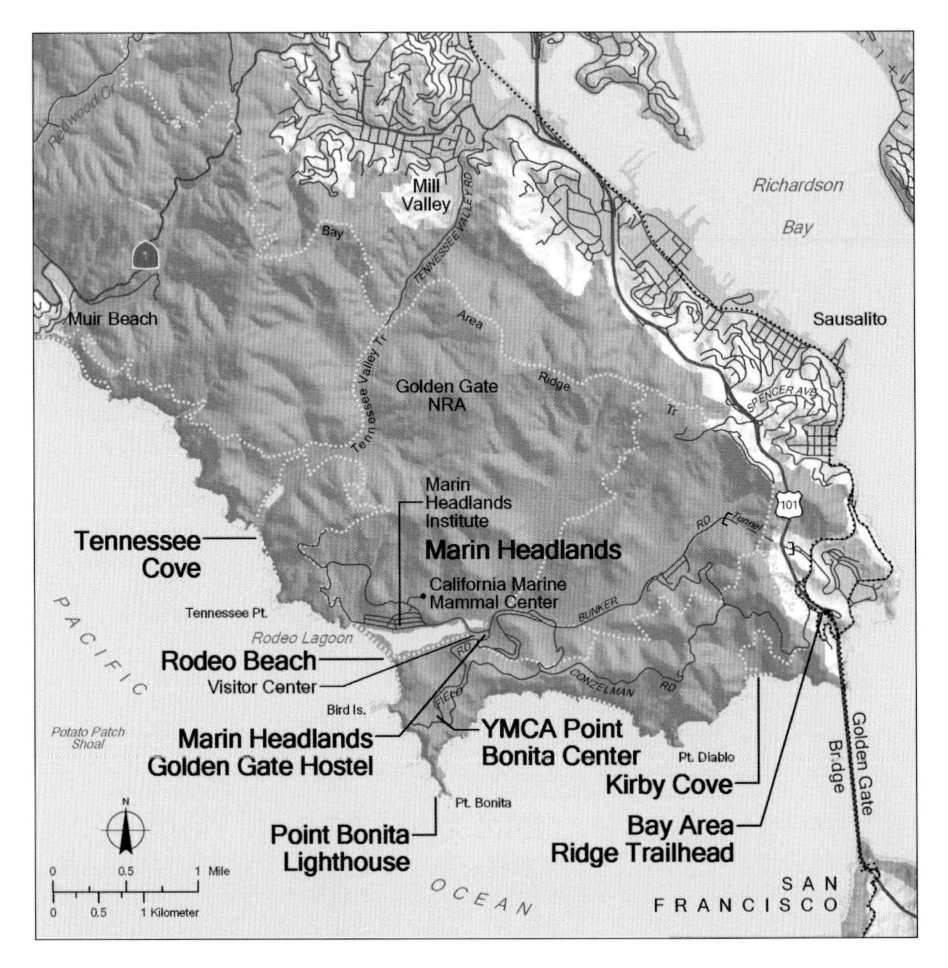

Caspian tern at Rodeo Lagoon

Marin Headlands

	Sandy Beach	Rocky Shore	Trail	Visitor Center	Campground	Wildlife Viewing	Historic Building	Fishing or Boating	Facilities for Disabled	Restrooms	Parking	Fee
Tennessee Cove	•	•	•		•			•	•	•	•	
Rodeo Beach	•	•	•	•					•	•	•	
Marin Headlands–Golden Gate Hostel			•				•		•	•	•	•
Point Bonita Lighthouse			•	•			•		•	•	•	
Marin Headlands	•	•	•	•	•	•	•		•	•	•	
The YMCA Point Bonita Center			•				•		•	•	•	•
Kirby Cove	•		•		•	•			•	•	•	
Bay Area Ridge Trailhead			•						•	•	•	

TENNESSEE COVE: *Tennessee Valley Rd., off Hwy. One in Mill Valley.* Gently sloping valley; hiking trails lead north to Muir Beach and south to Rodeo Beach. A two-mile-long trail leads from the end of Tennessee Valley Rd. to the beach at Tennessee Cove, where a coarse-sand beach is flanked by greenstone outcroppings to the north and south. Miwok Livery riding stables offers guided rides by appointment; call: 415-383-8048. Haypress hike-in campsite has five sites, each accommodating up to four persons; call: 415-331-1540.

RODEO BEACH: *Bunker Rd. off Hwy. 101 (Alexander Ave. exit).* A wheelchair-accessible bridge across a lagoon leads to the edge of pebbly Rodeo Beach, part of the Golden Gate National Recreation Area. A picnic area and restrooms are available. Trails lead up the bluffs and through Rodeo Valley, connecting to points in the headlands and beyond. Former barracks house several environmental education programs, including the California Marine Mammal Center and the Marin Headlands Institute. The Marine Mammal Center features a bookstore and is open 10 AM–4 PM daily; educational/interpretive programs by arrangement. For information, call: 415-289-7325. The Headlands Institute offers education programs

The Golden Gate National Recreation Area, located in San Francisco, Marin, and San Mateo Counties, reflects an effort by the National Park Service beginning in the 1960s to make its facilities more accessible to America's urban population. A number of unrelated events helped provide land for the recreation area, which was created in 1972, including the closure of the federal prison on Alcatraz Island and disposition of surplus military lands such as Fort Mason. Famed photographer Ansel Adams had proposed as early as the 1940s that the Golden Gate be designated as a national monument; today, the Golden Gate National Recreation Area (known as GGNRA) is one of the national park system's most popular units, drawing some 14 million visitors annually to its attractions that extend well beyond the Golden Gate. Visitor centers in Marin County are located at the Marin Headlands (open daily 9:30 AM to 4:30 PM; call: 415-331-1540) and at Muir Woods National Monument (open 8 AM to sunset; call: 415-388-7368).

about the outdoors, and a science building and coast laboratory that can be toured by prior arrangement, 8:30 AM–5 PM weekdays; call: 415-332-5771. For information on guided walks, call: 415-331-1540.

Rodeo Beach is popular with surfers, and the adjacent highlands offer elevated spots for observers. The beach includes weathered particles of greenstone, red jasper, and orange chalcedony, known as carnelians, from the nearby cliffs. Offshore near the southern end of the beach is Bird Island, a breeding place in springtime for pelagic cormorants. In summer, California brown pelicans roost on Bird Island. Rodeo Lagoon is a brackish water body where uncommon tufted ducks and harlequin ducks overwinter; in summer, California brown pelicans and Caspian terns may be sighted.

MARIN HEADLANDS–GOLDEN GATE HOSTEL: *Off Bunker Rd., near Rodeo Lagoon.* 60 beds, bicycle storage room; operated by Hostelling International. For information, write: Building 941, Fort Barry, Sausalito 94965; or call: 415-331-2777 or 1-800-909-4776, #62.

Rodeo Beach

POINT BONITA LIGHTHOUSE: *S.W. tip of Marin Headlands, end of Conzelman Rd.* Built in 1877, the lighthouse was manually operated until 1980. It is located on a treacherous point reached only via a tunnel hewn through solid rock and a suspension bridge over crashing surf. Open year-round from 12:30 PM to 3:30 PM, Saturday, Sunday, and Monday. Monthly full-moon walks and occasional sunset walks to the lighthouse are available by reservation only. For information and walk schedules, call: 415-331-1540.

The original lighthouse at Point Bonita was built in 1855 and proved to be an inhospitable place; seven different tenders were employed in the first nine months of operation. In 1856 an iron cannon was installed as a fog signal, the first on the Pacific Coast. The keeper of the fog signal found his task somewhat onerous, as he was required to fire the cannon every 30 minutes during continuous fogs, which last sometimes for days on end. The cannon was abandoned after a couple of years, and a foghorn was built in 1902.

MARIN HEADLANDS: *Off Hwy. One, N. of Golden Gate Bridge.* A series of grassy hills and valleys; former military lands, now part of the Golden Gate National Recreation Area. A wheelchair-accessible visitor center, located in the former Fort Barry chapel, is on Field Rd. off Conzelman Rd. The center features books, exhibits, maps, and information; open 9:30 AM–4:30 PM daily. Permits are required for camping anywhere in the headlands and are available at the visitor center. Bicentennial campsite has three sites; each accommodates one to two persons; for information, call: 415-331-1540.

The coastline from Tennessee Cove to the Golden Gate Bridge consists of steep sea cliffs with beaches located between rocky outcroppings. Rocks of the Franciscan Complex, formed as much as 150 million years ago, including red chert, greenstone, and a type of sandstone known as greywacke, can be observed in many exposed locations in the headlands.

Chert is a thin-bedded sedimentary rock found in uncharacteristically thick ribbons in the headlands, usually layered and folded in angular patterns. Red chert is exposed

Painting of the *Tennessee,* San Francisco Maritime National Historical Park

Before the advent of modern navigational aids, finding the entrance to the Golden Gate was a challenge for mariners. The *Tennessee* was an 1,194-ton steamship put into service by the Pacific Mail Steamship Company in late 1849 to transport gold-seekers from Panama to San Francisco. The *Tennessee* also carried mail and merchandise such as pistols, watches, jewelry, glassware, cigars, and playing cards, intended for sale in the fledging city of San Francisco, where gold dust in the pockets of miners created a ready market for luxury goods. But the *Tennessee*'s career was a short one: on March 6, 1853, as the ship felt its way through a dense fog toward San Francisco Bay, the crew saw surf ahead. Rather than strike the rocks that loomed up, the captain ordered full speed ahead toward a cove beach, now known as Tennessee Beach. There was no loss of life as the ship went ashore intact, although it later proved impossible to refloat the ship. Some passengers hiked over the hills to the Richardson ranch in Sausalito, where they hired small boats to ferry them to San Francisco, and others waited for rescue ships that removed the ship's cargo.

Only a month later, on April 9, 1853, the steamer *Samuel S. Lewis*, owned by Cornelius Vanderbilt, was on approach from Nicaragua when it overshot the entrance to San Francisco Bay and ran aground on Duxbury Reef. The *Samuel S. Lewis* broke up within a day, a total loss like the *Tennessee*. The 385 passengers, including William Tecumseh Sherman, made it safely to the beach, where bonfires were lit to dry their clothes and they awaited rescue. Within two years, the first light was constructed at Point Bonita, to help guide ships into San Francisco Bay.

Rodeo Beach

in road cuts along Conzelman Rd., which winds along the top of the headlands west of the Golden Gate Bridge. The basalt found in the headlands, called greenstone, is of volcanic origin, often with a greenish cast due to the presence of the minerals chlorite and pumpellyite. Bird Island, the large rock lying just offshore the south end of Rodeo Beach, is a greenstone formation. Cavities in the greenstone are filled with pillow lavas or with various minerals, such as red jasper, visible on the cliffs around the beach at Kirby Cove, or orange chalcedony. Greywacke is a type of sandstone made up of various-sized angular grains, probably formed originally in a deepwater environment. The rocky cliffs at the north end of Rodeo Beach are composed of greywacke, ranging in color from gray at the bottom to a weathered soft brown at the top.

THE YMCA POINT BONITA CENTER: *Off Field Rd., near Conzelman Rd.* The center offers overnight accommodations, by reservation only, for groups of 20 or more. Trails lead to a small beach also accessible by trail from Battery Alexander. For information, call: 415-331-9622.

KIRBY COVE: *Conzelman Rd., off Hwy. 101 (Alexander Ave. exit).* Follow Conzelman Rd. from Hwy. 101 one-half mile to locked gate; walk to the beach at the foot of the dirt road. Picnic facilities and firepits are available. Overnight camping at four sites, each of which can accommodate ten people; for reservations, call: 1-800-365-2267.

Wheelchair-accessible facilities include modified picnic tables, a bridge to the beach, and restrooms. Check with the Marin Headlands visitor center for road conditions, call: 415-331-1540. For information, call: 415-388-2070.

BAY AREA RIDGE TRAILHEAD: *Conzelman Rd., off Hwy. 101.* A link to the Bay Area Ridge Trail begins from the parking lot on Conzelman Rd. northwest of the Golden Gate Bridge. The Bay Area Ridge Trail system, planned to be 500 miles long connects Mount Tamalpais State Park, Muir Woods, and Samuel P. Taylor State Park in Marin County, as well as ridgetops, parks, and open space in ten counties around the perimeter of San Francisco Bay. The trail accommodates pedestrians, equestrians, and bicyclists. Call: 415-561-2595.

The separate San Francisco Bay Trail provides public access along the shoreline of the bay; some 230 miles of trail, about half of the planned system, have been completed. Call: 510-464-7900.

Golden Gate

The Golden Gate is the river outlet eroded by the Sacramento–San Joaquin Rivers during interglacial periods of the Pleistocene Epoch, around three million years ago. Although San Francisco Bay is relatively muddy and shallow, the floor of the Golden Gate is swept of sand and silt by strong tidal action; the depth of the sea at the gate is 341 feet. The name "Golden Gate" predates by far the world-famous bridge. Captain John C. Frémont, in an 1848 report to Congress on his explorations with the U.S. Topographical Engineers through Mexican California, called the opening of San Francisco Bay *Chrysopylae*, or "golden gate." Within the year, gold-seekers bound for the Sierra Nevada foothills from points around the globe began to pour into the bay, and they quickly adopted Frémont's uncannily appropriate label.

Afterword

In preparing this guide, we visited and researched all beaches, parks, and coastal accessways that are described here. We have incorporated comments and corrections supplied by various agencies and members of the public, who wrote to us regarding the information in previous editions of the *California Coastal Access Guide*. We also sought review of draft material by staff of parks departments, local governments, land trusts, and others. The book is accurate, to the best of our knowledge. Nevertheless, conditions on the coast are constantly changing, and there may be inaccuracies in this book. If you think something is incorrect or has been omitted, please let us know. The Coastal Commission intends to continue publishing revised guides in the future and would appreciate any additional information you can provide. Please remember, however, that this book includes only those beaches and accessways that are managed for public use.

Address all comments to:

Coastal Access Program
California Coastal Commission
45 Fremont Street, Suite 2000
San Francisco, CA 94105

or e-mail to: coast4U@coastal.ca.gov

Kelp

© 2004 Tom Killion

Acknowledgments

Tom Killion is an artist, teacher, historian, and humanist with roots in Marin County. His work and study have taken him to Europe, Eritrea, and the Sudan, among other places. Tom Killion produces landscapes using woodcut techniques strongly influenced by traditional Japanese methods. His striking prints have been exhibited at venues from San Francisco and Yosemite to New York and London. His work has been published in books including *The Coast of California*; *The High Sierra of California*; and *Walls: A Journey Across Three Continents*. We extend our deepest appreciation to Tom Killion for his stunning contributions to this book. To view more of his work, see www.tomkillion.com.

Researchers:
Melissa Cheung
Ben Hansch
Louisa Morris

Additional cartography:
Peter Tittmann
Doug Macmillan
Darryl Rance

Special thanks to Don Neuwirth and Pat Stebbins, former managers of the Coastal Access Program, for their imagination, perseverance, and inspiring leadership.

Principal Contributors, *California Coastal Access Guide* and *California Coastal Resource Guide*, from which selected material has been incorporated in this book:
Trevor Kenner Cralle
Linda Goff Evans
Stephen J. Furney-Howe
Jo Ginsberg
Christopher Kroll
Trish Mihalek
S. Briggs Nisbet
Sabrina S. Simpson
Mary Travis
Jeffrey D. Zimmerman
Gianmaria Mussio

Thanks to the following individuals and institutions for their invaluable assistance:
Dawn Adams
Marnie Atkins
Jim Baskin
Bodega Marine Laboratory
Loren Bommelyn
Michael Bowen
California Resources Agency
Susan Calla
Bruce Cann
Joel Canzoneri
Jeanne Coleman
Peter Connors
Su Corbaley
Cordell Bank National Marine Sanctuary
Jennifer Dare
Mark Delaplaine
John Dell'Osso
Shirley Eberly
Rebecca Ellin
Karren Elsbernd
Phyllis Faber
Thomas Gates
Rasa Gustaitis
Grey Hayes
Dan Howard
Susan Jordan
Chris Kern
Diane Knox
Karen Kovacs
Sylvie Bloch Lee
Lynne Mager
Moira McEnespy
Karl Menard
Bob Merrill
Ron Munson
National Oceanic and Atmospheric Administration
Eric Nelson
Office of Ocean and Coastal Resource Management
Brendan O'Neil
Steve Petterle

Chris Platis
Elizabeth Ptak
Richard Retecki
Ellen Rilla
Hawk Rosales
Rebecca Roth
Mary Jane Schramm
Gary Shannon
Larry Simon
Tony Smithers
Becky Smythe
Martha Spencer
State Coastal Conservancy
Jennifer Stock
Tolowa Tribe
Noah Tilghman
Ray Van de Water
Michael Walbrecht
Sarah Warnock
Stephanie Weigel
Wiyot Tribe
Yurok Tribe

And special appreciation to the Culture Committee, Yurok Tribe, for their help and hospitality.

Photo and Illustration Credits:

Dawn Adams, National Park Service, Pt. Reyes Natl. Seashore, 290

© 2002-2004 Kenneth & Gabrielle Adelman, 66, 72, 143, 239, 240, 261, 299 California Coastal Records Project, www.Californiacoastline.org

Scot Anderson, 165a

Sherry Ballard © California Academy of Sciences, 153c

Bancroft Library, University of California, Berkeley, reproduced by permission, 90, 146, 243b

Barry Breckling, 56b

Jeff Bright, 53a, 53b

Brousseau Collection, 55c, 125c

California Historical Society, 123, 147

Robert Campbell, 242, 260

Melissa Cheung, 262

Chris Christie, 126c

Jeanne Coleman, 119

Peter Connors, 126b, 195, 196

Gerald and Buff Corsi © California Academy of Sciences, 38a, 132b, 183b, 198a, 199a, 269a

Jeff Dao, 176

T.W. Davies © California Academy of Sciences, 151a

Brooke Delello, 104

Don DesJardin, 166b, 267a

Richard Doell, 139c

Jeremiah Easter, 55a

Kip Evans, 164c, 210c, 211a, 212a, 266b, 268b, 269b

Lesley Ewing/Jon Van Coops, 231

Jeff Gill, 220

Tom Greer, 197a

Thomas Grey, 229

Joyce Gross, 127b, 213a

Marlin Harms, 38b, 56a, 153b, 186a

Barbara Harris, 161

Roy Harrison, 99

Phoebe Apperson Hearst Museum of Anthropology and the Regents of the University of California – ID 15-3318 – reproduced by permission, 33

Scott Hess, 10, 11, 214, 224, 235, 283

William R. Hewlett © California Academy of Sciences, 198d

Carl Hoard II, 211c

Gary Hromada, 280, 296

Humboldt County Convention and Visitors Bureau, 74, 77

David Hurley, 227

Dr. Lloyd Ingles © California Academy of Sciences, 55b

Stephen Ingram, 139a

Russ Kerr, 151c

Tom Killion, 8, 52, 160, 194, 208, 264, 288, 300

Richard Lang, 183a

Dorothea Lange © The Dorothea Lange Collection, Oakland Museum of California, City of Oakland, Gift of Paul S. Taylor, 286a, 286b

Peter LaTourrette, 54b, 84c, 133c, 151b, 152c, 184c, 253b,

Ron LeValley, 112a, 112b, 132a, 167a, 167b, 184b, 197b, 289

Linda Locklin, 14, 269c

Glenn McCrea, 152b, 203

Art Mielke, 100, 105, 106a, 106b, 107, 137, 140, 148, 171, 172, 173, 174, 177, 178, 180

Monterey Bay Aquarium, 83c, 209a, 210a

Louisa Morris, 102, 103, 116, 118, 157, 158

Gianmaria Mussio, 26

Gary Nafis, 54c

National Oceanic and Atmospheric Admin. Cordell Bank Natl. Marine Sanctuary, 165b National Geophysical Data Center, 243a Robert Pitman, 185a

Courtesy of the National Park Service, Redwood National Park, 27, 36, 41, 42, 43, 50, 61, 62

Naval Historical Foundation, Washington, D.C., 81

Robert Orr © California Academy of Sciences, 252c

Bill Perry, 170

Gordon H. Phillip © California Academy of Sciences, 54a

Andrea Pickart, U.S. Fish and Wildlife Service, 127a

Steve Romaine, Fisheries and Oceans Canada, Institute of Ocean Sciences, Sidney, B.C., 83b

John Ryan, Monterey Bay Aquarium Research Institute, 162

San Francisco Maritime National Historic Park, reproduced by permission, 31, 297

Larry Sansone, 127c, 153a

Robert Schmieder, Cordell Expeditions, 166a, 200, 232

Steve Scholl, cover, 17, 20, 22, 23, 25, 28, 30, 32, 35, 37a, 37c, 39, 44, 46, 47, 48, 49, 51, 57, 60, 64, 67, 68, 70, 73, 78, 80, 82a, 82b, 83a, 84a, 84b, 85a, 85b, 86, 87, 89, 91, 92, 98, 110, 111a, 111b, 111c, 113, 120, 122, 124, 125a, 125b, 125d, 126a, 128, 129, 130, 131, 133a, 133b, 134, 136, 138a, 138b, 142, 144, 145, 150a, 150b, 150c, 152a, 154, 155, 156, 159, 164a, 164b, 168, 169, 175, 179, 181, 186b, 187, 190, 192, 193, 198b, 198c, 199b, 199c, 201, 204, 205, 206, 207, 209b, 210b, 211b. 212b, 213b, 213c, 216, 217, 218, 221, 222, 226, 228, 238, 241, 244, 245, 247, 248, 249, 250, 251, 252a, 252b, 253a, 253c, 254, 255, 256, 257, 258, 259, 265, 267b, 267c, 268c, 270, 271, 272, 273, 274, 276, 278, 279, 284, 287, 291, 292, 293a, 294, 298

Aaron Schusteff, 139b

Susan Spann, 12, 13

Rich Stallcup, 185b

Craig Stern, 263

Ian Tait, 38c, 293b

Peter Tittmann, 94

Guy Towers, 34

Beti Webb Trauth, 76

United States Geological Service, 108

Peggy Ushakoff, Washington State Department of Fish and Wildlife 266a,

Glenn and Martha Vargas © California Academy of Sciences, 184a

Larry Wan, 163, 268a

Warner Brothers Pictures, Inc., 149

Charles Webber © California Academy of Sciences, 37b

Wiyot Tribe, 88

Glossary

alluvial. Having stream deposits and sediments formed by the action of running water.

anadromous. Migrating from salt water to fresh water in order to reproduce.

annual. A plant that germinates, flowers, sets seed, and dies within one year or less.

arthropod. Invertebrate animal characterized by jointed limbs, segmented bodies, and chitinous exoskeletons; e.g., spiders, insects, and crustaceans.

barrier beach. A long, narrow beach separated from the mainland by a lagoon, bay, or rivermouth.

basalt. A dark igneous rock of volcanic origin. Basalt is the bedrock of most of the world ocean.

bay. A partially enclosed inlet of the ocean.

beach. The shore of a body of water, composed of sand, pebbles, or rocks.

benthic. Pertaining to the ocean bottom, and the organisms that inhabit the bottom.

bioluminescence. The production of visible light by biochemical reaction in living organisms, such as phytoplankton.

biomass. The total amount of living matter in a given area.

biota. The collective plant and animal life, or flora and fauna, of a region.

bivalves. Mollusks such as clams and oysters that have two-piece, hinged shells.

bluff. A steep headland or cliff.

bottom fish. Fish species that feed on the sea floor, such as rockfish and flounder; commercial fish harvested with trolling nets.

brackish. Water that contains some salt, but less than sea water (from 0.5 to 30 parts per thousand).

bunchgrass. Perennial grass that forms tufted clumps; includes many native California species, such as needlegrass, bentgrass, reedgrass, and hairgrass.

calcareous. Composed of or containing calcium carbonate; e.g., the shells of mollusks.

California Current. A cold-water ocean current in the North Pacific that flows southward along the west coast of North America.

cetaceans. A group of aquatic mammals including whales, dolphins, and porpoises.

chert. A fine-grained siliceous sedimentary rock; a source of flint used by California Indians to make spear and arrow blades.

closed-cone. Refers to coniferous trees having cones that remain closed for several or many years after maturing; e.g., bishop pine.

coastal scrub. A plant association characterized by low, drought-resistant, woody shrubs; includes coastal sage scrub and northern coastal scrub.

coastal strand community. A plant association endemic to bluffs, dunes, and sandy beaches, and adapted to saline conditions; includes sea rocket and sand verbena.

coastal terrace. A flat plain edging the ocean; uplifted sea floor that was cut and eroded by wave action. Also called marine terrace or wave-cut bench.

conifer. A cone-bearing tree of the pine family, usually evergreen.

continental shelf. The shallow, gradually sloping area of the sea floor adjacent to the shoreline, terminating seaward at the continental slope.

crustaceans. A group of mostly marine arthropods; e.g., barnacles, shrimp, and crabs.

current. Local or large-scale water movements that result in the flow of water in a particular direction, e.g. alongshore, or offshore.

delta. A fan-shaped alluvial deposit at the mouth of a river.

endangered. Those species designated by the California Department of Fish and Game or the U.S. Fish and Wildlife Service as "endangered" because of severe population declines.

endemic. A plant or animal native to a well-defined geographic area and restricted to that area.

erosion. The gradual breakdown of land by weathering, solution, corrosion, abrasion, or transportation, caused by action of the wind, water, or ice; opposite of accretion.

estero. Spanish for estuary, inlet, or marsh.

estuary. A semi-enclosed coastal body of water that is connected with the open ocean and within which seawater mixes with freshwater from a river or stream.

exotic. Any species, especially a plant, not native to the area where it occurs; introduced.

fault. A fracture or fracture zone along which visible displacement of the earth occurs, resulting from seismic activity.

forb. Any herbaceous (non-woody) plant other than a grass.

foredune. Dune closest to the seashore that is relatively unstable and subject to salt spray, wind erosion, and storm waves; generally sparsely vegetated by plants with special adaptations to harsh conditions.

Franciscan Complex. A group of sedimentary and volcanic rocks that occurs along much of the northern California coast and consists predominantly of sandstone, shale, and chert, with occasional limestone, basalt, serpentine, and schist.

groin. Wall or embankment, constructed at right angles to the shoreline, that projects out into the water to trap sand or to retard shoreline erosion; a shoreline protective device.

gyre. A large circular or spiral motion of currents around an ocean basin.

habitat. The sum total of all the living and nonliving factors that surround and potentially influence an organism; a particular organism's environment.

halophyte. A plant that is adapted to grow in salty soils.

haul-out. A place where pinnipeds emerge from the water onto land to rest or breed.

intertidal. The shoreline area between the highest high tide mark and the lowest low tide mark.

invasive species. Weedy, generally non-native plants or non-native wildlife species that invade and/or proliferate following disturbance or continued overuse.

invertebrate. An animal with no backbone or spinal column; 95 percent of the species in the animal kingdom are invertebrates.

jetty. A wall or embankment constructed at right angles to the coast at the mouth of a river or harbor to help stabilize the entrance; usually constructed in pairs on each side of a channel.

lagoon. A body of fresh or brackish water separated from the sea by a sandbar or reef.

longshore current. A current flowing parallel to and near shore that is the result of waves hitting the beach at an oblique angle.

marsh. General term for a semi-aquatic area with relatively still, shallow water, such as the shore of a pond, lake, or protected bay or estuary, and characterized by mineral soils that support herbaceous vegetation.

mollusks. Soft-bodied, generally shelled invertebrates; for example, chitons, snails, limpets, bivalves, and squid.

nearshore. The offshore area, extending an indefinite distance seaward of the intertidal zone, well beyond the breaker zone.

pelagic. Living in the open ocean rather than in inland waters or waters adjacent to land.

perennial. A plant that lives longer than a year.

pinnipeds. Marine mammals that have fin-like flippers, including seals, sea lions, and walruses.

plankton. Free-floating algae (phytoplankton) or animals (zooplankton) that drift in the water, ranging from microscopic organisms to larger species such as jellyfish.

prairie. Flat or rolling grassland.

predator. An animal that eats other animals; a carnivore.

raptors. Birds of prey, such as falcons, eagles, and owls.

reef. A submerged ridge of rock or coral near the surface of the water.

relict. In ecology, a genus or species from a previous era that has survived radical environmental changes resulting from climatic shifts.

revetment. A sloped retaining wall built of riprap or concrete blocks along the coast to prevent erosion inland and other damage by wave action; similar to a seawall.

rip current. A narrow, seaward-flowing current that results from breaking of waves and subsequent accumulation or build-up of water in the nearshore zone.

riparian. Pertaining to the habitat along the bank of a stream, river, pond, or lake.

riprap. Boulders or quarry stone used to construct a groin, jetty, or revetment.

rookery. A breeding site, such as an island, for seabirds or marine mammals.

sea stack. A tiny island of rock left standing after waves have eroded the shoreline.

seawall. A structure, usually a vertical wood or concrete wall, designed to prevent erosion inland or damage due to wave action.

seismic sea wave. A long-period wave generated by an underwater earthquake, landslide, or volcanic eruption; a tsunami; erroneously called a tidal wave.

serpentine. A green or black magnesium silicate mineral, or a rock composed principally of serpentine. Soils of serpentine origin support a unique association of endemic plants.

siliceous. Containing or consisting of silica.

slough. A small marshland or tidal waterway that usually connects with other tidal areas.

species. A taxonomic classification ranking below a genus, and consisting of a group of closely related organisms that are capable of interbreeding and producing viable offspring.

substrate. The surface on which an organism grows or is attached.

surf zone. The area affected by wave action, extending from the shoreline high-water mark seaward to where the waves start to break.

surge channel. A narrow gap in rocky intertidal areas through which waves surge; habitat for species of barnacles, mussels, and algae.

take. As defined by the Endangered Species Act, "to harass, harm, pursue, hunt, shoot, wound, kill, capture, or collect, or attempt to engage in any such conduct."

terrestrial. Living or growing on land, as opposed to living in water or air.

threatened. Those species designated by the California Department of Fish and Game or the U.S. Fish and Wildlife Service as "threatened" because of severe population declines.

tidal wave. The regular rise and fall of the tides; often misused for *tsunami*.

tide. The periodic rising and falling of the ocean resulting primarily from the gravitational forces of the moon and sun acting upon the rotating earth.

tidepool. Habitat in the rocky intertidal zone that retains some water at low tide.

tsunami. A large, sometimes destructive ocean wave caused by an underwater earthquake, submarine landslide, or volcanic eruption; inaccurately called a tidal wave.

understory. A layer or level of vegetation occurring under a vegetative canopy, such as the herbaceous plants growing under the taller trees of a riparian woodland.

uplifted. Pertaining to a segment of the earth's surface that has been elevated relative to the surrounding surface, especially as a result of tectonic activity.

upwelling. A process by which deep, cold, nutrient-rich waters rise to the sea surface.

waterfowl. Ducks, geese, and swans.

watershed. The land area drained by a river or stream system or other body of water.

wetland. Land periodically or permanently covered with shallow water or saturated in the root zone, including marshes, intertidal mudflats, sloughs, and vernal pools.

State and Federal Agencies, North Coast

California State Agencies:

California Coastal Commission
45 Fremont St., Suite 2000
San Francisco, CA 94105
415-904-5200

California Coastal Commission
710 E St., Suite 200
Eureka, CA 95501
707-445-7833

California Department of Boating and Waterways
2000 Evergreen St., Suite 100
Sacramento, CA 95815
916-263-1331

California Department of Fish and Game
1416 Ninth St.
Sacramento, CA 95814
916-445-0411

California Department of Fish and Game
619 Second St.
Eureka, CA 95501
707-445-6493

California Department of Fish and Game
1850 Bay Flat Rd.
P.O. Box 1560
Bodega Bay, CA 94923
707-875-4260

California Department of Fish and Game
19160 South Harbor Dr.
Fort Bragg, CA 95437
707-964-9078

California Department of Forestry and Fire Protection
Demonstration State Forest Program
1416 Ninth St.
P.O. Box 944246
Sacramento, CA 94244
916-653-9420

California Department of Parks and Recreation
1416 Ninth St.
Sacramento, CA 95814
1-800-777-0369

California Environmental Protection Agency
1001 I St.
Sacramento, CA 95812
916-323-2514

California Environmental Resources Evaluation System (CERES)
California Ocean and Coastal Environmental Access Network
900 N St.
Sacramento, CA 95814

California Resources Agency
1416 Ninth St., Suite 1311
Sacramento, CA 95814
916-653-5656

San Francisco Bay Conservation and Development Commission
50 California St., Suite 2600
San Francisco, CA 94111
415-352-3600

State Coastal Conservancy
1330 Broadway, Suite 1100
Oakland, CA 94612
510-286-1015

State Lands Commission
100 Howe Ave., Suite 100 South
Sacramento, CA 95825-8202
916-574-1900

Wildlife Conservation Board
1807 13th St., Suite 103
Sacramento, CA 95814
916-445-8448

Federal Agencies:

Bureau of Land Management
1695 Heindon Rd.
Arcata, CA 95521
707-825-2300

Bureau of Land Management
2550 North State St.
Ukiah, CA 95482
707-468-4000

National Park Service
Golden Gate National Recreation Area
Fort Mason, Building 201
San Francisco, CA 94123
Visitor Information: 415-561-4700

National Park Service
Point Reyes National Seashore
1 Bear Valley Rd.
Point Reyes Station, CA 94956
Headquarters: 415-464-5100
Visitor information: 415-464-5100

National Park Service
Redwood National Park
1111 Second St.
Crescent City, CA 95531
707-464-6101

Cordell Bank National Marine Sanctuary
1 Bear Valley Rd.
Point Reyes Station, CA 94956
415-663-0314

Gulf of the Farallones National Marine Sanctuary
Fort Mason, Bldg. 201
San Francisco, CA 94123
415-561-6622

National Oceanic and Atmospheric Administration
Office of Ocean and Coastal Resource Management
N/ORM 10th floor SSMC4
1305 East-West Hwy.
Silver Spring, MD 20910

U.S. Army Corps of Engineers
Civil Works Office
333 Market Street, Rm. 923
San Francisco, CA 94105
415-977-8600

U.S. Fish and Wildlife Service
911 N.E. 11th Ave.
Portland, OR 97232
503-231-6118

Bibliography

Alley, Bowen & Co. *History of Sonoma County*. Petaluma, CA: Veronda, republished 1973, originally published 1879.

Barron, Alan D. *A Birdfinding Guide to Del Norte County, California*. Crescent City, CA: Redwood Economic Development Institute and the Rose Foundation for Communities and the Environment, 2001.

Borhan, Pierre. *Dorothea Lange: The Heart and Mind of a Photographer*. Boston: A Bulfinch Press Book, Little, Brown and Co., 2002.

California Coastal Commission. *California Coastal Access Guide*. 6th ed. Berkeley: University of California Press, 2003.

———. *California Coastal Resource Guide*. Berkeley: University of California Press, 1987.

Carranco, Lynwood. *Redwood Lumber Industry*. San Marino, CA: Golden West Books, 1982.

Carranco, Lynwood, and John T. Labbe. *Logging the Redwoods*. Caldwell, ID: Caxton Printers, Ltd., 1996.

Cooper, Daniel S. *Important Bird Areas of California*. Pasadena, CA: Audubon California, 2004.

Delgado, James P. *To California by Sea: A Maritime History of the California Gold Rush*. Columbia, SC: University of South Carolina Press, 1990.

Elliott, W. Wallace & Co. *History of Humboldt County California*. San Francisco: Wallace W. Elliott & Co. Publishers, 1881.

Fix, David, and Andy Bezener. *Birds of Northern California*. Auburn, WA: Lone Pine Publishing, 2000.

Goodson, Gar. *Fishes of the Pacific Coast*. Stanford, CA: Stanford University Press, 1988.

Griffin, L. Martin. *Saving the Marin-Sonoma Coast: The Battles for Audubon Canyon Ranch, Point Reyes, & California's Russian River*. Sweetwater Springs Press, 1998.

Harris, Stanley D., Ph.D. *Northwestern California Birds*. Arcata, CA: Humboldt State University Press, 1996.

Hedgpeth, Joel W. *Introduction to Seashore Life of the San Francisco Bay Region and the Coast of Northern California*. Berkeley: University of California Press, 1962.

Hill, Mary. *Gold: The California Story*. Berkeley: University of California Press, 1999.

Hoover, Mildred Brooke, Hero Eugene Rensch, and Ethel Grace Rensch. *Historic Spots in California*. 3rd ed. Stanford, CA: Stanford University Press, 1966.

Kampion, Drew, ed. *The Stormrider Guide: North America*. Bude, Cornwall, UK: Low Pressure Ltd., 2002.

Kroeber, Alfred Louis. *Yurok Myths*. Berkeley: University of California Press, 1976.

Layton, Thomas N. *Gifts from the Celestial Kingdom: A Shipwrecked Cargo for Gold Rush California*. Stanford, CA: Stanford University Press, 2002.

———. *The Voyage of the Frolic: New England Merchants and the Opium Trade*. Stanford, CA: Stanford University Press, 1999.

Munz, Philip A. *Introduction to Shore Wildflowers of California, Oregon, and Washington*. Berkeley: University of California Press, 2003.

Ogden, Adele. *The California Sea Otter Trade*. Berkeley: University of California Press, 1941.

Partridge, Elizabeth, editor. *Dorothea Lange: A Visual Life*. Washington, D.C.: Smithsonian Institution Press, 1994.

Pearsall, Clarence E. et al. *The Quest for Qual-a-wa-loo (Humboldt Bay): A Collection of Diaries and Historical Notes Pertaining to the Early Discoveries of the Area Now Known as Humboldt County, California*. Oakland, CA: The Holmes Book Company, 1966.

Schmieder, Robert W. *Ecology of an Underwater Island*. Walnut Creek, CA: Cordell Expeditions, 1991.

Stallcup, Rich. *Ocean Birds of the Nearshore Pacific: A Guide for the Sea-Going Naturalist*. Stinson Beach, CA: Point Reyes Bird Observatory, 1990.

Stuart, John D., and John O. Sawyer. *Trees and Shrubs of California*. Berkeley: University of California Press, 2001.

Thompson, Thomas H. & Co. *New Historical Atlas of Sonoma County: Illustrated*. Oakland, CA: Thomas H. Thompson, 1877.

Waterman, Thomas T. *Yurok Geography*. Berkeley: University of California Press, 1920. Facsimile edition with new introductory material, Trinidad, CA: Trinidad Museum Society, 1993.

Webber, Bert, and Margie Webber. *Shipwrecks and Rescues on the Northwest Coast*. Medford, OR: Webb Research Group Publishers, 1996.

Suggestions for Further Reading:

California Coast & Ocean, a quarterly magazine covering trends, issues, and controversies shaping the California coast; see www.coastandocean.org.

Cralle, Trevor. *Surfin'ary: A Dictionary of Surfing Terms and Surfspeak*. Berkeley: Ten Speed Press, 1991. Surfers' lingo and technical terms, augmented by history of the sport.

Delgado, James. *Shipwrecks at the Golden Gate*. San Francisco: Lexikos, 1990. A thorough account of ocean shipwrecks.

Durham, David L. *Durham's Place Names of the California North Coast*. Clovis, CA: Word Dancer Press, 2001. A compendium of definitions and etiology of the names of topographical features, water bodies, towns, and other places.

Garth, John S., and J. W. Tilden. *California Butterflies*. Berkeley: University of California Press, 1986.

Groot, C., and L. Margolis. *Pacific Salmon Life Histories*. Vancouver, B.C.: University of British Columbia Press, 2003.

Hart, John. *Farming on the Edge*. Berkeley: University of California Press, 1991. The story of the Marin Agricultural Land Trust, first land trust of its kind, and family farms in Marin County.

House, Freeman. *Totem Salmon: Life Lessons from Another Species*. Boston: Beacon Press, 1999. The story of efforts by Humboldt

County residents to save a population of native salmon.

Humann, Paul. *Coastal Fish Identification: California to Alaska*. Jacksonville, FL: New World Publications, 1996.

Kozloff, Eugene N. *Seashore Life of the Northern Pacific Coast: An Illustrated Guide to Northern California, Oregon, Washington, and British Columbia*. Seattle: University of Washington Press, 1983.

Love, Milton. *Probably More Than You Want to Know about the Fishes of the Pacific Coast*. Santa Barbara: Really Big Press, 1996.

Lyons, Kathleen, and Beth Cuneo-Lazaneo. *Plants of the Coast Redwood Region*. Boulder Creek, CA: Looking Press, 1988.

Marin Agricultural Land Trust. *An Abundant Land: The Story of West Marin Ranching*. Point Reyes Station, CA: Marin Agricultural Land Trust. A 90-minute audio cassette driving tour narrated by actor Peter Coyote with oral histories, interviews, and music.

Mondragon, Jennifer, and Jeff Mondragon. *Seaweeds of the Pacific Coast: Common Marine Algae from Alaska to Baja California*. Monterey, CA: Sea Challengers, 2003.

Ricketts, Edward F., Jack Calvin, and Joel Hedgpeth. *Between Pacific Tides*. Stanford, CA: Stanford University Press, 1992. rev. ed. For generations, a classic in the field.

For younger readers:

Aykroyd, Clarissa. *Exploration of the California Coast*. Philadelphia: Mason Crest Press, 2002.

Fletcher, Susan. *Walk Across the Sea*. New York: Atheneum Books, 2001. Historical fiction set in Crescent City in 1886.

Kalman, Bobbie. *Life of the California Coast Nations*. New York: Crabtree Publishing Company, 2004. Day-to-day life along the pre-Columbian California coast.

St. Antoine, Sara. *The California Coast–Stories from Where We Live*. Minneapolis: Milkweed Editions, 2001. A collection of essays, letters, poems, and stories that celebrate and explore life on the California Coast.

Sobol, Richard. *Adelina's Whales*. New York: Dutton Children's Books, 2003.

Thalman, Sylvia Barker. *The Coast Miwok Indians of the Point Reyes Area*. Point Reyes, CA: Point Reyes National Seashore Association, 1993.

Index

Bold numeral indicates photograph.

St. Louis Community College
at Meramec
LIBRARY